Weed Control Methods for Recreation Facilities Management

Editor

Edward O. Gangstad
Aquatic Plant Control Specialist
U.S. Army Corps of Engineers
Washington, D.C.

CRC Press
Taylor & Francis Group
Boca Raton London New York

CRC Press is an imprint of the
Taylor & Francis Group, an **informa** business

First published 1982 by CRC Press
Taylor & Francis Group
6000 Broken Sound Parkway NW, Suite 300
Boca Raton, FL 33487-2742

Reissued 2018 by CRC Press

Library of Congress Cataloging in Publication Data
Main entry under title:

Weed control methods for recreation facilities
management.

Includes bibliographies and index.
1. Aquatic weeds—Control. 2. Recreation areas
—Management. 3. Aquatic sports facilities—Manage-
ment. I. Gangstad, Edward O.
SB614.W44 628.97 81-7748
ISBN 0-8493-5330-0 AACR2

A Library of Congress record exists under LC control number: 81007748

Publisher's Note
The publisher has gone to great lengths to ensure the quality of this reprint but points out that some imperfections in the original copies may be apparent.

Disclaimer
The publisher has made every effort to trace copyright holders and welcomes correspondence from those they have been unable to contact.

ISBN 13: 978-1-315-89860-5 (hbk)
ISBN 13: 978-1-351-07770-5 (ebk)

Visit the Taylor & Francis Web site at http://www.taylorandfrancis.com and the
CRC Press Web site at http://www.crcpress.com

PREFACE

The primary and secondary uses of a body of water determine the need, frequency, and kind of aquatic plant control required to meet the needs of a specific situation. That is to say, a given body of water may be used primarily, or even exclusively for such activities as fishing, boating, or swimming, and the water itself may be used for domestic (potable water), industrial production, and/or agriculture. These uses may seem to be incompatible in themselves, but it is incumbent upon management to supply the optimum conditions for total water use. It is the purpose of this volume to survey and assess the management problems and to present case problems from the field in which the technical data, published literature, and the operations mechanics are given in sufficient detail to provide a format for practical analysis and application. Special emphasis has been given to measures of control of Eurasian watermilfoil.

Edward O. Gangstad

THE EDITOR

Dr. E. O. Gangstad, Aquatic Plant Control Specialist, transferred to the Office of the Chief of Engineers, Washington, D.C. in October of 1966 from the Texas Research Foundation, Dallas, where he was employed as Principal Agronomist of the Hoblitzelle Agricultural Laboratory. Dr. Gangstad was graduated from the University of Wisconsin, Madison, with an M.A. in Biochemistry (1947) and from Rutgers University, New Brunswick, N.J., with a Ph.D. in Agronomy (1950) and minors in Plant Pathology and Plant Physiology.

During the past 25 years he has established an outstanding career in the environmental and biological fields and has over 80 scientific publications. He is a contributor to the *Journal of Biological Chemistry, Agronomy Journal, Crop Science, Journal of Heredity, Weed Science, Crops and Soils, Tropical Agriculture, Turrialba, Soil Science Society of Florida, Hoblitzelle Agricultural Bulletin, Wrightia, Hyacinth Control Journal, Environmental Quality Journal,* and the *Journals of the American Society of Civil Engineers.*

Dr. Gangstad's experience includes research and field studies on the physiology and pathology of plants at the Universities of Wisconsin and New Jersey; research and development studies on strategic long-vegetable fibers and related programs in Florida with the U.S. Department of Agriculture, Office of Naval Research, and Department of Defense; and agrobusiness studies to develop renewable natural resources of the State of Texas with the Texas Research Foundation and management of natural resources for the Department of the Army.

Dr. Gangstad is listed in *American Men of Science, Leaders in Science, American Men of Achievement,* and the *International Who's Who of Intellectuals.*

CONTRIBUTORS

Richard A. Cassidy
Hydraulic Engineer
U.S. Army Corps of Engineers
Wltham, Massachusetts

P. M. Dooris
Southwest Water Management District
Brooksville, Florida

A. K. Gholson, Jr.
Assistant Reservoir Manager
U.S. Army Corps of Engineers
Chattahoochee, Florida

Robert C. Hiltibran
Biochemist
Illinois Natural History Survey
Urbana, Illinois

D. F. Martin
University of South Florida
Tampa, Florida

Philip Phillips
Planning Division
North Atlantic Division
U.S. Army Corps of Engineers
New York, New York

Charles K. Rawls
Chesapeake Biological Laboratory
Natural Resources Institute
University of Maryland
Solomons, Maryland

Carl B. Schreck (Retired)
Assistant Professor of Fisheries
Virginia Polytechnic Institute and State University
Blacksburg, Virginia

Gordon E. Smith (Retired)
Biologist
Tennessee Valley Authority
Muscle Shoals, Alabama

IN MEMORIUM

Dr. Sigmond L. Solymosy

Dr. Sigmond L. Solymosy, retired Professor of Horticulture at the University of Southweatern Louisiana, passed away on Sunday, November 10, 1974 after a brief illness.

A native of Budapest, Hungary, Dr. Solymosy joined the U.S.L. Faculty in 1959. Immediately prior to coming to U.S.L., he was Research Director of Cornelius Nurseries, Inc., Houston, Texas. Prior to that, Dr. Solymosy was Director of the Hortus Botanicus, Hatzendorf, Austria.

Internationally renowned, he held active memberships in the International Society for Horticultural Science, International Association for Plant Taxonomy, Fellow of the Royal Horticulture Society, American Institute of Biological Science, American Society for Horticultural Science, Louisiana Academy of Science, American Society of Plant Taxonomists, American Horticultural Society, Holly Society of America, and the Louisiana Society for Horticultural Research.

TABLE OF CONTENTS

PART I: SPECIAL STUDIES FOR MANAGEMENT OF RECREATION WATERS

Part I: Special Studies for Management of Recreation Waters

Chapter 1

COSTS AND BENEFITS OF AQUATIC WEED CONTROL

INTRODUCTION

Recurring floods and droughts have been natural phenomena since prehistoric times. With the advent of man and his civilization of cities, cultivated farms, and industrial development, water storage and impoundment have become a vital necessity. While it is widely recognized that aquatic weeds can cause considerable damage in these water resource areas, thereby retarding economic growth, there has been little serious attempt to quantify, estimate, or even define the extent of the weed nuisance in terms of specific economic losses. Therefore, a sound basis for determining the priority among competing demands on limited resources that should be given to aquatic weed control programs has not been developed.

It is clearly not possible to define economic losses in exact dollar terms. Precise information does not exist nor could it readily be developed concerning the damage caused by specific weeds in specific settings, let alone the overall impact, including secondary effects, on an economy. Instead, an attempt is made to develop order-of-magnitude estimates for selected types of damage caused by weeds in specific settings. Such broad estimates should at least place the problem in better perspective and provide some guidance as to the priority that should be attached to corrective action in this field.

The damage and problems caused by weeds can have, in varying degrees, a negative effect upon a country's economy. Whereas a developed country, such as the U.S., can readily absorb such economic losses, as has happened in Florida, Louisiana, Mississippi, and Georgia, such losses cannot be easily accepted in a developing country. Countries which have only one, or a few, waterways or other fresh-water bodies can be more seriously affected than "wetter" countries with much fresh water and an extensive riverine network. In countries bordering the Niger, the Congo, or the Nile Rivers, for example, weeds can be a very serious potential threat to the overall economy. Even in the "wetter" countries such as Ceylon, India, Guyana, and Brazil with many ancillary waterways, massive weed infestation will, with time, have measurable economic consequences, especially where the major rivers are prime avenues for transportation, sources of food, or hydropower. The economic loss resulting from disease is particularly difficult to assess, even in societies where collection of complete and accurate demographic and social data is a well-established practice. In many of the larger semiarid countries roads, railroads, or air networks provide extensive transport linkages at reasonable cost. However, this approach is prohibitive for some of the delta or savannah countries where dense jungle or hills impose serious physical obstacles to land transport or airport construction.

ECONOMIC IMPACT

Some attempts have been made to assess the impact of weeds upon the economies of Florida and Louisiana which may give some baseline for estimating losses in developing countries. Even as far back as the turn of the century, when the embryonic hyacinth infestation plagued Florida with Lake George being blocked by hyacinths for 40 km, an annual loss to the lumber trade alone of $55,000 was estimated. Transport by water is still a vital factor in the economy of the Mississippi delta since corn, rice, cane, and cotton together with fish, citrus fruits, salt, oil, lumber, and other products are carried by water. The water hyacinth invasion has thus had serious consequences, and it has been estimated that the annual loss incurred in Louisiana through the effects of the weed reached $35 million annually in recent years from damage to agriculture, fish and wildlife, navigation, drainage, and public health.

Aside from such roughly estimated direct economic losses, there are additional costs of deliberate weed removal and prophylactic measures for growth prevention. Such costs, which may amount to millions of dollars per year, are incurred just to keep the weeds under some level of control and prevent further economic damage. Since 1959 the Corps of Engineers in Florida has received about $3.6 million annually for hyacinth eradication programs in Florida to which is added a 30% matching contribution by the state. That sum, however, is not all that is being spent. In addition, in 1969, the state of Florida spent $1.3 million for control efforts, and for fiscal year (FY) 1970, $2.5 million was requested by the Governor's Aquatic Weed Research and Development Committee because control attempts had not been very successful.

The economic losses estimated for the U.S. are, at best, rough estimates prepared by federal, state, and local organizations concerned with budgets. While these estimated dollar losses represent the principal "hard" data available for any country, they are difficult to apply directly to developing country situations, with different labor and other costs and differing methods of estimating costs. For example, with regard to water transport, the large motorized barge systems used in the U.S. are less sensitive to cooling-intake clogging and propeller fouling than are the small, individual, and relatively crude river steamers and small barges powered by outboard motors so commonly used for transporting goods in developing countries.

In developing countries considerable sums are spent for weed control. In most cases at least one half of the funds draw on scarce foreign exchange in paying for cutting apparatus and chemicals. Furthermore, needed labor for dam construction, flood control, fishing, and transportation is often idled as it must be used for weed harvesting or destruction as the control programs are carried out. The uranium and copper mining interests in the Congo Republic are known to have spent millions of dollars in this field. Egypt and the Sudan each spend about $1.5 million annually on control programs.

In the Sudan, an initial hyacinth control program involving technical advice, organization building, appropriate equipment and supplies, and control operations was supported by aid from 1960 to 1964. At the conclusion of the project, weed control activities employed 73 river craft of all types, 28 vehicles, 5 planes, 500 full-time employees and 500 seasonal laborers. Control was effected largely by 2,4-dichlorophenoxy (2,4-D) herbicide deployed from boats, land-based spray vehicles, and man-carried sprayers, as well as from aircraft under favorable weather conditions.[1-5]

Country Specific Situations

All available reports agree that aquatic weeds present "serious economic problems" for developing countries and that these problems will increase unless major control measures are undertaken. An attempt has been made to translate the weed-menace assessments into very broad and quasispeculative dollar-equivalent losses. Several developing countries were selected as reasonably representative of those suffering economic losses which seem to have a measurable impact upon their economies. Principal attention is directed to losses due to constraints on transportation and fishing which are the easiest aspects to address in a quantitative manner.

Zambia

The inadequate road and railroad network of this landlocked country forces dependence on short-haul water transportation on the Zambesi, Kaufue, Luangiva, and Chambesi Rivers and Lakes Tanganyika and Kariba for internal commerce. The presence of rapids and varying seasonal depths of water inhibit long-haul river transport although logs and other freight are transported on the Zambesi above the falls at Livingstone. Trade with Rhodesia across Lake Kariba is limited. For at least 8 months of the year, these bodies of water are infested to some extent with aquatic weeds restricting waterborne transportation.

The effect on internal commerce is significant. Considering agriculture, forestry, and fishing only, a 2 to 3% impact due to blockages of waterways and fishing areas means an internal loss of commerce of several million dollars since these sectors account for more than $75 million of the country's total commerce. Weed-clearance schemes, including reliance on imported equipment and materials, add another $750,000 to $1 million in costs annually. In addition, schistosomiasis has shown a 10% increase from 1965 to 1969, which may be related to the presence of aquatic weeds.

Thailand

The lower Mekong River Basin, as well as other riverine regions in Thailand, with their reservoirs, ponds, inlets, and channels, is infested with hyacinth, water lettuce, and water fern.

Very roughly, about 30% of the Thai labor force is engaged in activities related to fishing and river transport. Fresh-water fish serve as an important food source, with the rice fields and irrigation channels serving as good breeding grounds. There are important river and canal networks used for boat transportation around Bangkok and in the delta. There is evidence that weed concentrations are growing steadily and are beginning to have a noticeable influence on river transportation and fishing. In the Nam Pong and other reservoir areas the weeds are moving downstream where they accumulate in reservoir impoundments, and grow at an accelerated rate. While the economic consequences are difficult to quantify, order-of-magnitude estimates of the severity of the problem are presented below.

More than 80% of the rice crop, which earned $21 million in foreign exchange in 1970, moves over the many waterways. Similarly, about 30% of the tin ore, tungsten ore, corn, sugar cane, rubber, and other export products move by water. The approximate value of such water-shipped products is about $300 million/year. Assuming weed clogging reduces the mobility of boats and barges so that about 2% of the potentially marketable products do not arrive at their destination at all or too late, then an annual loss of several million dollars accrues, including a substantial foreign exchange component.

The annual fresh-water fish catch is valued at about $100 million. With the weeds interfering with spawning, a 1% loss could cost the economy $1 million/year. The actual loss is probably greater since the weeds interfere with fishermen's gear and boats, and they probably also harbor fish predators.

With regard to costs of preventive action, it is estimated that initial herbicide weed control in the Pa Mong dam site area would cost about $50/ha and would be required for 10% of the reservoir area and an additional $40/ha/year for 1% of the area to prevent weed encroachment. It is also estimated that a weed control research team costing $200,000 annually would be needed for 10 years to cover the delta region. If the same control costs were to accrue for other-than-delta waters, and assuming that about 1 million ha are threatened, then a probable initial cost would be several million dollars and annual expenditures would reach about $400,000.

Schistosomiasis is of growing concern in the Mekong region. The toll of this disease on the economy is not known. It is not unreasonable to estimate that the weeds could add 1% to the incidence of the disease which probably afflicts several million people. Assuming a 10% loss in productivity of the afflicted, it seems clear that from this point of view the weed menace is not insignificant.

Guyana

This small developing country suffers from both actual and potential aquatic weed infestations, mostly from the hyacinth. Guyana has 3700 mi of navigable waterways — the only transportation routes in most areas. There are several potential hydroelectric sites.

River fishing and forest products, which are heavily dependent on river transport, account for about 12% of the Gross National Product (GNP). The country, with a GNP of about $235 million, cannot afford any interference with the flow of export foods and materials amounting to $121 million, with about 20% moved by river. A 1% weed-caused loss in river-dependent exports seems conservative, given the extent of the weed problem. This means a $250,000/year loss — a modest but nevertheless serious item for a country with a trade deficit. Assuming that weed control would cost about $100,000 annually, a $350,000/year loss accrues — and most of it in foreign exchange.

Mosquito-borne diseases have been a long-time concern of the country. While malaria has been almost eliminated, weed infestations may cause a recurring control problem because the mosquitoes which breed in the weeds are very difficult to eradicate. Some schistosomiasis increase has been reported, and it is likely that the increase is related to weeds.

In the sense that aquatic weeds constitute a crop of potential use, much research has been done during the past 50 years to find an economic means of utilizing them as commercial products. These efforts have been summarized by an advisory committee of the Office of Science and Technology for the Agency for International Development in cooperation with the National Academy of Sciences. While it is possible to discuss these alternatives in terms of potential application, there are no practical applications that can be cited except herbivorous animals, and these have had very limited economic impact on the total problem.[6-8]

Economics of Aquatic Weed Control in the U.S.

An accurate assessment of the economic losses due to aquatic vegetation is not available. Apart from estimated losses due to aquatic weeds, it has been estimated that about $5.5 million is spent annually for treating lakes and ponds in the U.S., and this amount of money was for treatment of only approximately 10% of the weed-infested areas. Moreover, in 17 western states in 1957, 95,000 km of canals were treated for aquatic weeds which represented 54% of the aquatic weed-infested area.[9,10]

Rather elaborate calculations of the annual benefits of the Expanded Project for Aquatic Plant Control[11] in the states of Alabama, Florida, Georgia, Louisiana, Mississippi, North Carolina, South Carolina, and Texas have been compiled:

Flood and control and drainage (urban)	$ 580,000
Water and flood control (agriculture)	3,590,000
Fish and wildlife	3,310,000
Recreation	3,600,000
Mosquito control	170,000
Water supply	100,000
Pollution control	2,610,000
Total annual benefits	$13,960,000

In addition to these estimated benefits, which appeared to be very conservative, attention is called to the fact that if there had been no control program, the area of infestation would have grown from 90,000 ha in 1959 to 150,000 ha in 1963. The cost for treating 50,000 ha amounted to $187,600 or an average cost of $15.15/ha. A drainage engineer has estimated the hand labor costs on clearing a certain ditch at $600/km with the trimming out of regrowth a year later at $200/km. Chemical control in the same ditch costs $450/km the first year and the second spraying a year later costs only $150/km.[12]

AQUATIC WEED CONTROL IN FLORIDA

The original effort to control weeds by the U.S. government under the Corps of Engineers Aquatic Plant Control Program was initiated in 1899, and was accelerated and expanded in 1958 and in 1965 to achieve progressive control and eradication of primary weed infestations of greatest economic importance in the eight South Atlantic and Gulf Coastal States from North Carolina to Texas. Water hyacinth, alligator weed, Eurasian water milfoil infestations, and more recently, hydrilla, were the primary probem aquatic weeds. Currently, mechanical, chemical, and biological methods of control are being used. The benefit-to-cost ratio considerably exceeds the 5 to 1 estimate on the basis on which the program was authorized in 1958 and sustains the 14 to 1 ratio determined for the preceding project, in 1965.[13]

The state of Florida is somewhat unique to have flood control districts with operational costs and expenditures. A $396 million system of canals, pumping stations, water reservoirs, dams, and spillways is now nearly completed for central and southern Florida. This system is designed to control flooding and mitigate drought. Benefits are estimated at $82 million/year; however, they will be almost totally negated if aquatic weeds are left uncontrolled.

Drainage programs were launched in Florida in the early 1900s but they were largely underfinanced. Floods recurred, droughts were worsened by overdrainage, and hurricane-driven wind tides in 1926 and 1928 claimed a total of about 2500 lives near Lake Okeechobee. The feast-and-famine water cycles plagued Florida in the 1930s and 1940s, and in 1947 a gigantic flood inundated more than 5 million acres with damage running to $60 million.

It was clear that a large-scale water management program was needed, if Florida was to grow and prosper. Accordingly, in 1948 the Congress authorized such a program. The objectives were flood control, water conservation, water supply, preservation of desirable levels of fresh water in the ground, preservation and enhancement of fish and wildlife, navigation, and public recreation.[14] In 1949, the State Legislature created the Central and Southern Florida Flood Control District (FCD) to represent all local interests in the building and subsequent operation of the system. To take advantage of these primary facilities, there must also be secondary and tertiary works. Individuals, cities, farmers, drainge districts, and counties can use the primary canals to remove flood waters in rainy seasons and obtain needed water supplies in dry times. However, to do so they must construct their own ditches, dikes, and pumping stations.[15]

Weed Control Costs

The FCD in Florida uses both chemical and mechanical means of aquatic weed control. Floating aquatic vegetation, such as water hyacinth, are killed by chemical sprays usually applied from small boats. Submersed weeds are uprooted with steel A-frames dragged along canal bottoms by towboats, amphibious "ducks", and other equipment. The uprooted submersed weeds float to barriers and are removed by draglines. However, the job has to be done over and over again and grows more costly every year.

In the late 1950s, the annual cost of weed control climbed above $100,000 and in FY 1963 it exceeded $160,000. In 1964 it was $189,000; in 1965 and 1966 it was $230,000 each year. A breakdown of expenses indicates that $230,000 was spent on chemical weed control and $107,000 on mechanical weed control.[16]

Within the boundaries of the 18-county FCD, weed control programs are being conducted by other organizations. In the Everglades farming areas south of Lake Okeechobee, local drainage districts collect assessments that usually run between $5 to $5.25/acre/year. This money is used for local pumping operations, administration, and canal maintenance. Part of the local administration costs and most of the canal-maintenance expenses go for aquatic weed control. It is conservatively estimated that drainage districts are spending at least $2 an acre for these programs. There are now about 360,000 acres in agricultural production between Lake Okeechobee and the conservation areas, and weed control in secondary canals and ditches exceeds $700,000/year.

Local counties, such as Dade, Broward, Palm Beach, and Lee, are heavily committed to aquatic weed control. The U.S. Fish and Wildlife Service, the Lake Worth Drainage District in southeastern Palm Beach County, and many other organizations are also spending considerable sums to check aquatic growth. Estimated costs for the FY 1978 are given in Tables 1 and 2.

Weed Control Benefits

With the warm sunshine and year-round growing season, aquatic weeds would soon clog all primary and secondary canals, reservoirs, and most waterways if left uncontrolled. The FCD would thus be unable to remove flood waters, which would inundate vast acreages. Large pumping stations operated by the FCD would be unable to pump very long before weeds were sucked into intake gratings so tightly that a head of water would be created, causing enough of a drawdown of the water level to stall the pumps.

In testimony prepared for Congress, it is reported that average annual benefits from the FCD total $82,169,000/year. This is broken down as follows: (1) flood damages prevented — $30,467,300, (2) increased land use — $49,498,100, (3) recreation — $1,794,100, (4) fish and wildlife — $359,100, and (5) navigation — $51,000. It is also reported that the water control project has a benefit-to-cost ratio of 4.9 to 1, which means that $4.90 in benefits are being realized (in the above categories) in return for every $1.00 of initial cost. This is probably one of the most generous benefit-to-cost ratios for any large public works project in America.[17]

Other benefits of the existing project include the municipal water supply and prevention of salt-water intrusion; however, these have never been evaluated in dollars. Municipal supplies and farm irrigation supplies would be seriously affected if the waterways were filled with aquatic weeds. Pumping stations in cities such as West Palm Beach would soon experience difficulty in drawing water supplies from the District's canal if that canal were choked with vegetation. In fact, the city's canal (which brings fresh water directly into the 12,000-acre city water catchment area) could become blocked to the flow of water.

Navigation could be brought to a halt almost everywhere, and waterways used by fresh-water fishermen would be blocked. Irrigation supplies for farmers could be drastically cut down, if not cut off. Lack of aquatic weed control would spell a nightmare of economic chaos for central and southern Florida and it is doubtful if an accurate estimate of all damages could be made. However, taking into consideration the benefits of water control, those that have been evaluated in dollars plus those that are

Table 1
WATERSHED WORK AREAS AND ESTIMATED EXPENDITURE FY 1978

Area	Total cost ($)	Hydrilla ($)	Water hyacinth ($)
Board of County Commissioners of Citrus County			
Lake Tsala Apopka	81,000	61,000	20,000
Crystal River Canals	130,000	130,000	
Crystal River	269,000	259,000	10,000
Total	450,000	450,000	30,000
Game and Fresh Water Fish Commission			
St. Johns Basin	10,000		10,000
Oklawaha River	2,000		2,000
Oklawaha Lakes	110,000	100,000	10,000
Suwannee	75,000	15,000	60,000
Alafia-Manatee	79,000	44,000	35,000
Lake Istokpoga	53,000	8,000	45,000
Lake Trafford	106,000	4,000	102,000
Peace River	54,000	20,000	34,000
Myakka River	54,000	25,000	29,000
Nassau	7,300		7,300
Aucilla-Wacissa	15,000	10,000	5,000
West Coast Basin	65,000	40,000	25,000
East Coast Basin	1,000		1,000
Total	631,000	266,000	365,000
Board of County Commissioners of Highlands County			
Lake Istokpoga (70—30%)	77,000	72,000	5,000
Hillsborough County Mosquito Control Department			
Hillsborough River (70—30%)	52,000	22,000	30,000
Lake County Water Authority[a]			
Oklawaha River (70—30%)	8,000		8,000
Oklawaha Basin Lakes (70—30%)	85,000	75,000	10,000
Total	93,000	75,000	18,000
Lee County Hyacinth Control District			
Caloosahatchee River (100%)	75,000	50,000	25,000
Board of County Commissioners of Orange County			
St. Johns Basin Lakes (70—30%)	354,000	349,000	5,000

Table 1 (continued)
WATERSHED WORK AREAS AND ESTIMATED EXPENDITURE FY 1978

Area	Total cost ($)	Hydrilla ($)	Water hyacinth ($)
Board of County Commissioners of Polk County			
Lake Pierce (70—30%)	165,000	150,000	15,000
Polk County Lakes (70—30%)	193,000	119,000	74,000
Totals	358,000	269,000	89,000
South Florida Water Management District			
Kissimmee River Tributaries (70—30%)	350,000	230,000	120,000
Okeechobee Basin Minor Tributaries (70—30%)	189,000	176,000	13,000
Kissimmee River Main Channel (100%)	600,000	450,000	150,000
Lake Okeechobee (100%)	266,000	266,000	
Total	1,405,000	1,122,000	283,000
Southwestern Florida Water Management District			
Gulf Coast (70—30%)			
Anclote River	4,000	4,000	
Lake Tarpon	4,000	1,000	3,000
Medard Reservoir	6,000	2,000	4,000
Homosassa River	41,000	38,000	3,000
Chassahowitzka River	4,000	3,000	1,000
Weekiwachee River	20,000	18,000	2,000
Total	79,000	66,000	13,000
Withlacoochee (100%)			
Lake Panasoffkee	20,000	12,000	8,000
Withlacooche River	486,000	344,000	142,000
Total	506,000	356,000	150,000

[a] Includes the Oklawaha Basin Reclamation and Water Control and Conservancy Authority.

From Bureau of Aquatic Plant Research and Control, U.S. Department of Natural Resources, Tallahassee, Fla., FY 1978.

known to be large and significant but are unevaluated, values far in excess of $100 million/year are at stake. With weed control as a part of water management, these are benefits; without control, the benefits would become damages.[18,19,23]

Table 2
WATER WORK AREAS AND ESTIMATED EXPENDITURE

Agency	Local funds ($)	State funds ($)
Lee County Hyacinth Control District	184,030	55,209
Melbourne-Tillman Drainage District	65,747	19,725
Sumter County	9,615	2,885
Indian Trail Water Control District	78,777	23,633
Collier County	165,000	49,500
Greenacres City	2,031	609
Coral Springs Improvement Districts	49,682	14,905
Sunshine Drainage District	77,185	23,156
North Lauderdale Water Control District	25,592	7,677
Seminole Water Control District	20,580	6,174
Broward County	66,667	20,000
Brevard County	101,351	30,405
Charlotte County	50,000	15,000
Indian River Farms Water Management District	106,001	31,800
Manatee County Mosquito Control	38,555	11,567
Highlands County	41,690	12,507
Bailey Drainage District	7,692	2,308
Hollywood Reclamation District	91,442	27,433
West Lauderdale Water Control District	26,154	7,846
Margate	103,260	30,978
Orlando	45,708	13,713
Palm Beach County	86,445	25,933
Central Broward Drainage District	115,806	34,742
Altamonte Springs	40,342	12,103
West Palm Beach	41,546	12,464
Spring Lake Improvement District	20,070	6,021
North Springs Improvement District	13,755	4,126
Central County Drainage District	21,222	6,367
Fort Pierce	21,896	6,569
Tamarac	36,781	11,034
Gerber Groves	33,077	9,923
Sebastian River Drainage District	25,572	7,672
Dade County	1,067,678	320,303
Coconut Creek	5,077	1,523
Palm Beach Gardens	47,475	14,242
Martin County	20,506	6,152
Fort Pierce Farms Water Control District	17,803	5,340
Loxahatchee Sub-Drainage District	29,522	8,857
Juno Beach	1,492	448
Dunedin	10,517	3,155
Pinellas County	78,151	23,445
Lauderdale Lakes	58,266	17,480
Hobe-St. Lucie Conservancy District	47,885	14,365
Fellsmere Water Control District	65,070	18,622
St. Johns River Water Management District	54,615	16,385
Volusia County	36,496	10,949
Lake Worth Drainage District	398,178	119,453
Acme Improvement District	33,043	9,913
Barron Water Management District	34,000	10,200
City of Atlantis	2,942	883
St. Petersburg	322,684	96,805
Miramar	4,231	1,269
Oakland Park	13,192	3,958
Indiantown Drainage District	6,769	2,031
Troup-Indiantown Water Management District	10,769	3,231

Table 2 (continued)
WATER WORK AREAS AND ESTIMATED EXPENDITURE

Agency	Local funds ($)	State funds ($)
Polk County	27,636	8,291
Cape Coral	32,000	9,600
East County Water Control District	123,077	36,923
North Port Water Control District	49,231	14,769
Sarasota County	192,729	57,818
Orange County	116,667	35,000
Hastings Drainage District	23,346	7,004
Bay County	—	200,000
Lake Clark Shores	4,073	1,222
Manatee County Board of Commissioners	136,358	40,907
South Florida Water Management District	—	184,661
Southwest Florida Water Management District	—	78,893

From Bureau of Aquatic Plant Research and Control, U.S. Department of Natural Resources, Tallahassee, Fla., FY 1978.

Conservation Districts

Florida's small lakes and ponds are owned and far overshadowed in importance and use by the larger bodies of water. Numbering into the tens of thousands, they are of considerable economic and recreational importance to Florida. Economically, they provide farmers, ranchers, and growers with water for irrigation, livestock watering, and commercial fishing. Their recreational importance lies in providing fishing, swimming, and other water-based outdoor recreation.

Florida's 60 soil and water conservation districts are local subdivisions of the U.S. government whose boundaries ordinarily correspond to county lines. Organized by referendum and chartered by the state, they are governed by five locally elected supervisors.[20] The U.S. Department of Agriculture's Soil Conservation Service (SCS) provides organized districts with technical assistance in soils, engineering, agronomy, and woodland and wildlife conservation.

The SCS began to show peninsular Florida how to grow fish crops in natural pond waters 32 years ago. After some experience, it was possible to take ponds that had not produced good fishing in 48 years, and develop such fishing in 18 months. By chemically removing existing fish populations and counting and weighing all fish removed, it was found that natural ponds could produce about 1000 kg/ha. The lowest weight was taken from a pond in poor sandy soil; the highest weight came from a very fertile farm pond. It was concluded that the total number of fish a pond would support was dependent upon the water's fertility and watershed.

As a result of this work, farmers, ranchers, and growers in Florida's soil and water conservation districts stocked and have under management more than 8000 farm ponds; several of those in the 1-acre size class have provided good fishing for 20 years.

Most of these farm ponds had or have aquatic weed problems, a situation which generally leads to poor fish ponds. The weeds form such a tangled mass of vegetation that it is difficult or impossible to fish or even paddle a boat. Additionally, small fish (particularly bluegill) are protected from the larger fish. The pond then becomes overcrowded with small bluegill, which are stunted and never grow to any size. Partial or total elimination of the weeds can significantly change this picture.

Floating plants, such as water hyacinth (*Eichhornia crassipes*), water lettuce (*Pistia stratiotes*), salvinia (*Salvinia rotunfifolia*), and the duckweeds (*Lemna minor*) exclude sunlight from the water, thus causing a chain reaction that eliminates oxygen from the water. Desirable fish, such as the large-mouth black bass and bluegill bream, need adequate oxygen while garfish and mudfish require very little. Timmer and Weldon[21] have described this situation well and cited the evapotranspiration rate of water hyacinth as being 3.7 times greater than the evaporation rate of water. Pond weeds also clog up and damage irrigation equipment and machinery.

Since the early 1950s, technical assistance in wildlife has been more involved with waterfowl management on privately owned lands. Although there are hundreds of thousands of wetland acres in Florida, many are useless because of aquatic weeds — few of which are of any value to waterfowl. This does not seem to be commonly understood, but it is a fact. Therefore, the SCS is involved more in the control of noxious aquatics than in trying to grow more of them.

In north Florida, wild ducks and geese have abandoned lakes and marshes containing aquatic weeds and are using man-made duck fields; the planted choice food is corn and millets flooded by water control measures to a depth suitable for feeding.

Seeking more efficient use of land devoted to the production of food and fiber, Florida's farmers, ranchers, and growers have sought to equalize the vagaries of the annual rainfall pattern. SCS engineers have assisted them in the design of V-type field ditches intended to control water flow. Subsurface irrigation is practical and effective in porous sandy soils where a high water table exists. Annually, more than 1000 mains and laterals have been constructed by district cooperators with the technical guidance of SCS engineers; aquatic weeds are a constant problem in these channels.[22]

The Watershed Protection and Flood Prevention Act (Public Law 566, 83rd Congress) was signed by President Eisenhower on August 4, 1954. To date, 16 small watershed projects have been authorized in Florida. They cover more than 500,000 ha with an estimated total cost of more than $29 million.

It does not seem to be commonly understood that small watershed projects are locally initiated and sponsored by Soil and Water Conservation District Boards of Supervisors, Boards of County Commissioners, or special purpose districts. A watershed planning party of the SCS assists the local sponsors in the preparation of a work plan, each of which contains the stipulation that the sponsors of each small watershed project will maintain the improvements and control the water weeds.[12-20]

AQUATIC WEED CONTROL IN LOUISIANA

Control Cost

The cost of operations for control of 1 acre of aquatic vegetation in Louisiana has been arrived at by using two separate approaches. The first involved a summation of all estimated detailed costs necessary to support a crew performing control operations. The accuracy of these detailed costs range from "a good estimate", such as known costs of labor, equipment or supervision and administration, to "a guess estimate", limited by uncontrollable factors such as climatic conditions or accessibility. This control operations cost is for the chemical method which is the most widely used and the most economical. The second approach was an accumulation from 13 years of actual previous records of money expended in the Expanded Project and the number of controlled acres resulting. The detailed cost-analysis estimate concluded in a per acre cost of $25.18 (1977 cost) while the actual record of accumulated cost per acre was $28.12. The detailed cost estimate and the actual 13-year record are given as Tables 3 and 4. The treatment per acre cost difference between estimate and actual is easily explained by: (1) a decrease in controlled area during inclement weather, (2) performance of noncontrol types of activities, and (3) less efficient methods of control such as mechanical destruction.

Table 3
ACTUAL OPERATION COST PER CONTROLLED ACRE[a]

FY	FY allotment for control operations[a] ($1000)	Cumulative FY allotment for control operations[a] ($1000)	Acres controlled per year	Cumulative acres controlled	$ per acre for control operations	Cumulative $ per acre for control operations
1965	283.5	283.5	5,848	5,848	48.48	48.48
1966	253.3	536.8	4,374	10,222	57.91	52.51
1967	276.6	813.4	7,497	17,719	36.89	45.91
1968	320.7	1,134.1	9,472	27,192	33.86	41.71
1969	219.1	1,353.2	4,309	31,500	50.85	42.96
1970	284.4	1,601.6	8,399	39,899	29.57	40.14
1971	280.1	1,881.7	8,400	48,299	33.34	38.96
1972	283.3	2,165.0	12,532	60,831	22.61	35.59
1973	245.7	2,410.7	10,540	71,371	23.31	33.78
1974	248.8	2,669.5	10,873	82,244	23.80	32.46
1975	241.9	2,911.4	11,873	94,117	20.37	30.93
1976	180.9	3,092.3	11,371	105,488	15.91	29.31
1977	222.3	3,381.3	10,357	120,244	21.46	28.12

[a] Does not include research, engineering, or design.

From General Design Memorandum, Corps of Engineers, New Orleans District, Louisiana, FY 1978.

Table 4
1977 ESTIMATED COST PER CONTROLLED ACRE[a]

Item	Daily cost ($)
Salaries, field people	
Aquatic growth controlled — as required ($6.17/hr) (8 hr) (1.2 overhead factor)	59.23
Launch operator — 1 required ($5.52/hr) (8hr) (1.2 overhead factor)	52.99
Aquatic growth foreman — 1/6 required ($8.12/hr) (8 hr) (1.2 overhead factor)/(6 crews)	12.99
Field-office coordinator — 1/11 required ($9.00/hr) (8 hr) (1.2 overhead factor)/(11 crews)	7.85
Chemicals sprayed ($5.10/gal) (1 gal/acre) (11.36[a] acre/day)	57.99
Equipment ($84,000/year rental)/(260 paid days/year) (11 crews)	29.34
Supervision and administration — 1/11 required ($113,400/year office supervision and administration)/ (260 pdaid days/year) (11 crews)	29.65
Subtotal	260.04
Maintenance, repairs, project equipment, shops, etc. (10% of subtotal)	26.00
Daily cost of one crew	286.04
Estimated cost/controlled acre ($286.04/11.36[a] acres)	25.18

[a] $$\text{Acre/day} = \frac{\text{program acres controlled in 1977}}{\text{(work days in 1977)} \times \text{(crews)}} = \frac{27253}{218(11)} = 11.36 \text{ acre/day}$$

From General Design Memorandum, Corps of Engineers, New Orleans District, Louisiana, FY 1978.

Benefits

The control of aquatic plant life has been an important factor in the expansion of economic development in Louisiana, particularly along the Gulf of Mexico coastline and in wetland areas. The control of water hyacinths, alligator weed, and other problematic infestations has improved navigation, agriculture, commercial and sport hunting and fishing, drainage, and irrigation; the net effect has been increased productivity, reduced costs, and newly stimulated economic activity. The presence of aquatic vegetation in the waters of Louisiana results in losses and interferences of various degrees, not only with respect to the twofold responsibility of the Corps of Engineers — navigation and flood control — but also with respect to agriculture, fish and wildlife, and the recreational uses of these water areas by tourists and the native populace. Although obviously adding up to a large total and affecting many aspects of Louisiana's life and economy, these losses are not readily susceptible to precise determination; therefore, the estimates presented herein should be considered as approximations of the nature and extent of the losses.

Agricultural losses (i.e., benefits to agriculture) that would be experienced by the half-billion-dollar agricultural industry in Louisiana without adequate aquatic plant control would primarily be losses due to drainage interference. The deterioration of local drainage systems prior to project control in 1948 caused thousands of acres of rich soil, formerly devoted to agricultural production, to degenerate into unimproved pasture or to go out of production entirely and revert to swamp. Related to the agriculture industry are the numerous irrigation systems which supply water to 700,000 acres of cropland in the state. These are seriously impaired when infested by aquatic vegetation. Additionally, agriculture is adversely affected by water shortages which are caused by high transpiration losses due to the presence of aquatic vegetation. Estimated annual crop damages were computed by comparison of previously furnished data from the U.S. Department of Agriculture relating to agricultural conditions and losses on average annual yields, crop distributions, agricultural prices, and agricultural damage produced by aquatic plants prior to project control.

Since the inception of the water hyacinth control project in the late 1940s, soybean production has become the number one agricultural endeavor in Louisiana, the basis for crop evaluation has changed from the use of seasonal averages to Water Resources Council current normalized agricultural prices, and yields have increased through advanced research and technology. It is estimated that the annual agricultural losses that would be incurred with the discontinuance of the aquatic plant control project would approximate 2% of the value of all farm products sold, or some $8.1 million.

Estimated Agricultural Damages Prevented (1975)

Parish	Total value of specified crops harvested[a]	Annual crop damage
Acadia	$ 31,050,000	$ 170,000
Ascension	6,290,000	110,000
Assumption	18,230,000	370,000
Calcasieu	21,220,000	2,100,000
Iberia	16,000,000	920,000
Jefferson Davis	33,400,000	100,000
Lafourche	6,340,000	430,000
Plaquemines	50,000	20,000
St. Landry	12,740,000	630,000
St. Martin	14,030,000	660,000
St. Mary	20,030,000	470,000

Estimated Agricultural Damages Prevented (1975) (continued)

Parish	Total value of specified crops harvested[a]	Annual crop damage
Tangipahoa	280,000	70,000
Terrebonne	8,520,000	1,130,000
Vermilion	32,000,000	920,000
Total	$220,180,000	$8,100,000

[a] Includes cotton, rice, corn, soybeans, and sugarcane.

Navigation losses, which occur under the influence of wind, tide, and current, masses of floating vegetation that move about through the swamps, marshes, and open waterways which abound throughout Louisiana, are extensive. Dense mats of water hyacinths under a foot in height cause only moderate difficulties for large vessels and tug and barge tows. However, even a small patch of water hyacinths or alligator weed increases navigation hazards for all traffic by masking snags or floating logs; these small patches also seriously interfere with the progress of all small and moderately sized commercial and pleasure craft. Large plants in large mats seriously handicap all classes of vessels and can completely block all navigation. The plants accumulate in hugh rolls at the bow of the vessel or the leading barge, forward movement becomes retarded or completely stopped, rudders become jammed, and control of the craft is lost. Under the impact of wind and tide, hyacinth gorges may form against some obstruction in the waterway and completely block all traffic. Small boats have been overturned by the pressure of the massed vegetation against their sides. Some water hyacinth mats become sufficiently dense to permit men to walk atop the massed vegetation while cutting the pressed hyacinths into blocks so that they can be removed.

Without an effective plant control program, waterborne traffic would largely disappear in Louisiana as the low costs which presently allow successful competition with rail and other transportation no longer would be obtainable. Under the condition of uncontrolled infestation, smaller craft would be afffected first and then the larger craft, although perhaps better able to fight their way through in some fashion, would also give up the uneconomical struggle. Soon, no traffic would move with any degree of dependability in any Louisiana waters except on the Mississippi, Atchafalaya, Ouachita, Black, Pearl, Calcasieu, and Sabine Rivers.

Information data of waterborne traffic in selected streams located in the southern portion of the belt of maximum infestation in Louisiana are given below. These data provided the basis for determining the losses to navigation if all aquatic plant control operations were suspended. The streams listed are the main-traveled waterways which would be most severely affected for which accurate commercial statistics are available.

Waterway Traffic

Selected Waterways within Belt of Greatest Aquatic Growth Infestations

Waterway	Ton-Mi
Amite River and Bayou Manchac	50,400
Atchafalaya River	531,918,623
Barataria Bay Waterway	60,225,041
Bayou Bonfuca	488,628
Bayous Grand Caillou, Dulac, and Lacarpe	5,606,484
Bayou Grosse Tete	17,724
Bayou Lacombe	155,015
Bayou Lafourche	23,006,175

Waterway Traffic (continued)

Selected Waterways within Belt of Greatest Aquatic Growth Infestations

Waterway	Ton-Mi
Bayou Segnette	150,150
Bayou Teche	26,587,653
Bayou Terrebonne	14,511,594
Bayou Vermilion	19,792,507
Big and Little Pigeon Bayous	420,490
GIWW-Mississippi River to Sabne River	8,677,533,503
GIWW-Morgan City to Port Allen	970,581,229
Mermentau River, Bayous Nespique, and Des Cannes	19,551,773
Pass Manchac	4,224,983
Tchefuncte and Bogue Falaya Rivers	452,258
Tickfaw, Natalbany, Pontchatoula, and Blood Rivers	145,200
Total	10,355,419,430

On the basis of approximately 10.4 billion ton-mi of traffic for the waterways listed above and the lowest known operating cost per ton-mile of $0.003, the operating loss for navigation would approximate $31.1 million. Additionally, traffic carried on largely by small craft and individual owners on these secondary streams as well as other streams is an estimated 200 million ton-mi/year with a normal operation cost of $0.015/ton-mi. The actual loss for this unreported traffic due to aquatic growth would be an estimated 50% of the operating cost, or $1.5 million/year.

However, the figures for the common carrier and commercial towing cannot fairly be taken as an indication of the benefits of the present program. If control operations were abandoned, all business would cease long before any such increase in costs had actually taken place. It is considered that a maximum increase of 25% in operating costs would eliminate profitable operations and bring about a suspension of traffic. Therefore, the losses incurred without an aquatic growth control program would be 25% of $31.1 million (the computed cost of handling the traffic shown above), or approximately $7.8 million per annum. This, with the $1.5 million loss to small boat operators on secondary waterways, gives a conservatively estimated annual loss of $9.3 million prevented by the project.

Dense mats of water hyacinth and, to a lesser extent, alligator weed constitute a pest and a hindrance and alter the environment. Areas covered by extensive rafts of water hyacinths undergo a change in aquatic fauna composition; often the rafts render areas incapable of supporting large populations of sport fish and waterfowl. These plants interfere with surface water mixing (wave formation) and phytoplankton primary productivity due to shading.

The extent of these adverse effects by water hyacinths and alligator weed depends to a large degree on such factors as location, water current, water depth, availability of nutrients, seasonal temperature extremes (especially low temperatures), and the extent to which the water surface is covered by these plants.

Water hyacinth roots provide an excellent substrate for many species of macroinvertebrates and cover for larval and juvenile fish. The roots filter sediment from the water, reducing the sediment level downstream and retaining nutrients for recycling. Spawning areas in shallow water can be ruined bywater hyacinth. Roots of this plant extend 12 to 18 in. below the water surface, and large mats drifting about can easily make a spawning area unusable. Dense growth of water hyacinth on water bodies will exclude light from the water and eliminate submersed plant growth. This not only eliminates a source of food for ducks and coots, but makes the water unsuitable for usage by the birds.

Losses to fishing- and trapping-related activities are primarily based on the interference with the exploitation of these resources produced by their inaccessibility. Under the worst circumstances, waterways are completely blocked and, therefore, these activities are stopped; under the best circumstances, the plants produce only a minor hindrance to navigation.

If total coverage by aquatic plants of all secondary and other fresh-water streams occurred within Louisiana, an estimated 30% of the commercial fisheries catch would be lost annually. In 1973, commercial fishermen landed over 22 million lb of fresh-water fish and shellfish in Louisiana valued at approximately $4.5 million. Thus, the annual benefit to the commercial fisheries industry is an estimated $1.35 million.

The coastal swamps and marshes are responsible for the state's leadership in the raw fur industry. During the 1969 to 1970 season, the state's meat and pelt production, led by nutria, muskrat, and raccoon, was valued at $6.9 million. However, without the aquatic plant control program, channels and trails would be blocked and this industry would be threatened. In areas such as the marshes of Terrebonne, Lafourche, and Plaquemine Parishes, many trapping areas are only accessible by pirogues. It is estimated that continuation of the aquatic plant control program would keep open 26,000 acres which are used for waterfowl and game hunting. Per acre benefits are estimated to be $1.15 for small game hunting and $2.56 for large game and waterfowl hunting. Therefore, the total benefit accrued to hunting activities from continued aquatic plant control is $96,460.

The annual beneficial impact to vocational and avocational activities related to Louisiana's fish and wildlife resources from continuation of the aquatic plant control project approximates $1.45 million as is shown in the tabulation below:

Activity	Annual benefit
Commercial fishing	$1,350,000
Hunting	100,000
Total	$1,450,000

Flood control on large rivers and main waterways is not affected by aquatic vegetation, but other flood control areas can be hampered by altering flow capacities and velocities in floodways, by forming aquatic vegetation dams with subsequent local flooding, or by creating violent erosion, eddies, and dangerous soil deposits. The effectiveness of flood control efforts would be decreased if any aquatic vegetation program in Louisiana did not exist. General drainage throughout the low flat lands of Louisiana is especially susceptible to the abundant growth of aquatic vegetation. Natural drainage can be neutralized by repeated buildups or dams of aquatic vegetation, thereby causing local flooding. Additionally, drainage projects in Louisiana, initially costing millions of dollars, would require continuous maintenance to remain effective.

Water Recreation

Water recreation includes sport fishing and hunting, all water sports, wildlife appreciation activities, and camping. Some $2.2 million was expended during 1973 to 1974 for sport fishing and hunting licenses and camping fees at the ten available camping areas in Louisiana. Far outshadowing this figure are the indirect economic benefits derived from monies spent for related equipment, supplies, and living accommodations, and the development of these businesses. However, losses which would result from the discontinuance of the aquatic plant control program are not readily measurable. The magnitude of the benefits to be derived would depend on locality, accessibility, and annual temperature cycles. As water hyacinths are killed by temperatures below freezing, mild winters allow for more severe infestations.

A B

PLATE 1. (A) Water hyacinth growing in the St. Johns River, Florida, August 1973, before application of herbicide; (B) clear water, 3 months after application of herbicide.

A B

PLATE 2. (A) Eurasian watermilfoil growing in Currituck Sound, North Carolina, September 1967, before application of herbicide; (B) clear water, 1 year after application of herbicide.

A B

PLATE 3. (A) Egeria, July 1973, growing in the Chickahominy Reservoir, Virginia; (B) clear water, 1 year after application of herbicide.

PLATE 4. (A) Alligator weed growing in the Ortega River, Florida, September 1966, before introduction of the alligator weed flea beetle; (B) clear water, 1 year after the introduction of the flea beetle.

A

B

PLATE 5. (A) Alligator weed growing in the Cross Bayou Canal, Louisiana, April 1972, before introduction of the alligator weed flea beetle; (B) clear water, 1 year after the introduction of the flea beetle.

A

B

PLATE 6. (A) Alligator weed growing in Steinhagen Lake, Texas, April 1968, before introduction of the alligator weed flea beetle; (B) clear water 2 years after the introduction of the flea beetle.

In the past, aquatic plant control was of concern to public health authorities primarily from the standpoint of mosquito infestation and water pollution control. However, research over the last several decades has changed the conception of this problem to the point that the adverse impacts in this area which were previously accounted for are now either eliminated or considered to be offset by the beneficial effects which aquatic plants can offer.

When light is excluded by a blanket of floating plants, phytoplankton growth is reduced, and a basic link in the aquatic food chain is affected. Dissolved oxygen in the water is reduced because of reduced growth of submersed plants and because the blanket cover prevents the free exchange of oxygen at the water surface. Since aquatic plants are able to absorb more nutrients than they can utilize, they have been recommended for usage on ponds draining livestock feedlots. It should be noted that this is an exceptional condition found in Louisiana; in most waters nutrients range from deficient to normal.

During previous decades when malaria and encephalitis epidemics threatened and mosquito control was less sophisticated than at present, the benefits obtained from the reduction of mosquito breeding areas were more viable. Research has now shown that some mosquito larvae do thrive, or often even survive, in dense aquatic plant mats. The *Mansonia* larvae obtain their oxygen directly from the roots of aquatic plants and never surface. The only way to control these mosquitoes is to remove these aquatics. The species *M. titillans* and *M. indubitans* show a marked preference for the water hyacinth and water lettuce. Therefore, control of areas heavily infested with aquatic plants will substantially reduce the mosquito population.

Thus, any adverse impacts on public health created by aquatic plants are offset by the beneficial effects they produce. Therefore, no benefits are attributed to this project for elevating the public health climate.

Other benefits which cannot be monetarily defined, but which would result from aquatic plant control, are an increase in waterfront land values because of elimination of unsightly and inaccessible adjoining areas, a decrease in the cost of water supplies because of less required purification treatment and more readily available water areas, and the reduction of effort, time, and cost required to regain control of a reinfested aquatic area.

The following tabulation presents the total quantifiable benefits to be derived from continued aquatic plant control within the state of Louisiana.

Benefit category	Annual benefit
Agriculture	$ 8,100,000
Navigation	9,300,000
Commercial fishing and trapping	1,450,000
Total	$18,850,000

Comparison of Benefits and Costs

The annual federal appropriation in FY 1974 for the Aquatic Plant Control Program, based on operations costs for minimum control, was $370,000; quantifiable benefits due to all aquatic vegetation control operations were $18.85 million. The Aquatic Plant Control Program was supplemented by $600,000 of Operations and Maintenance (O & M) funds in 1974, for a total of $970,000 in obligated expenditures. Since these O & M funds are only authorized to control water hyacinths to benefit navigation, the Aquatic Plant Control Program is responsible for some 40% of the nagivation benefits ($4.45 million) plus all other benefits ($9.55 million) for a total of some $14 million in average annual benefits. Therefore, the benefit-to-cost ratio for the Aquatic Plant Control Program is 37.8 to 1.

AQUATIC WEED CONTROL IN TEXAS

Texas has had water hyacinths in Lake Corpus Christi, a lake of 10,000 ha near the southern Gulf Coast, since 1935. Here plants were introduced by well-meaning but uninformed people who thought these plants were an attractant for waterfowl. Aquatic plant control work on Lake Corpus Christi began in 1952. This first work to eradicate water hyacinths was erratic, piecemeal, and restricted to the warm months of the year. Consequently these efforts served only to reduce the plants prolific reproduction to some extent, without making headway into large-scale plant removal. In 1961 spraying with aquatic herbicides was prohibited by federal regulations and work on this important aspect unfortunately had to be halted. From 1961 to 1969 this area of infestation grew from the remaining 300 ha to approximately 5000 ha of water hyacinths. These green masses prohibited boat traffic, furnished breeding grounds for mosquitoes, reduced lakeside land values, increased potable water losses by transevaporation, and brought staggering economic losses to camp operators.

Since March 1970, when the Parks and Wildlife Department of the state of Texas entered into a contract with the U.S. Corps of Engineers, over 7000 ha of water hyacinths have been removed from the Nueces Watershed, principally from Lake Corpus Christi. This plant removal was performed with a government approved formulation of 2,4-D for use in potable waters. Part of the job was accomplished by aerial application, shoreline truck units, and shallow watercraft, but the major portion of the work was performed by using specially designed deep-water boat spraying units. It became mandatory that some plant eradication work be done in this area because boat traffic was not possible over many parts of the lake, and floating water hyacinth mats prevented any type of trotline fishing. Many good black bass areas were closed in by plants, lakeside home owners could not launch their boats, and the plants transevaporation processes were reducing the available drinking water supply to a precarious level.

The preliminary phase of the eradication work encompassed 5000 ha of plants. These plants reduced fish populations by competing for available water space, which resulted in the overabundance of small, undesirable fish. Consequently this led to diminished use of public fishing and recreational waters. Recreational values based on fisherman expenditures and usage for this amounts to $1.24 million/year. Recreational values on the basis of $0.50/day for 80,000 man-days per year amount to an additional $40,000 for the year. Hunting revenue for this area is low because there is not much waterfowl activity present, but this phase does add $6250/year. Also the usage of the lake and surrounding area by sightseers, bird watchers, and hikers adds another $10,000 to the benefits. We seldom include the benefits to real estate and land values in economics cost or benefits analysis; however, plant removal is not an easy matter and money must be spent to realize that goal. In the Nueces Watershed Project, the property owners of 200 mi of shoreline benefited by at least $50/ha. This can be computed as a benefit of $3.2 million/year.

The conservation of water is rarely considered. The water saved from transevaporation on Lake Corpus Christi is enough to furnish 20,000 gal of water per month to 108,900 households for 1 year. Assuming savings of $1,361,125/year or a grand total of $4,632,375/year benefits from the control program would be realized.

Environmental Quality

Effective control of water hyacinth and alligator weed reduces impediments to navigation, flood control and drainage, irrigation, and recreation. A balanced ecosystem with a diversity of fresh-water species occurs in waters freed from heavy growths of these aquatic plants. Removal of favorable mosquito habitat would aid vector control

Table 5
ESTIMATED DAILY COST PER CREW FOR AQUATIC PLANT CONTROL PROGRAMS IN THE STATE OF TEXAS[a]

Item	Daily cost ($)
Salaries	
2 fish and wildlife technicians and project leader ⅓ time	55.50
Per diem	19.00
Equipment rental	
Boat, motor, trailer, sprayer, etc.	10.00
Mileage ($0.08/mi av 100 mi/day)	8.00
Chemical cost ($3.25/gal at 0.75 gal/acre)	24.00
Total	117.00

[a] Based on treating 10 acre/day.

From General Design Memorandum, Corps of Engineers, New Orleans District, Louisiana, FY 1978.

operations. Water supply systems would be aided by freeing intake structures and supply canals from infestations. It is evident that many factors in nature may affect the fish and wildlife reaction to the chemicals, such as water temperature, rainfall, size, and maturity of the affected organism, and others. In addition, plants that have been destroyed by herbicides remain in the water and, in decomposing, deplete the oxygen supply and release stored nutrients for support of new growths. Oxygen depletion would temporarily subject fresh-water species to physiological stress; however, by treating the lake or body of water in sections at different times, the motile species move to areas of more favorable conditions. The same low-oxygen condition would be present if a large mat of vegetation were to remain untreated, as some plants continually die and deplete the oxygen, while the heavy mat prevents aeration at the surface. The dying vegetation from either natural or chemical causes produces temporary noxious conditions.

The chemical 2,4-D, the proposed herbicide for the control of water hyacinths, is extremely toxic to many desirable broadleaf plants. However, application of the herbicide will be controlled at all times to prevent the drifting of spray onto cultivated lands, ornamentals, or desirable aquatics. No spraying will be attempted when high winds make such operations hazardous.

Biological and integrated control, including herbivorous fish, are a part of the research program, but are not operational in Texas at this time.

Estimated Costs

The estimated federal and nonfederal costs based on current information on work required in Texas to eradicate as thoroughly as possible obnoxious aquatic growth in the waterways of Texas are shown in Tables 5 and 6. Thereafter the program is a continuous patrol and inspection of the work areas that might have been missed in the initial eradication operations.

Estimate of Benefits

Benefits creditable to the control of aquatic plants in Texas waters are considered to be derived from the restoration of these waters to their former uninfested state. Benefits are derived from: (1) restoration of water surface for recreational activities,

Table 6
ESTIMATED FEDERAL AND NONFEDERAL COSTS FOR AQUATIC CONTROL PROGRAM IN THE STATE OF TEXAS (1972-1981)[a]

	Acres of infestation		Cost					
Work area no.	Hyacinth	Alligator weed	Total first cost	Annual charges	Annual O&M[b]	Annual charges plus O&M[b]	Annual benefits	Benefit-to-cost ratio
1	8,000	0	184.5	6.25	0.80	7.05	96.1	13.6
2	2,000	0	47.0	1.59	0.50	2.09	10.0	4.8
3	3,000	2,000	134.0	4.54	0.75	5.29	38.6	7.3
4	5,000	1,500	164.2	5.56	0.75	6.31	28.4	4.5
5	2,500	3,000	83.0	2.81	0.50	3.31	8.5	2.6
6	800	400	34.0	1.15	0.50	1.65	4.3	2.6
7	100	0	2.3	0.08	0.05	0.13	0.6	4.6
8	580	0	11.2	0.38	0.13	0.51	3.3	6.5
9	700	1,500	64.0	2.17	0.50	2.67	4.2	1.6
10	5	0	0.5	0.02	0.02	0.04	0.04	1.0
Total	22,605	8,400	724.7	24.55	4.5	29.05	194.04	6.7

[a] All costs in thousands of dollars.
[b] O & M = operations and maintenance.

From General Design Memorandum, Corps of Engineers, New Orleans District, Louisiana, FY 1978.

(2) restoration of waterfront land values that have been depressed by reduced access to water resulting from aquatic plant obstructions, (3) restoration of drainage courses to their capacity, (4) restoration of normal costs of water supply by eliminating extra costs caused by plant growth, and (5) reduction of existing aquatic plant control costs by implementing a comprehensive control program in lieu of the existing sporadic control measures.

The evaluation of benefits is based on data obtained from field surveys and interviews with members of the Texas Parks and Wildlife Department, managers of private and public lakes, operators of pump stations and irrigation canals, real estate developers, and private citizens concerned with boating, fishing, and other recreational activities. Although specific information on aquatic plant infestation and its economic effects over a long period is somewhat limited, the known benefits in relation to estimated costs of control under existing conditions are high. It is doubtful if future benefits would significantly enhance the feasibility of the aquatic plant control project.

Nueces River Basin Work Area No. 1

The infestation of water hyacinths is limited to the vicinity of Lake Corpus Christi. The evaluation is based on complete infestation of Lake Corpus Christi by 1975. The average annual benefits are based on the prevention of the loss of recreation activities on the 15,500-acre lake and are estimated at $74,700. Annual benefits are also evaluated for the restoration of land values to 3800 acres of waterfront lots surrounding the lake and are estimated at $21,000. The total annual benefits for eradicating water hyacinths in the Nueces River Basin are as follows:

Restored recreation activities	$74,700
Restored land values	21,400
Total benefits	$96,100

Guadalupe River Basin Work Area No. 2

The Guadalupe-Blanco River Authority has been engaged in planning and promoting water resource development since 1936. Numerous small reservoirs for power generation situated on the Guadalupe River between the cities of Gonzales and New Braunfels have been infested continually with water hyacinth. These reservoirs are well established as popular recreation areas. Homes valued from $7000 to $40,000 line the shores of these reservoirs. The total water surface area of these reservoirs is about 2400 acres. Most of these reservoirs were overcrowded on weekends by fishermen, boatmen, and skiers until the construction of Canyon Lake in 1964, which is a relatively large lake in the same general vicinity. The visitation to these lakes is approximately 16,800 visitor-days per year. With approximately 330 acres of infestation in the reservoirs, it is estimated that recreational use has been reduced by about 2300 visitor-days per year. This reduction in recreational use, valued at $1.00 per visitor-day, is estimated at $2300/year. The landowners of approximately 83,000 ft of shoreline presently have restricted access to the reservoirs because of the water hyacinth obstruction in front of their property. In the past, landowners in some areas have attempted to control the infestation along the shorelines; however, these attempts have been generally ineffective. Land depreciation is considered to be at least equal to the cost to the landowners for their local control operations. The cost for the localized control has been about $0.75/ft of shoreline. Without any plant control program the average land value would depreciate from $75 to about $74.25/ft of shoreline. The total land depreciation is estimated at $62,250. Reduction to an annual basis and application of standard long-term financing rate of 8% result in a depreciation value of $3700. Benefits are derived from returning infested water areas to recreational use and from restoring the land values; the annual benefits are 3800 and 3700, respectively. Minor infestation has been reported on the San Marcos River. The water hyacinths float downstream on the Guadalupe River and accumulate in various distributary channels near the mouth of the river. Approximately 600 acres of water hyacinths are located in the wide bottomland and marsh areas between the Guadalupe River and the Gulf Intracoastal Waterway Channel to Victoria. A pumping plant near the channel to Victoria is used to supply industrial and irrigation water in Calhoun and Refugio counties. The annual cost for cleaning the trash racks at the pumping plant to remove water hyacinths is estimated at $1000. Annual benefits for reducing this operational cost is estimated at $1000. Statistics are not available on the fishing activities in bayous near the mouth of the river. Adjacent areas such as the channel to Victoria and the mouth of Guadalupe River appear to be popular fishing areas, and it is assumed that fishing in the bayous would be similar if the water hyacinths were removed. It is estimated that about 300 out of the 600 acres will be made accessible for fishing when the area has been cleared of water hyacinths. The annual visitation is estimated at about 1500 visitor-days for the additional 300 acres made accessible. The annual benefits for the restored recreation activity in this area is estimated at $1500 (300 acres × 5 visitor-days per year per acre × $1.00 per visitor-day). The total annual benefits for eradicating water hyacinths on the Guadalupe River are as follows:

Restored recreational activities	$ 5,300
Restored land values	3,700
Reduced cost of operation of pump station	1,000
Total benefits	$10,000

North Coastal Area Work Area No. 3

Water hyacinth and alligator weed impair the recreation activities in Taylors Bayou, Oyster Bayou, Double Bayou, Lake Anahuac, Turtle Bayou, Oyster Creek, San Bernard River, Cedar Lake Creek, and Caney Creek. Water hyacinths are also found in small, privately owned lakes which are located near the streams. All of these areas are popular for boat and bank fishing. Fish generally caught in these areas are catfish, mullet, bass, sunfish, and crappie. The north coastal area is well known for the migratory waterfowl hunting activity which occurs for about 2 months during the fall. There are no visitor-day counts available for these areas except for a few small private lakes. The statistics developed from the private lake and bait camp operators indicate that there are about 7 visitor-days per acre of water area per year by fishermen. On larger streams or lakes the average visitation is about 5 visitor-days per acre per year. Information available on duck hunters indicates a usage of 1 visitor-day per 3 acres of water area per year. Access to most of these water areas is through private property where fees are usually paid, or from boat ramps provided by the state, county, or bait camp operators. It is estimated that water hyacinth and alligator weed infestations prevent recreation usage in the amount of 25,000 visitor-days annually for fishing and 1700 visitor-days annually for duck hunting. Based on a value of $1.00 per visitor-day, the reduction in recreational use caused by the existing infestation of 5000 acres of water hyacinth and alligator weed is estimated at $26,700. The annual benefits for returning these infested areas to recreational use are estimated at $26,700. Summer home developments are found along the banks of the coastal streams, but they are not as extensive as the developments around the large lakes in other areas. Much of the area along the banks of the coastal streams is marsh area or is in use for grazing cattle. Summer homes are located on the banks of Taylors Bayou, Double Bayou, Oyster Creek, San Bernard River, and Caney Creek. It is estimated that about 20,000 lineal ft of shoreline in the summer home development areas is presently affected by the infestation of water hyacinths. Assuming the same values of land depreciation used in the Guadalupe River analysis, the total land depreciation is estimated at $15,000. Reducing this amount to an annual basis and applying an average long-term mortgage financing rate of 6% results in reduction of the depreciated value to $900 annually. Annual benefits for preventing this land depreciation are estimated at $900.

Sections of Linnville Bayou, Caney Creek, and Cedar Lake Creek are entirely covered with water hyacinths. The drainage in these watersheds is poor, partly because of the topography which has a very gentle slope and partly because of choking or obstruction of drainage courses by water hyacinths. The area affected is used extensively for grazing of cattle. The average annual net return is about $2.20/acre. The net return will increase by about 10% with the improved drainage that will result from the eradication of the water hyacinths. About 50,000 acres are affected; therefore, the restored drainage will result in annual benefits of about $11,000.

Irrigation canals and drainage ditches near Elm Bayou, East Bay Bayou, and Oyster Bayou are infested with water hyacinth and alligator weed. These bayous are located in the southern part of Chambers County. In 1969, there were approximately 54,000 acres in rice production in Chambers County. The irrigation and drainage systems are quite extensive. The water supply is principally furnished by Chambers-Liberty Counties Navigation District and Trinity River Authority. The drainage is maintained by the Trinity Bay Soil Conservation District. These agencies have continually attempted to eradicate the water hyacinth and alligator weed in the irrigation canals and drainage ditches. The flow of water in the irrigation canals is impeded by the dense growths of these plants. Growth in the drainage ditches obstructs flow causing low areas, some of which are in rice production, to be flooded. The local agencies are continually eradicating the heaviest infestations; the aggregate cost is about $11,000 annually. A program of repeated eradication in all areas is needed, but the program is more expensive than the local agencies can afford. The annual benefits for eradicating the water hyacinth and alligator weed are assumed to be at least equal to the local agencies' maintenance costs, which will be greatly reduced by the more comprehensive eradication program proposed in this report. The annual benefits for restoring the irrigation and drainage canals to original condition in lower Chambers County are estimated at $11,000.

J. D. Murphree Wildlife Management Area is located between the mouths of Taylors Bayou and Big Hill Bayou. The area, which covers about 8400 acres, is operated by the Texas Parks and Wildlife Department. There have been periods when as much as 60% of this area has been covered with water hyacinth. The area is also heavily infested with alligator weed. The area is one of the primary feeding grounds for migratory waterfowl. The predominant native feeds are millet, sedges, and eleocharis. The encroachment of the water hyacinth and alligator weed reduces the amount of feed available to the wildfowl, which in turn affects the migration into the area. Detailed statistics are not available on the effects noxious plants have on the wildlife management area; therefore, benefits from eradication of water hyacinth and alligator weed have not been evaluated.

The total annual benefits for eradicating water hyacinth and alligator weed in the North Coastal Areas are as follows:

Restored recreational activities	$26,700
Restored land values	900
Restored irrigation and drainage canals	11,000
Total	$38,600

Sabine River Basin Work Area No. 4

The Toledo Bend Lake is currently the major problem area with water hyacinth infestation in the Sabine River Basin. The lake is operated by the Sabine River Authority of Texas and the Sabine River Authority, state of Louisiana. The project was completed in 1968, and its popularity as a recreational area has steadily increased. Sport fishing for bass and catfish is particularly popular in the Toledo Bank Lake. It is estimated that about 1 million visitor-days of recreational use will occur per year. The surface area of the lake is about 186,000 acres. About 750 of the total 1265 mi of shoreline is in Texas. Approximately 200 mi of the Texas shoreline are owned by the U.S. Forest Service and on which no private land development is allowed. About 40 commercial establishments are located on the remaining 550 mi of shoreline in Texas to service the needs of fishermen and recreationists who visit the lake. Presently, there are about 1100 registered waterfront lots that have been developed. About 5000 acres of water hyacinth infestation are estimated to be located on the Texas portion of the lake. Since the recreation usage of the lake is considered to be approaching the saturation level, any coverage of the lake by water hyacinth would reduce the recreational use by the proportionate value of recreation usage per acre of the lake. The current usage is about 5.5 visitor-days per acre per year. Therefore, the annual loss in recreation activity is estimated at 27,500 visitor-days (5000 acres of water hyacinths × 5.5 visitor-days per acre per year). Based on a value of $1.00 per visitor-day, the annual benefits for eradicating the water hyacinths and restoring the area for recreational activities is estimated at $27,500. The infestation is concentrated in the middle of the lake where the uncleared tree areas are located; however, there are several coves which have infestations. It is estimated that the accessibility to the lake from 20,000 ft of shoreline is restricted by the water hyacinth coverage. Some of the landowners have taken steps to remove the plants. The cost to the landowners for spraying the infestation is about $0.75/ft of shoreline. This activity is usually ineffective because other water hyacinth plants float back into the same areas and multiply until the access to the water is cut off again. It is estimated that the average land value will depreciate from $50 to $49.25/ft of shoreline because of the infestation. The total depreciation that will be prevented by the authorized eradication program is about $15,000. Reducing this amount to an annual basis and applying a standard long-term financing rate of 8% results in reduction of the depreciation value to $900 annually which is considered to be an annual benefits value.

The infestation of the marsh lands and swamps below Toledo Bend Lake with water hyacinths and alligator weed downgrades the marine habitat in the area. The detrimental effect cannot be measured at this time because supporting statistical data are not available. The areas are wooded and covered with heavy brush vegetation. Accessibility for recreation or any other purpose is almost impossible. There are a few areas around the fringes of the swamps which are accessible for treatment. The water hyacinths in these fringe areas will be eradicated and suppressed under the proposed maintenance program in order to eliminate the transportation of the plants to other areas. The benefits from the eradication of the water hyacinth and alligator weed in portions of the marsh lands and swamps are derived from restoring the marine habitat and from extending the effectiveness of the overall eradication program; however, no monetary value of these benefits has been estimated. The total annual benefits for eradicating water hyacinth and alligator weed in the Sabine River Basin are as follows:

Restored recreational activity	$27,500
Restored land values	900
Total benefits	$28,400

Trinity River Basin Work Area No. 5

Lake Livingston is operated by the Trinity River Authority and is located adjacent to the Big Thicket in southeast Texas. The surface area of the lake at normal pool elevation is about 83,000 acres. Construction of the dam and reservoir area was completed in November 1969. There are already over 75 marinas, bait camps, and camping suppliers located on the perimeter of the lake. The lake is used for fishing, sailing, water skiing, and swimming. The popular sport fish in the lake are black bass and channel catfish. The lake is frequented by pelicans, spoonbills, and sea gulls. Ducks and geese are found on the lake in the winter. Approximately 50 acres of water hyacinths have been located on the lake. If not controlled the plants will reproduce rapidly and cause serious problems on the lake. At the estimated peak-level visitation, full benefits for protecting recreational activities are evaluated for the eradication of the 50 acres of water hyacinths. Using an average peak visitation as experienced on similar lakes, the annual benefits for returning the usage of 50 additional acres to recreational activities in Lake Livingston are estimated at $275 or $300 (50 acres × 5.5 visitor days per acre per year × $1.00 per visitor-day).

About 730 acres of water hyacinth and alligator weed are located on small creeks, marshes, and wide river bottom areas on the Trinity River between Lake Livingston and Wallisville Lake which is now under construction. Bank fishing for catfish is popular in many of these areas. Since statistics are not available, it is estimated that half of this area will be returned to recreational use at the normal rate of about 5.5 visitor-days per acre per year. The annual benefits for returning 365 acres of the streams between Lake Livingston and Wallisville Lake back to recreational activities is estimated at $2000 (365 acres × 5.5 visitor-days per acre per year × $1.00 per visitor-day).

Approximately 4720 acres of water hyacinth and alligator weed are presently located in the Wallisville Lake site. The navigation locks and reservoir are presently under construction by the Corps of Engineers. The benefits for the project are as follows:

Salinity control	$ 413,000
Navigation	670,000
Water supply	60,000
Fish and wildlife	157,000
Recreation	257,000
Total benefits	$ 1,557,000

The presence of such an abundant amount of water hyacinth will reduce all of these benefits to some degree. Consequently, an Aquatic Plant Control Program has been included with the maintenance of the project. As long as the area is being fed with these aquatic plants from upstream sources, perpetual maintenance will be needed. The annual benefits for eradicating water hyacinth and alligator weed located upstream from Wallisville Lake are estimated at $6200, which is the estimated annual cost of the aquatic plant control maintenance. The total annual benefits for eradicating water hyacinth and alligator weed infestations on the Trinity River Basin are as follows:

Restored recreational activities	$2300
Reduction of aquatic plant control maintenance in Wallisville Lake	6200
Total benefits	$8500

Neches River Basin Work Area No. 6

The major infestation of water hyacinth has been in Dam B Lake, a Corps of Engineers project. The 15,700-acre lake is a very popular recreational area because of its close proximity to Beaumont and Port Arthur. The infestation of water hyacinth has never exceeded more than 250 acres, primarily due to the aquatic plant control maintenance conducted at the lake. New infestations are fed into the lake by small streams flowing directly into the main stream of the Neches River above Dam B Lake. Annual benefits are estimated at $2100 for eradicating the sources of the water hyacinth infestation, thereby eliminating the annual maintenance cost of the same amount at Dam B Lake.

Approximately 800 acres of water hyacinth are located on small creeks, marshes, and wide river bottom areas on the Neches River between Dam B Lake and the mouth of the river. Numerous summer homes and camps are located in these areas. The river is wide enough to facilitate extensive boating and fishing activities. Statistical information on the recreational activities is not available; however, it is estimated that half of the 800 acres now covered by water hyacinth and alligator weed will be returned to recreational activities with the eradication of the plants. Using the normal rate of 5.5 visitor-days per acre per year for the recreational activity in the area, the annual benefits for improving the recreational activities in the lower reach of the Neches River below Dam B Lake are estimated at $2200 (400 acres × 5.5 visitor-days per acre per year × $1.00 per visitor-day). Water hyacinths have not been located on the Angelina River, a tributary of the Neches River, or on Sam Rayburn Lake which is on the Angelina River. The total annual benefits for eradicating water hyacinth and alligator weed in the Neches River Basin are as follows:

Restored recreational activities	$2100
Reduction of existing aquatic plant control maintenance	2200
Total benefits	$4300

Cypress Creek Basin Work Area No. 7

The surface area of Caddo Lake is 32,700 acres, divided almost equally between Texas and Louisiana. Boating and fishing for bass, crappie, and sunfish are very popular. The annual visitation in the Texas area is about 96,000 visitor-days; the average visitor-day per acre per year is about 6.0. Water hyacinth infestation was severe between 1950 and 1955 when about 10,000 acres of the lake surface was covered by the plants. The water hyacinths were completely eradicated, and the lake remained clear of water hyacinth until 1967, when several boat channels were dredged. Apparently, water hyacinth seeds uncovered by the dredging in 1967 germinated, and there now exists about 100 acres of water hyacinth on the lake. The annual benefits from eradicating the water hyacinth infestation and returning the area to recreational use is estimated at $600 (100 acres × 6.0 visitor days per acre per year × $1.00 per visitor-day). The lake now has a severe problem with submerged vegetation consisting of approximately 5000 acres each of milfoil and coontail. There is no apparent economical solution to the submerged aquatic vegetation problem at this time.

South Coastal Area Work Area No. 8

Irrigation canals, drainage ditches, and resacas near San Benito, Harlingen, Los Fresnos, and Brownsville are infested with water hyacinths. The resacas are used as a source of water supply for irrigating citrus orchards, cotton fields, and truck farms. The water hyacinths reduce the flow of water in the irrigation canals and cause flooding in the lower areas when the water cannot drain through the ditches. The total average expenditure by Water Improvement Districts 2, 10, and 11 for partially controlling the water hyacinth has been about $3,300 annually. The local program is ineffective because the eradication operation is limited by the amount of funds available and the operation needs to be repeated every 2 or 3 years. The annual benefits from the more intensive water hyacinth control programs covered in this design memorandum will be derived from the reduction of practically all of the local maintenance cost for water hyacinth control, which is estimated at $3300. Occasionally, water hyacinths appear on two resacas located within the city of Brownsville. When this happens, the local citizens and city employees remove the plants manually. The existing problem is very minor; however, the plants need to be removed to prevent the development of a severe infestation and to prevent anyone from transporting the plants to some other area. The existing infestation on the two resacas in Brownsville is less than an acre; therefore, benefits have not been evaluated. The annual benefits for water hyacinth control in the South Coastal Area are estimated at $3300.

San Jacinto River Basin Work Area No. 9

There are large infestations of alligator weed and water hyacinth in Lake Houston and the water supply canals below the lake which are operated by the city of Houston as part of the municipal water supply system. The annual visitation on the 12,800 acre lake is about 77,000 visitor-days. The recreational activities include boating, fishing, skiing, and picnicking. Little has been done to reduce the acreage of alligator weed in the area. The periodic eradication program that the city has conducted has cost an average of about $4200 annually. The annual benefits will be a least equal to the cost of maintenance which will be reduced or eliminated by the more intensive aquatic plant control program authorized in this design memorandum; therefore, the annual benefits for providing this program in the San Jacinto Basin are estimated at $4200.

Rio Grande Basin Work Area No. 10

Only 5 acres of water hyacinth have been located in the river basin. The infestation is located on San Felipe Creek, near Del Rio, Texas. A major problem could arise if flood waters could cause the infestation to flow down San Felipe Creek into the Rio Grande and Falcon Lake. Falcon Lake, which has a surface area of 98,960 acres, is used for water supply, flood control, and recreation. The resulting infestation of water hyacinth would reduce the recreation activities on the lake. However, benefits have not been estimated for this contingency. The annual benefits for eradicating water hyacinth in the Rio Grande Basin are considered equal to the estimated annual cost of eradication in the small areas of infestation. The estimated amount is less than $100 annually and is therefore negligible.

Summary of Benefits

The total average annual benefits for the eradication of water hyacinth and alligator weed in the ten work areas in the state of Texas are estimated at $194,040 and are summarized as follows:

Work area	Location	Annual benefits
1	Nueces River Basin	$ 96,100
2	Guadalupe River Basin	10,000
3	North Coastal Area	38,600
4	Sabine River Basin	28,400
5	Trinity River Basin	8,500
6	Neches River Basin	4,300
7	Cypress Creek Basin	600
8	South Coastal Area	3,300
9	San Jacinto River Basin	4,200
10	Rio Grande Basin	40
	Total benefits	$194,040

Comparison of Benefits and Costs

The average annual benefits are estimated at $194,040. Federal and nonfederal benefits are estimated at $724,700. The annual charges, based on 3¼% interest rate for amortization and project life of 100 years, are estimated at $24,550. The annual operation and maintenance cost is $4500 which makes the total annual charges $29,050. The benefit-to-cost ratio is 6.7.

REFERENCES

1. **Sculthorpe, C. D.**, The problem of aquatic weeds, in *The Biology of Aquatic Vascular Plants*, Edward Arnold, London, 1967.
2. **Little, E. C. S.**, Report to the Government of Indonesia on the Control of Aquatic Vegetation in the Lake of Rawa Pening, Central Java, TA 2534, Food and Agriculture Organization, Rome, 1968.
3. **Holm, L. F., Weldon, L. W., and Blackburn, R. D.**, Aquatic weeds, *Science*, 166, 699, 1969.
4. **Gangstad, E. O.**, Herbicidal control of aquatic plants, *J. Sanit. Eng. Div. Am. Soc. Civ. Eng.*, 98, 397, 1972.
5. **Gangstad, E. O.**, *Weed Control Methods for River Basin Management*, CRC Press, Boca Raton, Fla., 1978.
6. USAID, Economic Damage Caused by Aquatic Weeds, TA/OST 71-5, Office of Science and Technology, U.S. Agency for International Development, Washington, D.C., 1971.
7. Board on Science and Technology for International Development, *Making Aquatic Weeds Useful; Some Perspectives for Developing Countries*, National Academy of Sciences, Washington, D.C., 1977.
8. **Vietmeyer, N.**, *The Large Scale Production of Feed Stuffs from Aquatic Vegetation*, Board on Science and Technology for International Development, National Academy of Sciences, Washington, D.C., 1978.
9. **Zur Burg, F. W., Foret, J. A., Solymosy, S. L., and Hayes, S. A.**, *Annual Report of the Control of Alligatorweed and Other Aquatic Plants*, University of Southwestern Louisiana, Lafayette, 1967.
10. **Timmons, F. L., Bruns, V. F., Lee, W. O., Yeo, R. R., Hodgson, J. M., Weldon, L. L., and Comes, R. D.**, Studies on the control of common cattail in drainage channels and ditches, *U.S. Dep. Agric. Bur. Reclam. Tech. Bull.*, 1286, 1963.
11. U.S. Department of the Army, Expanded Project for Aquatic Plant Control, House Document No. 251, 89th Congress, Washington, D.C., 1965.
12. **Abramson, S. C.**, The importance of proposed programming in industrial weed control as it pertains to government agencies, *South. Weed Conf. Proc.*, 18, 362, 1965.

13. Aquatic Plant Control Program in Florida, U.S. Army Corps of Engineers, Jacksonville District, Florida, 1971.
14. Comprehensive Report on Central and Southern Florida for Flood Control and Other Purposes, House Document No. 643, 80th Congress, 2nd Session, 1948.
15. Annual Report, Central and Southern Florida Flood Control District (FCD), 1967.
16. **Grant, Z. C.**, Aquatic weed control program of the Central and Southern Florida Flood Control District, *Hyacinth Control J.*, 1, 24, 1962.
17. Current Estimates of Annual Benefit-Cost Ratio, Army Corps of Engineers, Jacksonville District, Florida, 1968.
18. Public Works Appropriations, hearings before a subcommittee of the Committee on Appropriations, House of Representatives, 90th Congress, 1st Session, Part I, 1968.
19. **Huser, T. E.**, Economics of aquatic weed control in the Central and Southern Florida Flood Control District, *Hyacinth Control J.*, 7, 16, 1968.
20. **Bissland, H. H.**, Aquatic weed control in Florida's soil and water conservation districts, *Hyacinth Control J.*, 7, 14, 1968.
21. **Timmer, C. E. and Weldon, L. W.**, Evapotranspiration and pollution of water by waterhyacinth, *Hyacinth Control J.*, 6, 34, 1967.
22. **Weldon, L. W. and Blackburn, R. D.**, Suggested Control Measures for Common Aquatic Weeds in Florida, U.S. Department of Agriculture, Washington, D.C., 1963, CR 56.
23. **Blanchard, J. L.**, Economic aspects of weed control in the lakes of Winter Park, Florida, *Hyacinth Control J.*, 6, 21, 1967.
24. **Anon.**, Removal of Waterhyacinth and Aquatic Plant Control Program, U.S. Army Engineer District, New Orleans, La., 1977.
25. **Kniffen, F. B.**, *Louisiana, Its Land and People*, Louisiana State University Press, Baton Rouge, 1968.
26. **Lagler, K. F.**, *Freshwater Fishery Biology*, William C. Brown, Dubuque, Iowa, 1956, 421.
27. **Lambou, V. W.**, Fish populations of the streams on the Red Dirt Game Management Area, Natchitoches Parish, Louisiana, *Proc. La. Acad. Sci.*, 22, 46, 1959.
28. **Lambou, V. W.**, Louisiana impoundments, their fish populations and management, *Trans. 24th N. Am. Wildl. Conf.*, 187, 1959.
29. **Lambou, V. W.**, Fish populations of backwater lakes in Louisiana, *Trans. Am. Fish. Soc.*, 88(1), 7, 1959.
30. **Lambou, V. W.**, Fish populations of Mississippi River oxbow lakes in Louisiana, *Proc. La. Acad. Sci.*, 23, 62, 1960.
31. **Lambou, V. W. and Geagan, D. W.**, Fish populations of alluvial flood plain lakes in Louisiana, *Proc. La. Acad. Sci.*, 24, 95, 1961.
32. **Lantz, K. E.**, *An Ecological Survey of Factors Affecting Fish Production in a Louisiana Natural Lake and River*, Louisiana Wild Life and Fisheries Commission, Baton Rouge, 1971, 92.
33. **Lantz, K. E., Davis, J. T., Hughes, J. S., and Schafer, H. E., Jr.**, Water level fluctuation — its effect on vegetation control and fish population management, *Proc. 18th Annu. Conf. S.E. Assoc. Game Fish Comm.*, 18, 483, 1964.
34. **Lawrence, J. M. and Lyle, W. W.**, Identification of aquatic weeds, *Hyacinth Control J.*, 4, 6, 1965.
35. **Lee, G. F.**, *Eutrophication*, University of Wisconsin, Madison, 1970.
36. **Little, E. C. S.**, The world distribution of the water hyacinth, *Hyacinth Control J.*, 4, 30, 1965.
37. Aquatic Plant Control and Eradication Program, U.S. Army Engineer District, Galveston, Tex., 1972.
38. **DeVaney, T. E.**, Chemical Vegetation Control Manual for Fish and Wildlife Management Programs, U.S. Department of Interior, Washington, D.C., 1967.
39. **Gangstad, E. O.**, Integrated Control of Alligator Weed and Water Hyacinth in Texas, Tech. Rep. 9, U.S. Army Engineer Waterways, Experiment Station, Vicksburg, Miss., 1975.
40. **Hughes, J. S. and Davis, J. T.**, Variations in toxicity to bluegill sunfish of phenoxy herbicides, *Weeds*, 11, 50, 1963.
41. **Hughes, J. S. and Davis, J. T.**, Toxicity of selected herbicides to sunfish, *Proc. La. Acad. Sci.*, 25, 86, 1962.
42. **Khanna, S. and Fang, S. C.**, Metabolism of ^{14}C-labeled 2,4-dichlorophenoxyacetic acid in rats, *J. Agric. Food Chem.*, 14, 500, 1966.
43. **Little, E. C. S.**, The world distribution of the water hyacinth, *Hyacinth Control J.*, 4, 30, 1965.
44. **Maddox, D. M.**, Bionomics of an alligatorweed flea beetle, *Agasicles* sp. in Argentina, *Ann. Entomol. Soc. Am.*, 61, 1299, 1968.
45. **Maddox, D. M. and Hambric, R. N.**, Use of alligatorweed flea beetle in Texas: an exercise in environmental biology, *Proc. S. Weed Sci. Soc.*, 23, 283, 1970.

46. **Maddox, D. M., Andnes, L. A., Hennessey, R. D., Blackburn, R. D., and Spencer, N. R.,** Insects to control alligatorweed, an invader of aquatic ecosystems in the United States, *Bioscience*, 21, 985, 1971.
47. **Mulrennan, J. A.,** The relationship of mosquito breeding to aquatic plant production, *Hyacinth Control J.*, 1, 6, 1962.
48. **Raynes, J. J.,** Aquatic plant control, *Hyacinth Control J.*, 3, 2, 1969.
49. Bureau of Aquatic Plant Research and Control, FY 1978 to 1979, U.S. Department of Natural Resources, Tallahasee, Fla., 1979.
50. General Design Memorandum, Corps of Engineers, New Orleans District, Louisiana, 1978.

Chapter 2

USING HERBICIDES EFFECTIVELY IN LAKES AND PONDS*

INTRODUCTION

Bodies of water in the Midwest range from small farm ponds of 1 acre or less, constructed for the control of soil erosion and as a source of water for farm activities; intermediate size lakes from ten to several hundred acres for domestic water supply, recreation (such as state-owned lakes), and/or residential developments and irrigation; lakes from several hundred to several thousand acres primarily as sources of cooling water for electric power generation, flood control, and water supply. In addition to the specific uses many, if not all, also have water-related recreational uses such as fishing, swimming, and boating. Many lakes have relatively large areas of water less than 15 ft deep and many small farm ponds have a maximum water depth of less than 15 ft. The shallow water areas and water areas less than 15 ft deep can become inhabited with aquatic plants, which may seriously interfere with the intended water uses and the associated water-related recreational activities.

If the water is sufficiently clear submersed aquatic plants may grow in water areas deeper than 15 ft, but they will not reach the surface of the water. However, in shallow water (water up to 3 or 4 ft deep) emersed, i.e., rooted aquatic plants with foliage extending above the water surface, may inhabit these water areas. Much of the shallow water is around the pond or lake shoreline, but the emersed aquatic plants can inhabit large areas if the water is not too deep. These marginal aquatic plants may not have a nuisance level, but they can be problems in some situations.

AQUATIC PLANTS

The aquatic plants inhabiting the water areas can be placed into two major groups: the nonvascular aquatic plants, i.e., aquatic plants without a vascular system and more commonly referred to as algae, and vascular aquatic plants, i.e., plants that have a well-developed vascular system. Many of the vascular group have roots in the hydrosoil.

There are three basic groups of nonvascular plants, or algae, based on their growth characteristics. The filamentous algae are single filament-type cells, however these filaments may cling together to form mats of cells and/or may occupy the water column from the hydrosoil to the water surface; the large mats of filamentous algae cells will float on the water surface and drift with the wind and water currents. The filamentous algae adhere to rocks, debris in the water, and the vascular plants. These mats of filamentous algae, superimposed upon submersed aquatic plants, are responsible for the unsightly appearance of many bodies of water in late summer and early fall.

The unicellular phytoplankton algae are single algae cells suspended in the water, however some algae cells will clump together or form distinct patterns of cells. Probably all waters in the world contain some phytoplankton algae. The unicellular algae cells are green, blue-green, red, and brown; when the density of the algae cells in the water is sufficient the algae will impart a color to the water, and the algae "blooms" that develop in many bodies of water during the growing season.

The third group of algae is the higher-branched algae, and, although they do not

* Contribution of Robert C. Hiltibran, Biochemist, Natural History Survey, Urbana, Ill., May 3, 1978.

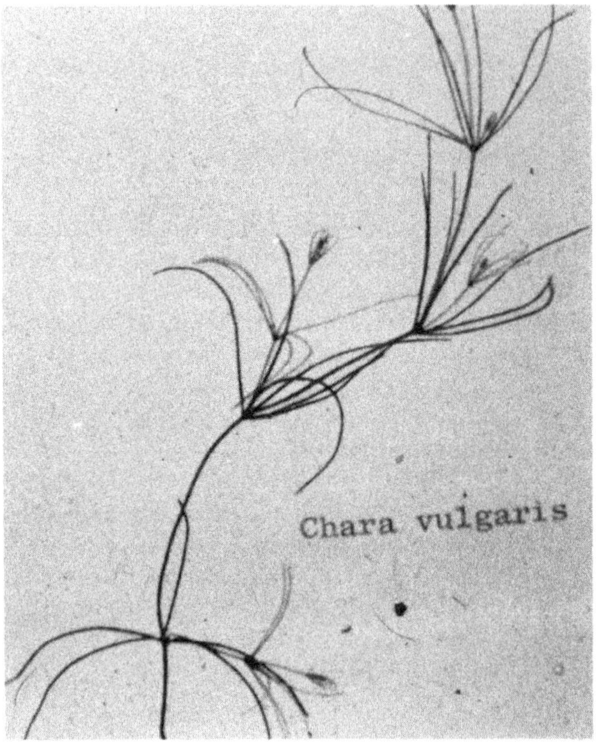

FIGURE 1. Higher branched algae; chara, *Chara vulgaris*.

have a vascular system, the cells take distinct forms and in appearance resemble, and are frequently confused with, the submersed aquatic plants. These algae also produce reproductive bodies. Two common examples are chara, *Chara* spp. (Figure 1), and *Nitella* spp.[8,32] *Chara vulgaris* has a very pungent odor resembling the odor of musk, which is the origin of one common name — muskgrass. *Chara* is rough to the touch, and as the growing season progresses marl, a salt deposit, forms on the algae surface and in late summer it will have a gray appearance. *Nitella*, on the other hand, is smooth and apparently does not develop marl on the algae surface.

The vascular aquatic plants were divided into groups based on their distribution in water and their growth characteristics. The common groupings are emersed, free-floating, submersed, and floating-leaved aquatic plants.

Emersed Aquatic Plants

The aquatic plants of this group are rooted in the hydrosoil and the foliage extends above the water surface. Usually members of this group of aquatic plants inhabit the shallow water around ponds and lakes and provide nesting areas for birds. They usually have a root rhizome system and new growth develops from the root rhizome system, but most of these aquatic plants also produce flowers and fruiting bodies and/or seeds that are small enough to be blown by the wind or large enough to be consumed by birds. The aquatic plants of this group can occupy large areas of shallow water and some members, such as arrowhead, provide a source of food for waterfowl and birds. The aquatic plants of this group are generally not very troublesome, but can prevent or reduce bank erosion.

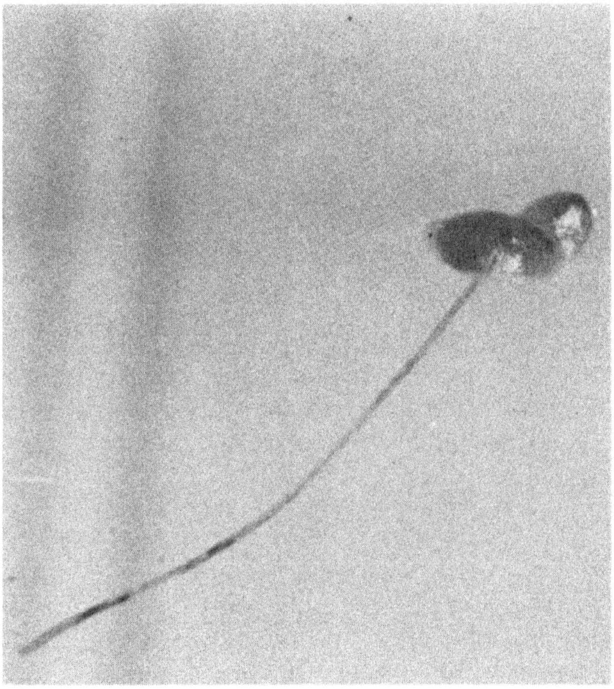

FIGURE 2. Common duckweed, *Lemna minor* L.

Free-Floating

These very small aquatic plants are not rooted in the hydrosoil but float in or on the water surface. The plants such as duckweed, *Lemna minor* L. (Figure 2), consists of two or more leaf-like structures called fronds and have a rootlet attached to each frond, or a series of rootlets attached to the fronds. Other plants, such as watermeal, *Wolffia columbia* Karst (Figure 3), are small granular plants. Watermeal is a flowering plant and produces seeds. The common duckweed produces from 3 to 7 seeds per plant and will grow from a seed to a seed-producing plant in 3 days. These aquatic plants possess great reproductive potential, and, also because of their size, will form layers on the water surface making their control difficult. [Common coontail, *Ceratophyllum demersum* (Figure 4), is also a free-floating aquatic plant, but because of its form will be discussed in another group.]

Submersed Aquatic Plants

This group of aquatic plants, many of which are rooted in the hydrosoil and occupy the water volume to the water surface, are the aquatic plants that seriously interfere with water uses such as fishery production, swimming, and boating, and reduce the aesthetic appearance of the pond or lake. Aquatic plants of this group are subdivided into three groups based on the attachment of the leaves to the center stem. The center stems of this group of plants are not strong enough to support the plant, and leaves floating in the water provide the support for the plant.

Alternate leaf attachment — The characteristic of these plants is easily seen when the plant is floating in water, as one leaf is attached on one side of the stem and the adjacent leaves are on the opposite side of the stem. Most of the alternate leaf aquatic plants, represented by curly-leaf pondweed, *Potamogeton crispus* L. (Figure 5), in the Midwest bodies of water are in the Potamogeton family of plants. Each has characteristic flowers and fruiting bodies. Flowers of some aquatic plants extend above the

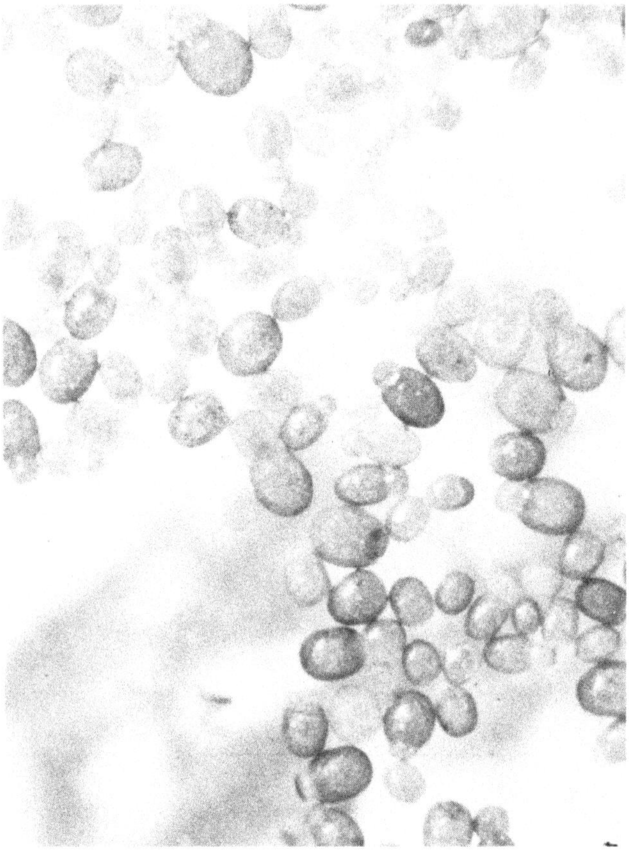

FIGURE 3. Watermeal, *Wolffia columbiana* Karst.

water surface for pollination, then the developing fruit floats on the water surface. In other species the flowers do not extend above the water surface. The membranous leaves vary: thin, broad, wavy leaves; narrow grass-type leaves; and round, pointed leaves.

Opposite leaf attachment — In this group of aquatic plants two leaves are attached to the same node of the center stem. The leaves may be simple grass-type leaves as possessed by the Najas family as represented by slender naiad, *Najas flexilis* (Figure 6), or the leaves may be high-branched, flat leaves as in cabomba, *Cabomba caroliniana* Gray, or round and pointed as shown by horned pondweed, *Zanchilla palustris* L. (Figure 7). Members of this group also have flowers and fruits, many of which have specific characteristics which frequently contribute to their common names, such as horned pondweed. The fruits of slender and southern naiad, *N. guadalupensis* (Spreng.) Magnus, are borne at the juncture of the leaves with the center stem as shown in Figure 6.

Whorled-leaf attachment — The aquatic plants of this group have four or more leaves attached to the center stem at the same node, represented by coontail (Figure 4). The leaves may be simple, branching, and/or finely divided. Many of the plants in this group have flowers and/or fruiting bodies that extend above the water surface for pollination and later they float on the water surface. (Coontail, because of its leaf attachment and structure, is listed with this group of plants as it has more characteris-

FIGURE 4. Common Coontail, *Ceratophyllum demersum* L.

tics in common with this group. It is not a rooted aquatic plant, but some plants remain anchored in the hydrosoil throughout the season, and large mats of this highly branched plant will float on the surface of the water.)

Floating-Leaved Aquatic Plants

The main characteristic of this group of aquatic plants is the large leaves floating on the water surface, represented by American pondweed, *Potamogeton nodosus* (Figure 8); they may also have submersed leaves of different texture and shape. The floating leaves of these plants vary from long to small elliptical leaves or to the very large, round leaves as shown by the water lily, *Nymphaea ordorata* Cut. The plants of this group are from several different families. While usually not as troublesome as the submersed group of aquatic plants, members of this group can occupy large areas of water. Usually these aquatic plants inhabit relatively shallow water, i.e., water 3 to 6 ft deep.

FIGURE 5. Alternate leaf attachment; curlyleaf pondweed, *Potamogeton crispus* L.

Of the various groups of aquatic plants discussed above, the submersed group are the most troublesome, have high nuisance levels, and may seriously interfere with water uses. Also this group of aquatic plants, although below the water surface, will catch debris and frequently are the cause of unaesthetic appearances of bodies of water in the Midwest. In late summer these submersed aquatic plants form a substrate to which filamentous algae can adhere thus adding to the unsightly appearance. It is this group of aquatic plants for which control is required or usually desired.

HERBICIDES

The control of aquatic plants in the U.S. was initiated in late 1884 following the accidental introduction of water hyacinth, *Eichonnia crassipes* (Mart.) Solms, in the U.S.[42] Mechanical methods, such as crushing, and other physical removal techniques were used. In 1904 copper sulfate was found to be effective for the control of algae in potable water,[31] and about the same time the use of sodium arsenite was introduced for the control of water hyacinth.[42] The combination of mechanical and chemical control methods was very effective for the control of water hyacinth. However, some of the water areas cleared of water hyacinth soon were invaded by alligator weed, *Alernanthera philoxeroides* (Mart.) Griseb., which was not susceptible to sodium arsenite. The metallic salts of copper sulfate and sodium arsenite were widely employed for the control of algae and aquatic plants.[33]

FIGURE 6. Opposite leaf attachment; slender naiad, *Najas flexilis*(Willd.) Rotsk. and Schmidt.

During the early 1940s, the herbicidal properties of chloro-substituted organic compounds, primarily (2,4-dichlorophenoxy) acetic acid* (2,4-D), were discovered, and by 1946 water hyacinth was reported to be susceptible to 2,4-D; soon it was widely used for the control of water hyacinth.[11] The success of 2,4-D for the control of other aquatic plants[34] and water hyacinth led to an investigation of the aquatic herbicidal activities of additional organic chemical agents.

By the mid 1950s and early 1960s, the aquatic herbicidal activities of endothall (7-oxalicyclo[2.2.1] heptane-2,3 dicarboxylic acid),[12] diquat (6,7-dihydrodipyrido[1.2-α:2′,1′-C] pyrazindiium ion),[13] dichlobenil (2,6-dichlorobenzonitrile), Fenac® (2,3,6-trichlorophenyl acetic acid),[14] and simazine (2-chloro-4,6 bis[ethylamino]-s-triazine)[10,16,38] were being investigated and their effectiveness as aquatic herbicides was being reported.[4,15,17,20,39,40]

The early investigations of the aquatic herbicidal activities of chemical agents were summarized by Lawrence.[29]

Flood control channels, particularly in Florida[4,37] and irrigation canals in the western states were developed and these water systems were invaded by aquatic plants,[1,2] which severely reduced the effectiveness of these water systems. The presence of the aquatic plants in these water systems led to the development of techniques and aquatic herbicides that were effective under flowing water conditions.[3,7,9,37]

To be effective as aquatic herbicides, the effects of possible aquatic herbicides against fishes, fish production, and fertilized fish eggs were investigated and reported.[19,28] To be an effective aquatic herbicide requires that the chemical agent must not have a direct toxicity or adverse effects on fishes at the rates which were found to be effective against aquatic plants.[6,15,27,35]

* Chemical names obtained from *Herbicide Handbook* of the Weed Science Society of America, 4th ed., Champaign, Ill., 1979.

FIGURE 7. Opposite leaf attachment; horned pondweed, *Zannichellia palustris* L.

Experimental Procedures

Many of the early test areas used in estimating the effects of the various aquatic herbicides against a large number of aquatic plants were conducted in 20-ft² areas.[15] The surrounding aquatic vegetation served as a barrier to contain the experimentally applied aquatic herbicides, and permitted observation on the drift of the aquatic herbicide from the application site. In a former coal-strip mine converted into a fishing club, 50- to 100-ft test areas in fingers of water from 25 to 50 ft wide were used. Some test areas were established by measuring a 100-ft length of aquatic vegetation and applying the herbicide to 20-ft areas on each side of the midline. Thus a 4000-ft test area was easily established. Frequently 100 to 200 ft of shoreline areas were used, where the vegetation would extend from 10 to 100 ft from the shoreline. Frequently 4000- to 6000-ft areas in bays of ponds were utilized as test areas. Other test areas of various sizes were utilized depending on the aquatic plant growth within the pond. In some situations total pond applications were made where the aquatic vegetation occupied the total pond area. Test areas were established only in the areas of the pond invaded by aquatic plants and the deeper-water areas usually devoid of aquatic plants were ignored.

Preemergence 20-ft² or larger test areas were established on exposed pond hydrosoil in late fall prior to the winter freeze, or in early spring, when possible, between the spring thaws and the pond filling with water. Some preemergence test areas were established after pond filling but prior to the growth of the aquatic plants. However, best results were obtained when the preemergence aquatic herbicides were applied to the exposed pond hydrosoil.

Long-term control of curlyleaf pondweed was initiated in 1960 in Mansion Pond, Allerton Park, Monticello, Ill., and the control of cattail, *Typha latifolia* L., in Allerton Lake, Allerton Park, Monticello, Ill., in 1960.[18,20] Long-term control of small pondweed, *Potamogeton pusillus* L., leafy pondweed, *P. foliosus* Raf., in Whetzel

FIGURE 8. American pondweed, *Potamogeton nodosus* Poir.

Pond, Dewey, Ill., and small and leafy pondweed in Giffin Pond, Charleston, Ill., began in 1964.[22,25]

Beginning in 1968 the long-term control of curlyleaf pondweed in Big, Turtle, and Southeast ponds in Utterback Complex, Gibson City, Ill., was also initiated. The long-term control of sago pondweed, *P. pectinatus* L., in Yankee Ridge Lake, Urbana, Ill., and sago and leafy pondweeds in Miller Pond, Monticello, Ill., was initiated in 1970 and 1971, respectively, and the aquatic plant control activities have been continuous. Later the long-term control of aquatic plants in North and Hedge Pond, Utterback Complex, and Goldfish Pond, Allerton Park, was also initiated. Long-term observations of the aquatic plants, complexes, and frequently aquatic plant control research activities resulted in the elimination of year stands of several aquatic plants in other ponds, such as Brunhow Pond, Pasture, Orchard Pond, Miller Complex, and Allerton Lake. The results of many of these studies have been reported.[20,22,24,25]

Aquatic herbicides were applied to test areas in a variety of water bodies. These include former gravel and coal-strip mines, and typical farm ponds formed by the construction of an earthen dam on an existing waterway. Many had diverse uses such as farm recreation, fishing, swimming, residential development, 4-H club camp activities, reflection pools, and water supply. The water bodies ranged in size from a circular pond of approximately 25 ft in diameter to a 5000-acre cooling water supply reservoir. No attempt was made to compare the water quality in these bodies of water used in these studies. Recently, however, an investigation of the interaction of aquatic plants and aquatic herbicides on water quality from 30 water sources was initiated,[23,26] although we had long been interested in these interactions.[21] Duplication of test areas was not usually undertaken in the same body of water, as duplication was undertaken by applying the same aquatic herbicide to the same aquatic plant in one or more bodies of water, or by follow-up studies conducted on the same aquatic plant in the same water body in succeeding years. The effects of the aquatic herbicides on the aquatic plants within each test areas were estimated by visual observations, and all observations of the test areas were recorded on 35-mm colored slides.

Rates of Application

The object was to determine the appropriate amounts of aquatic herbicides which applied to a test area would eliminate all of the existing aquatic vegetation within the test area.

Aquatic herbicides are available as granular, liquid, or wettable powder formulations. Diquat is available only as a liquid formulation, whereas only the granular formulations of 2,4-D are registered by the Federal Environmental Protection Agency (EPA) for use as aquatic herbicides. Aquatic herbicides available as either liquid or granular formulations were equally effective as long as the amounts of active ingredient (ai) were the same. However, it should be mentioned that not all aquatic plants are susceptible to all the aquatic herbicides. Therefore, it is necessary to identify the aquatic plants, at least to the family. The application of an aquatic herbicide to an aquatic plant not susceptible to that aquatic herbicide will not produce suitable results, i.e., the complete elimination of the existing aquatic plants within the areas treated.

Field rates of herbicides for terrestrial applications are usually given as pound per acre, and similar rates were suggested for the control of emersed and submersed plants. Field experience indicated that the application of aquatic herbicides on an area basis may not give satisfactory results, particularly in deeper water. Currently there are rates of application of aquatic herbicides suggested on an area basis; however, such rates of applications have resulted in the failure of the aquatic herbicide to remove the existing stand of aquatic vegetation from several bodies of water.[26a]

Emersed Plants

To eliminate the need to estimate irregular marginal areas within bodies of water, the development of rates of application based on the dilution of a volume of liquid or wettable powder formulation to 1 gal of water was initiated. The resulting solution was to be applied to the emersed foliage to the point of runoff. Suitable rates were developed and the estimation of the irregular areas was discontinued.

If a granular formulation is used an estimate of the size of the areas to be treated is necessary. However, this does not complicate the situation since the emersed aquatic vegetation usually extends out into the water to a depth of 3 to 4 ft. Thus, the application of a granular formulation over these areas is not difficult, even though the density of the aquatic plant growths within the area vary. It was the variation in the density of the stands that made the liquid application difficult. Also the application of additional solution to the emersed foliage beyond the point of runoff did not provide increased effectiveness of the herbicide.

Free-Floating Rates of Application

Suitable rates of application are given in either parts per million or volume of liquid formulation diluted to 1 gal with water. In the latter case, the herbicide solution is applied to the foliage floating on the water surface as a fine spray, particularly if a contact herbicide is being used. Diquat can be applied in either rate expression.

Submersed Aquatic Plants

Early rates of application suggested for liquid formulations of Silvex® and others were also based on pound per surface acre. Submersed aquatic plants inhabit shallow water, water up to 15 ft in depth, and when the water is very clear even in greater water depths, although the plants may not reach the surface. Due to the consideration of the toxicities of some aquatic herbicide formulations to fishes, the application of some herbicide formulations, based on pound per surface acre, could exceed toxic levels for fishes. Further, since the deeper water frequently contained greater quantities of aquatic plants, the distribution of the herbicide throughout the water mass would be necessary; therefore, the rate of application which took into account water depth would be more precise. The ratio of pounds of herbicide per million pounds of water (abbreviated ppm) appeared to be a suitable rate expression that would consider water depth. The quantities of aquatic herbicides in the formulations are expressed as percent, and are expressed in pounds per gallon or pounds per hundred pounds of the granular formulation.

When the test areas were established, usually the 20-ft^2 areas, the water depth was measured at the four corners of the test area. In test areas, such as 50-ft^2 areas, water depth measurements of the four corners of the test area were adequate. In larger test areas additional measurements are necessary to obtain adequate water depth values. A stick, metal pole, or a weighted chain, calibrated in feet, made an excellent device to measure water depth. Then it was only necessary to use the following formula to estimate the water volume infested by aquatic plants.

$$\text{Average length} \times \text{average width} \times \text{average depth} \times 62.4 \text{ lb} = \text{lb water (1 ft}^3 \text{ of water weighs 62.4 lb)}$$

To determine the amount of active ingredient that would be applied to the water volume, it is necessary to divide by 1 million, and the formula has the following form:

$$\frac{\text{Average length} \times \text{average width} \times \text{average depth} \times 62.4 \text{ lb}}{1 \text{ million}} = \begin{array}{l}\text{lb ai that must be added}\\ \text{to 1 million lb water to}\\ \text{give a rate of 1 ppm}\end{array}$$

One surface acre with a water depth of 1 ft (1 acre-ft) contains 2.718 million lb of water. Therefore, 2.7 lb of active ingredient applied to 1 acre-ft of water would give a rate of 1 ppm. If the desired rate is greater than or less than 1 ppm, the amount of active ingredient can easily be calculated. The remaining calculation is determining the amount of formulation that would contain the necessary amount of active ingredient. The content of the active ingredient is given on the label.

For example, using the data for a rate of 1 ppm in 1 acre-ft of water (2.7 lb), the liquid formulation contains 2-lb active ingredient per gallon. Dividing the pounds required, 2.7 by 2 lb/gal, indicates that 1.35 gal of formulation are required. Here it would be convenient to use the metric system since 1 gal contains 3785 mℓ, thus 0.35 gal would be the equivalent of 1325 mℓ. For a granular formulation using 20 lb/100 lb, 1 lb of granular formulation would contain 0.2 lb of active ingredient. Thus, to determine the amount of granular formulation would be to divide 2.7 by 0.2 lb which gives 13.5 lb of granular formulation and would contain 2.7 lb of active ingredient. Applying these calculations the rate of application and amount of formulation required to contain the appropriate amount of aquatic herbicide can be easily determined.

Floating-Leaved Aquatic Plants

Suitable rates of application are either parts per million or volume of liquid formulation diluted to 1 gal with water. Liquid formulations are applied to the floating leaves, whereas a granular formulation would be applied on a water volume basis. It is always necessary to apply to any aquatic plant an aquatic herbicide to which the plant is susceptible. The appropriate herbicides are listed below. Do not use a herbicide that is not listed and apply only at the suggested rate. Do not increase nor reduce the rate. In certain situations, such as along the margins of a bay where wind and water currents are prevalent, it may be necessary to increase the rate to compensate for losses of the herbicides due to drift and diffusion.

It is always advisable to apply liquid for the control of submersed aquatic plants below the water surface, since the top layer of water skims and if the aquatic herbicide acts as a wetting agent, much of the herbicide may drift out of the treated area.

The primary reasons for aquatic herbicide failures have been applying a herbicide that is not effective against that aquatic plant, not using a suitable rate, or misapplication of the aquatic herbicide.

Herbicide Activities

The aquatic herbicides can be divided into two main groups: (1) contact and (2) systemic which are based on how they affect the plant. A contact herbicide kills the plant or plant parts, such as leaves, on direct contact with the plant, hence a rapid killing action is observed. Contact herbicides include endothall and diquat, and copper is a contact algicide.

Systemic herbicides are usually slower acting since the plant must absorb and translocate the herbicide within the plant to the site of activity. Systemic herbicides are 2,4-D, Silvex®, and simazine. Before any effects are noted, 1 to 2 weeks may pass, albeit the actual contact time of the herbicide with the aquatic plant may be relatively short.

Aquatic Herbicide Formulations

Aquatic herbicides are available in three general formulations: (1) liquids which are the solutions of the active ingredient in water or some other suitable carrier, (2) granular in which the active ingredient is impregnated into or onto a pellet which is then dispersed into the water (various types of pelleting materials have been used and are listed as inert ingredients on the label), and (3) wettable powders which are finely divided powders containing some type of solubilizing agent and are distributed as a dry powder which is mixed with water for dispersion into the aquatic environment.

Some aquatic herbicides are available only as a liquid, such as diquat, whereas others are available as liquid or granular formulations. Some are available only as wettable powders. The chemistry, solubility, and other factors dictate, in part, the formulation available. The amount of active ingredient is clearly stated on all packages.

Experimental formulations of herbicides are currently being developed in which the active ingredient is mixed with a suitable matrix that provides a controlled release of the active ingredient over a period of time. As yet these formulations are not on the open market.

Application Techniques

Liquid or wettable powder formulations are more easily applied into the aquatic environment than are the granular formulations.

The conventional hand garden sprayer can be used to apply the herbicide solution to the emersed, free-floating, and floating-leaved aquatic plants. This sprayer can also be used to apply aquatic herbicides for the control of submersed aquatic plants and algae in small areas such as home water frontages and around boat docks. For large applications, suitable pressure pumps are available and the liquid or wettable powder herbicide solution can be mixed into the water flow.

One suitable arrangement is the use of a boat bailer. The bailer is attached to an outboard motor and the bailer hose connected to a suitable container which contains the herbicide solution. This is usually the least expensive, but must be used before the aquatic plants reach the water surface. This method is very useful in total pond applications for contact or systemic herbicides.

Suitable application equipment for small-scale applications of granular formulations into the aquatic environment has not been developed. Cyclone-type grass seeder or similar devices have been mounted on boats and driven by electric motors; however, there are some drawbacks with this application technique.

A hand scoop is frequently satisfactory for small-scale application; however, the operator should wear protective equipment such as a mask and rubber gloves.

Mist blowers have been used satisfactorily for the application of granular formulations in large areas.

Preemergence Aquatic Herbicides

Two aquatic herbicide formulations have been registered by the Federal EPA for preemergence use: dichlobenil (2,6-dichlorobenzonitrile) developed by the Thompson-Hayward Chemical Company and sold under the tradename of Casoron® (only the granular formulation is registered for aquatic use), and Fenac® (2,3,6-trichlorophenyl acetic acid) developed by the Amchem Products, Inc. (liquid and granular formulations are available). Fenac® must be applied to the exposed pond hydrosoil as is it not effective when applied through the water.

Dichlobenil, applied to the exposed pond hydrosoil in late November or early December before the winter freeze, at rate of 4-lb active per acre, provided season-long control of the higher-branched algae, chara, sago pondweed, and small pondweed; at 8-lb active per acre, it gave season-long control of southern naiad. Rates of dichlobenil, 12-lb active per acre, did not provide season-long control of leafy pondweed. Suitable results were not readily obtained when the dichlobenil was applied at higher rates and through the water.

Fenac®, 10-lb active per acre, applied to the exposed pond hydrosoil in late November or early December prior to the winter freeze prevented the growth of sago pondweed and leafy pondweed. However, leafy pondweed later, in mid-July, infested the area treated with 10 lb of Fenac®. Higher rates of Fenac® (20, 30, and 40 lb/acre) did not provide season-long control of leafy pondweed. Ten lb of Fenac® applied to exposed pond hydrosoil reduced, but did not completely prevent, the slender naiad. Higher rates of 20 and 40 lb of Fenac® further reduced, but did not completely prevent, the growth of slender naiad. The growth of small pondweed was prevented by 40 lb of Fenac®, but it would be presumed that Fenac® would be effective at a lower rate. Rates of Fenac® as high as 60 lb/acre did not provide season-long control of leafy pondweed. In central Illinois, sago and small pondweed appear to develop only one stand, early in the growing season, whereas leafy pondweed will develop stands all season long. Thus, it is presumed that the levels of Fenac® in the hydrosoil are rapidly reduced below the level necessary to prevent the growth of leafy pondweed.

Efficacy of Herbicides

Emersed Aquatic Plant Control

For the most part, the emersed group of aquatic plants are not very troublesome although all or most of the plants can and will infest shallow water areas. Liquid ester formulations of 2,4-D applied to the foliage at a rate of ¼ cup of liquid (4 lb/gal of acid derivative) ester or amine derivatives diluted with water to 1 gal eliminated the creeping water primrose, *Jussiaca repens* L. var *glabrescens* Ktze.; waterwillow, *Justicia americana* (L.); Vahl hardstem bulrush, *Scirpus acutus* Muhl; and the common arrowhead, *Saggittaria latifolia* Willd. These are the common emersed species found in central Illinois, but are widely distributed throughout the Midwest and the U.S. They were eliminated by the application of 1-lb granular (20% acid equivalent) 2,4-D/440 ft^2.

Common cattail, *Typha latifolia* L., and narrowleaf cattail, *T. augustifolia* L., were controlled by application of liquid ester formulations of 2,4-D; however, high rates were required. Applications of 2,2-dichloropropionic acid, dalapon (4 oz/gal of water), 3-amino-s-triazole, amitrole (2 oz/gal of water) were very effective on both species. The addition of a wetting agent or liquid household detergent added to the water solution greatly increased the effectiveness of the herbicides. Very low rates of diquat (10 m*l*/gal of water), plus either a nonionic wetting agent (X-77) or common household liquid detergent Joy®, gave excellent results. It was possible, by making repeated applications of the herbicides followed by the hand pulling of newly developed cattail shoots, to completely eliminate cattail from a body of water. Many of these emersed species help prevent bank erosion. A bed of cattails is being maintained in an area of severe bank erosion while keeping the remainder of the shoreline free of cattails.

Another common representative of this group of aquatic plants is the American lotus, *Nelumbo lutea* (Willd.) Pers. The large, 20-cm diameter bluish-green leaves extend above the water surface. The petiole is attached to the center of the leaf. Some leaves may float on the water surface. In late summer, fruits are contained in a light, spongy, circular top-shaped structure that extends above the water surface. American lotus is a member of the water lily family, Nymphaea, and has a rootstock buried in the hydrosoil. It is susceptible to 2,4-D or liquid Silvex®.

Free-Floating Plants

Two common representatives of this group of plants are the common duckweed, *Lemna minor* L., and watermeal, *Wolffia columbiana* Karst. The control of these two species is complicated by the large number of plants on the surface of the water, and the duckweed plants will form layers. Thus, it is difficult to apply herbicides, such as diquat or endothall derivatives, in contact with the duckweed; however, duckweed has been substantially reduced in many test areas by the application of diquat, disodium or dipotassium salts of endothall. Repeat applications may be necessary every 1 to 2 weeks to reduce the duckweed below nuisance levels. By repeated applications of the above mentioned herbicides, duckweed can be brought under control throughout the growing season; however, the effort must be continuous because of the great reproductive potential of the duckweed. During the normal growing season a single duckweed plant will produce from 3 to 7 seeds and grow from a seed to a seed-producing plant in 3 days. Duckweed can also divide vegetatively.

The stand of duckweed can be reduced by applying an aquatic herbicide early in the growing season, and can, by repeated applications, keep the duckweed stand below nuisance levels. In this manner, the water surface could be kept relatively free of duckweed. Duckweed produces overwintering structures which sink to the pond hydrosoil in late fall and rise in the spring to develop into duckweed plants. Reducing the duckweed population in the fall might also prevent the development of a severe stand of duckweed the following spring.

Simazine, 0.5 ppm, is very effective against duckweed.

Bodies of water infested by watermeal would appear very similar to bodies of water infested by duckweed. However, the watermeal on the surface would not seriously interfere with fishing. Frequently watermeal and duckweed occur in the same body of water. Some aquatic herbicides used to control duckweed do not affect watermeal. Without the competition from the duckweed, watermeal will develop an extensive stand.

Simazine at a rate of 1 ppm will reduce substantially the stand of watermeal. Higher rates of simazine have not been used against watermeal.

There are additional members of the Lemnaceae, or duckweed, family in Illinois, such as *Spirodella* and *Wolffiella*, but they usually are not serious problems; however, they can develop severe infestations.

Submersed Aquatic Plants with Alternate Leaf Attachment

Most, if not all, of the aquatic plants with the alternate leaf attachment to the center stem comprise members of the Potamogeton family. Sago pondweed, *Potamogeton pectinatus* L.; leafy pondweed, *P. foliosus* Raf.; small pondweed, *P. pusillus* L.; and curlyleaf pondweed, *P. crispus* L., have been controlled in many bodies of water by the application of disodium endothall, 1 ppm, or diquat, 0.5 ppm. Recently these four species have been found to be more susceptible to the dipotassium salt of endothall, and have been eliminated from bodies of water or in large-scale application (4000 ft^2) following the application of 0.3 ppm dipotassium endothall. However, in several bodies of water, these pondweeds have been removed each year, and the numbers of plants have decreased in succeeding years; the few plants that developed each year have been removed by the use of the granular formulation of dipotassium endothall. In these bodies of water pondweeds have not been serious problems. However, the control or removal activities must be continuous to obtain the reduction in the growth of these aquatic plants.

Submersed Aquatic Plants with Opposite Leaf Attachment

The plants of this group are characterized by two leaves emerging from the center stem at the same point or node. Several families are represented. Two of the most ubiquitous and widely distributed throughout the U.S. are southern naiad, *Najas guadalupensis* (Spreng.) Magnus, and slender naiad, *N. flexilis* (Willd.) Rotsk. and Schmidt. These two species are readily controlled by diquat at 1 ppm or by dipotassium salt of endothall at 3 ppm.

The control of cabomba, *Cabomba caroliniana* Gray, has not yet been achieved. The only herbicide which was found to be effective against cabomba was a granular isooctyl ester formulation of 2-(2,4-dichlorophenoxy) propionic acid (dichlorprop) at a rate of 5 ppm; however, no damage to the cabomba was evident until 8 weeks after the application and the cabomba was not completely removed from the treated area until 12 weeks following the application. Applications of the liquid formulation of dichlorprop at the same rate were not effective. Applications of granular 2,4-D or Silvex®, at extremely high rates, were not effective; neither were high rates of liquid diquat, 2,4-D, and Silvex® nor the disodium and dipotassium salts of endothall. Combinations of diquat and chelated-copper formulations, at rates as high as 3 ppm diquat and copper, were not effective. Combinations of dipotassium salt of endothall and dipotassium salt of Silvex® in small- and large-scale applications were not effective. The liquid and granular formulations of the mono- or di-(N,*N*-dimethylalkylamine) salts of endothall at a rate of 2 ppm endothall content caused severe top burn of the cabomba, but by the end of the growing season the cabomba had recovered. Application of either the mono- or di-(N,*N*-dimethylalkylamine) salt of endothall to small areas could provide some temporary relief from cabomba. Larger-scale applications of these herbicide formulations for the control of cabomba are not possible due to the toxicities of these herbicide formulations to fishes.

Horned pondweed, *Zannichellia palustris* L., has the opposite leaf attachment to the center stem, but because of the thin, delicate, rounded, pointed leaves, it can easily be confused with sago pondweed. It is easily distinguished from the sago pondweed by the characteristic horned fruiting body; sago pondweed has leaves that are larger and coarser and has the cluster of fruiting bodies at the end of the peduncle. Horned pondweed can be controlled by dipotassium salt of endothall and by diquat at 1 ppm. Horned pondweed was not widely distributed within the state of Illinois and recently patches have been observed in several bodies of water in central Illinois, but horned pondweed has remained a minor component in these bodies of water. However, in 1975, an extensive stand of horned pondweed developed in Fox Pond which was eliminated by the application of 1 ppm of dipotassium salt of endothall.

Simazine was applied to Fox Pond for the control of filamentous algae in 1976 and 1977. Extensive stands of horned pondweed did not develop in Fox Pond during 1976 and 1977.

During 1977 0.5 ppm simazine was applied to Crappie Pond in the Pollywog Association complex Oakwood, Ill., primarily for the control of cabomba; however, northern watermilfoil and coontail were also present. The coontail and the existing submersed aquatic plants had not disappeared after 4 weeks, whereas the emersed aquatic plants, the waterwillow, and the creeping waterprimrose were eliminated and the water smartweed severely damaged. An additional 0.5 ppm of simazine was applied and the coontail disappeared in about 2 weeks, and by late summer the small amount of cabomba present at the time of application could not be found. The small quantity of northern watermilfoil persisted throughout the summer, but the stand did not increase.

Submersed Aquatic Plants with Whorled-Leaf Attachment

The plants of this group are characterized by four or more leaves attached to the center stem at the same node. The leaves may be relatively simple or they may be highly branched. Several families of plants are also represented by members of this group. Probably the most common of this group is watermilfoil, such as northern watermilfoil, *Myriophyllum exalbescens* Fernald, and Eurasian watermilfoil, *M. spicatum* L. Also present, but not quite as common, is the broadleaf watermilfoil *M. heterophyllum* Michx. All three species are susceptible to granular 2,4-D at 2 ppm free-acid equivalent, and to liquid Silvex® at 2 ppm free-acid equivalent. Diquat cation and potassium salt of endothall at 1 ppm are also effective against these species.

Another species that has wide distribution throughout the U.S. and is very common in Illinois and the Midwest, is the common coontail, *Ceratophyllum demersum* L. Common coontail is susceptible to 2,4-D and Silvex®, 2 ppm, but is also susceptible to dipotassium salt of endothall at 2 ppm. The disodium salt of endothall is not as effective against the common coontail as is the potassium salt of endothall. Diquat cation, 1 ppm, is also very effective against the common coontail. The common coontail is a free-floating aquatic plant, but because of the leaf attachment and other characteristics is usually discussed with this group of plants. Common coontail, however, can remain anchored throughout the growing season, or the highly branched plant will break loose and float with the wind and water currents.

The control of elodea, *Elodea canadensis* Michx., is possible by the application of diquat at a rate of 1 ppm. Other investigators have reported that elodea can be controlled by the application of a combination of the dipotassium salt of endothall and the dipotassium salt of Silvex®; however, the author was not successful in obtaining any control of elodea following the application of this combination of aquatic herbicides in small plot applications.

Floating-Leaved Aquatic Plants

The primary characteristic of this group of aquatic plants is the various sizes and shapes of the leaves that float on the water surface. All these aquatic plants are rooted in the hydrosoil.

Probably the most well-known and widespread representative of this group is fragrant or white water lily, *Nymphaea odorata* Art. or *N. tuberosa* Paine, respectively. The petiole of the water lily is attached to the leaf at the summit of a deep notch in the nearly circular leaf.

Another member of the water lily family with floating leaves is water shield, *Braseni schreberi* Gmel. The dark brown circular floating leaves of water shield are 10 cm or less in diameter and frequently the underside of leaves has a heavy coating of gelatinous slime; the flowers when submersed in water also have the gelatinous slime covering. Water shield is not very common, but when present, will form dense beds in ponds or lakes.

Another member of the water lily family that could be placed in this group is the spatterdock, *Nuphar advena* (Ait.) Ait. f. The leaves are longer than wide, with some leaves floating on the water surface, whereas many of the leaves extend above the water surface. In spatterdock the petiole is attached at the summit of a less severe notch in the leaf than in water lily. The spatterdock has a yellow flower and a large mass of seeds extending above the water surface.

The control of water lily and water shield is possible by the application of granular 2,4-D or liquid Silvex® to the floating leaves. To control or eliminate these species, the spreading rootstock in the hydrosoil must be killed, which might make the use of granular 2,4-D more effective.

Spatterdock is not readily controlled by 2,4-D, although the populations can be reduced following the use of the granular 2,4-D formulations. Dichlobenil applied postemergent at a rate of 5 lb/acre is very effective against spatterdock.

Another common representative of this type of plant is the floating-leaved *Potamogeton*. The floating leaves vary from the small oval leaves of *P. vaseyi* Robbins to the long, narrow leaves of American pondweed, *P. nodosus* Poir, and floating pondweed, *P. natans* L.

American pondweed is very common in Illinois and generally is not a serious problem. However, in some ponds it can become a serious problem. *P. vaseyi* is less common, but can form dense stands in some bodies of water; it has finely divided submersed leaves. American pondweed has thin, broad, membranous submersed leaves. Floating pondweed is not common in Illinois.

Liquid dipotassium endothall (¼ cup/gal of water) will control American pondweed and *P. vaseyi*. Also 1 ppm of granular dipotassium endothall will control American pondweed and *P. vaseyi*. Diquat is not effective against these species.

Miscellaneous Submersed Aquatic Plants

The attachment of the leaves to the center stem in the aquatic plants in this group do not readily fit any of the types discussed previously.

The finely divided leaves of *Ranunculus trichophyllus* Chaix, white water buttercup, are borne singly along the center stem and appear to be borne on one side of the stem. The plants has a white flower that extends above the water surface in early spring. The plant is usually found in shallow water, but it can grow in deeper water.

Diquat at a rate of 1 ppm will control white water buttercup.

Water stargrass, *Heteranthera dubia* (Jacq.) MacM., has long ribbon-like leaves trailing through the water. If seen in the water early in the growing season, water stargrass closely resembles some of the Potamogeton group of aquatic plants; however, it is easily distinguished from the *Potamogeton* by the complete lack of a midvein in the leaves. Later in the growing season the yellow flower with fives points, resembling a star, develops.

Diquat, 1 ppm, controls water stargrass. The dipotassium salt of endothall also will control water stargrass, but a rate of 3 to 5 ppm will be required.

Many bodies of water contain a few water smartweed, *Polygonum amphibium* L., plants. These plants are usually anchored at the shoreline and slender stems float on the water surface supporting the leaves, flowers, and fruits. Usually these are not serious problems since only a few plants are present and very little control activity has been required.

In a new water supply lake, another amphibious polygonum, *P. coccineum*, developed a dense stand after the lake had been filled and was existing in 12 ft of water.

Granular 2,4-D was not effective. Liquid applications of 2,4-D and Silvex® could severely damage the emersed foliage but did not provide any degree of control, as the slender underwater stem remained viable. Introduction of liquid 2,4-D into the water containing the dense bed of *P. coccineum* and application of liquid 2,4-D to the emersed foliage substantially reduced the stand, and by the following spring the amphibious water smartweed had disappeared.

Control of Aquatic Plants

Control of aquatic plants is usually construed to mean the removal of an existing stand of aquatic vegetation by the application of an aquatic herbicide. Pond owners appear frequently to hope that the problem will disappear following one application of an aquatic herbicide.

To learn about the long-term control of aquatic plants, in 1960 we initiated the removal of curlyleaf pondweed from Mansion Pond, near Monticello. The growth cycle of curlyleaf pondweed in central Illinois begins late in July or August, a dense stand will develop by late fall, overwinter, develop flowers and fruiting bodies early in the summer, and disappear during June or July. The stand of curlyleaf pondweed present in Mansion Pond in the spring of 1960 was removed by the application of 1 ppm of the disodium salt of endothall. Only the water infested with curlyleaf pondweed was treated with endothall. A stand developed in August and it also was removed following the application of endothall. It was hoped that by the removal of the fall stand, further work during the following spring would not be necessary. However, another stand of curlyleaf pondweed developed in the spring of 1961, and it also was removed as well as the stand that developed in the fall. The spring and fall stands of curlyleaf pondweed were also removed in 1962. In spring of 1963 the stand of curlyleaf pondweed was less dense; this stand was removed using endothall. A dense fall stand did not develop, but there might have been a few plants. All applications of endothall were made as early as possible to prevent the development of viable fruiting bodies and/or turions and reduce potential reinfestation in subsequent years. During some years only a few plants of curlyleaf pondweed developed while in other years more curlyleaf pondweed plants were present, presumably from tubers or viable seeds present in the hydrosoil. Very few curlyleaf pondweed plants developed in 1977; those that did were removed. Curlyleaf pondweed has not been eliminated from Mansion Pond, but since 1962 it has not been a nuisance. Fortunately we have not had an invasion of another aquatic plant in Mansion Pond.

Chara, which was present in 1960 as a minor component, was not affected by the sodium endothall and became abundant in Mansion Pond. A natural die-off of the chara occurred in 1963, but in 1964 the existing stand of chara was removed. Nuisance levels of chara have not been present in Mansion Pond since 1964. From time to time it has been necessary to apply algicides to reduce the filamentous algae content in Mansion Pond, but, in general, filamentous algae has not been at nuisance levels.

In 1964, the control of leafy pondweed and small pondweed in Whetzel Pond, Dewey, Ill., was undertaken and the existing stands of these pondweeds were eliminated following the application of diquat, 0.5 ppm. During the intervening years, the pondweed plants that developed were removed using granular endothall as soon as possible after their presence was noted. The object was not to let Whetzel Pond become seriously infested with either leafy or small pondweed. Leafy pondweed plants can develop throughout the growing season, thus periodic checks of Whetzel Pond were necessary. When leafy pondweed plants were found they were eliminated. Thus, the continuous removal of the leafy pondweed plants has been very effective, and since 1964 Whetzel Pond has been relatively free of leafy and small pondweed.

Control of filamentous algae has been necessary, but the stands were removed before they became dense. Filamentous algae also has not been at nuisance levels in Whetzel Pond.

Beginning in 1968, the control or elimination of curlyleaf pondweed from Big, Turtle, Southeast, and North Ponds of the Utterback Complex, Gibson City, Ill., has been undertaken each year. Potassium endothall, 0.3 ppm, has been used in these ponds. Nuisance levels of filamentous algae have developed and were removed using an algicide. The stands of curlyleaf pondweed in Big, North, and Southeast Ponds have been less, in part due to the herbicidal activity of the algicide. The sago pondweed in Yankee Ridge Lake, Urbana, also has been under control since 1971 using potassium endothall at 0.3 ppm.

The aquatic plant complex in the 6-acre Miller Pond, Monticello, Ill., has been controlled since 1971. Simazine, 0.5 ppm, has been used primarily for the control of the dense stands of filamentous algae, but simazine also has herbicidal activity so that sago and small pondweed and slender and southern naiad have been controlled as well. Diuron, 0.8 lb/acre (equivalent to 0.05 ppm), has been used in Miller Pond and has given reasonably good control of the pondweeds, naiad, and filamentous algae. The use of simazine or diuron (diuron has not been granted registration for use in the aquatic environment) has some advantages over similar work since these two agents control both aquatic plants and algae. In Big and Southeast ponds, two applications were made to obtain the same results.

SUMMARY OF RESULTS

This work does indicate that control of aquatic plants and algae are possible; however, once initiated the control activities must be undertaken each year to reduce the aquatic plants below nuisance levels. The pond owner can obtain similar results without severly damaging the aquatic environment. Results of these studies are summarized in Table 1.

Table 1
CHEMICAL CONTROL OF SOME AQUATIC PLANTS

Group and species	Chemical, ai[a], or free-ae[a]	Rate of application	Remarks
Emergent plants			
Arrowhead (*Sagittaria* spp.)	Use one of the following		
	2,4-D		
	Ester (20% granules)	1 lb/440 ft^2	Spread on water
	Ester (4 lb/gal)	¼ cup/2 gal	Wet foliage
	Amine (4 lb/gal)	¼ cup/2 gal	Wet foliage
	Silvex®[b]		
	Ester (4 lb/gal)	¼ cup/2 gal	Wet foliage
	Potassium salt (6 lb/gal)	¼ cup/2 gal	Wet foliage
	Potassium salt (20% granules)	1 lb/440 ft^2	Spread on water
	Diquat cation (2 lb/gal)	¼ cup/gal	Wet foliage
Bulrush (*Scirpus acutus*)	Use one of the following		
	2,4-D		
	Ester (20% granules)	1 lb/440 ft^2	Spread on water
	Ester (4 lb/gal)	½ cup/2 gal	Wet stems
	Diquat cation (2 lb/gal)	2 tbsp/3 gal and 1 tsp nonionic wetting agent	Wet foliage to point of run-off
	Dichlobenil (aquatic granules 4%)[c]	100 lb/acre	Apply in March to exposed bottom soil
Cattails (*Typha* spp.)	Use one of the following		
	Dalapon	4 oz/gal and 3 caps detergent	Wet foliage
	Amitrole	2 oz/gal and 3 caps detergent	Wet foliage
	2,4-D Ester (4 lb/gal)	½ cup/gal and 3 caps detergent	Wet foliage
	Diquat cation (2 lb/gal)	2 tbsp/ 3 gal and 1 tsp nonionic wetting agent	
Creeping water primrose (*Jussiaea repens* var. *glabrescens*)	Use one of the following		
	2,4-D		
	Ester (20% granules)	1 lb/440 ft^2	Spread on water
	Ester (4 lb/gal)	¼ cup/2 gal	Wet foliage
	Amine (4 lb/gal)	¼ cup/2 gal	Wet foliage
	Silvex®[b]		
	Ester (4 lb/gal)	¼ cup/2 gal	Wet foliage

	Potassium salt (6 lb/gal)	¼ cup/2 gal	Wet foliage
	Potassium salt (20% granules)	2 lb/440 ft^2	Spread on water
	Diquat cation (2 lb/gal)	¼ cup/2 gal	Wet foliage
Spatterdock (*Nuphar advena*)	Dichlobenil	6 lb ai/acre 3 lb 4% granules per 440 ft^2	Spread on Water
Waterwillow (*Justicia americana*)	Use one of the following		
	2,4-D		
	Ester (20% granules)	1 lb/440 ft^2	Spread on water
	Ester (4 lb/gal)	¼ cup/2 gal	Wet foliage
	Amine (4 lb/gal)	¼ cup/2 gal	Wet foliage
	Silvex®[b]		
	Ester (4 lb/gal)	¼ cup/2 gal	Wet foliage
	Potassium salt (6 lb/gal)	¼ cup/2 gal	Wet foliage
	Potassium salt (20% granules)	1 lb/440 ft^2	Spread on water
	Diquat (2 lb/gal)	¼ cup/2 gal	Wet foliage
Submersed plants with alternate leaf attachment			
Curlyleaf pondweed (*Potamogeton crispus*)	Use one of the following		
	Endothall (potassium salt, 4.23 lb/gal or 10% granules)	0.3 ppm (total or large-scale application) 1.0 ppm (marginal application)	Apply on or below surface
	Diquat (2 lb/gal)	0.5 ppm or 1 gal/surface acre	Same as above
	Dichlobenil (aquatic granules 4%)	200 lb/acre	Preemergent application
	Fenac®	See manufacturer's directions	Must be applied to exposed pond bottom
	Diquat/copper-triethanolamine complex	0.25 ppm diquat plus equal volume of copper-triethanolamine complex	Apply on or below water surface
	Endothall [di (N, N-di-methylalkylamine)salt] (liquid) (10% granules)	0.5 ppm (endothall content) 100 lb/acre	Apply on or below water surface Spread on water
	Simazine (80-WP)	0.5 ppm	Apply to total water volume
Leafy pondweed (*P. foliosus*)	Use one of the following		
	Endothall (potassium salt, 4.23 lb/gal or 10% granules)	0.3 ppm (total or large-scale application)	Apply on or below water surface

Table 1 (continued)
CHEMICAL CONTROL OF SOME AQUATIC PLANTS

Group and species	Chemical, ai[a], or free-ae[a]	Rate of application	Remarks
		1.0 ppm (marginal application)	
	Diquat cation (2 lb/gal)	0.5 ppm, or 1 gal/surface acre	Same as above
	Dichlobenil (aquatic granules 4%)	300 lb/acre	Preemergent application[d]
	Fenac® (10% granules)	See manufacturer's directions	Must be applied to exposed pond bottom
	Endothall [di(N, N-di-methyl-alkylamine) salt] (10% granules)	100 lb/acre	Spread on water Apply to total water volume
	Simazine (80-WP)	0.5 ppm	
Sago pondweed (*P. pectinatus*)	Use one of the following		
	Endothall (potassium salt, 4.23 lb/gal or 10% granules)	0.3 ppm (total or large-scale application) 1.0 ppm (marginal application)	Apply on or below water surface
	Diquat cation (2 lb/gal)	0.5 ppm, or 1 gal/surface acre	Same as above
	Dichlobenil (aquatic granules 4%)	100 lb/acre	Preemergent application
	Fenac® (10% granules)	See manufacturer's directions	Must be applied to exposed pond bottom
	Simazine (80-WP)	0.5 ppm	Apply to total water volume
Small pondweed (*P. Pusillus*)	Use one of the following		
	Endothall (potassium salt, 4.23 lb/gal or 10% granules)	0.3 ppm (total or large-scale application) 1.0 ppm (marginal application)	Apply on or below water surface
	Diquat cation (2 lb/gal)	0.5 ppm	Same as above
	Dichlobenil (aquatic granules 4%)	200 lb/acre	Preemergent application
	Fenac® (10% granules)	See manufacturer's directions	Must be applied to exposed pond bottom
	Endothall [di(N, N-di-methyl-alkylamine)salt] (10% granules)	100 lb/acre	Spread on water
	Simazine (80-WP)	0.5 ppm	Apply to total water volume
Waterstar grass (*Heteranthera dubia*)	Diquat cation (2 lb/gal)	1 ppm, or 2 gal/surface acre	Apply on or below water surface

	Endothall (potassium or 10% granules)	5 ppm	Same as above
Submersed aquatic plants with opposite leaf attachment			
Buttercup (*Ranunculus* spp.)	Diquat cation (2 lb/gal)	0.55 ppm	Apply below water surface
Cabomba (*Cabomba caroliniana)*	Endothall [di(N, *N*-di-methyl-alkylamine)salt] (liquid)	2ppm	Same as above, small-scale application only
Slender naiad (*Najas flexilis)*	Use one of the following		
	Diquat cation (2 lb/gal)	1 ppm, or 1.5 gal/surface acre	Same as above
	Endothall (potassium salt, 4.23 lb/gal or 10% granules)	3 ppm (total or large-scale application) 4 ppm (marginal application)	Same as above
	Endothall [di(N, *N*-di-methyl-alkylamine)salt] (liquid)	2 ppm (endothall content)	Apply on or below water surface
	Dichlobenil (aquatic granules 4%)	200 lb/acre	Preemergent application
Southern naiad (*N. guadalupensis)*	Use one of the following		
	Diquat cation (2 lb/gal)	1 ppm, or 1.5 gal/surface acre	Apply below water surface
	Endothall (potassium salt, 4.23 lb/gal or 10% granules)	3 ppm (total or large-scale application) 4 ppm (marginal application)	Same as above
	Endothall [di(N, *N*-di-methyl-alkylamine)salt]	2 ppm (endothall content)	Apply on or below water surface
	Dichlobenil (aquatic granules 4%)	200 lb/acre	Preemergent application
Submersed aquatic plants with whorled-leaf attachment			
Coontail (*Ceratophyllum* spp.)	Use one of the following		
	Endothall (potassium salt, 4.23 lb/gal or 10% granules)	2 ppm	Spread on water
	2,4-D ester (20% granules)	2 ppm	Spread on water
	Silvex®[b] ester (4 lb/gal)	2 ppm	Spread on water
	Diquat cation (2 lb/gal)	1 ppm, or 2 gal/surface acre	Apply below water surface
	Diquat/copper-triethanolamine complex	0.5 ppm diquat plus equal volume of copper-triethanolamine complex	Apply on or below water surface

Table 1 (continued)
CHEMICAL CONTROL OF SOME AQUATIC PLANTS

Group and species	Chemical, ai[a], or free-ae[a]	Rate of application	Remarks
Elodea (*Elodea canadensis)*	Diquat cation (2 lb/gal)	1 ppm, or 2 gal/surface acre	Apply below water surface
Watermilfoil (*Myriophyllum* spp.)	2,4-D ester (20% granules)	2 ppm	Spread on water
	Silvex®[b]		
	Ester (4 lb/gal)	2 ppm	Apply below water surface
	Potassium salt (6 lb/gal)	2 ppm	Apply below water surface
	Potassium salt (20% granules)	2 ppm	Spread on water
	Endothall (potassium salt, 4.23 lb/gal or 10% granules)	3 ppm 3 ppm	Apply below water surface Spread on water
	Diquat cation (2 lb/gal)	1 ppm	Apply below water surface
	Dichlobenil (aquatic granules 4%)	240—375 lb/acre or 2.5—3.8 lb/400 ft^2	Spread on water
	Fenac® (10% granules)	See manufacturer's directions	Must be applied to exposed pond bottom
Floating-leaved aquatic plants			
American pondweed (*Potamogeton nodosus)*	Use one of the following		
	Endothall (10% granules)	1 ppm	Spread on water
	Endothall (4.23 lb potassium salt/gal)	½ cup/gal	Apply to leaves
Water lilies (*Nymphaea* spp.)	Dichlobenil	5 lb ai/acre	Spread on water
Free-floating aquatic plants			
Duckweed (*Lemna minor*)	Use one of the following		
	Endothall (potassium salt, 4.23 lb/gal or 10% granules)	1 cup/4 gal	Apply to leaves
	Diquat cation (2 lb/gal)	1 cup/4 gal	Apply to leaves
	Simazine (80-WP)	0.5 ppm	Apply to total water volume
Watermeal (*Wolffia* spp.)	Simazine (80-WP)	1 ppm	Apply to total water volume

Chara (has cylindrical, whorled branches and resembles, in form, some of the plants mentioned above) (*Chara* spp.)	Use one of the following		
	Dichlobenil (aquatic granules 4%)	100 lb/acre	Preemergent application only
	Copper sulfate[e]	1 ppm	Postemergent
	Endothall [di(N, *N*-dimethyl-alkylamine)salt] (10% granules)	0.3 ppm	Postemergent
	Mariner	15—25 lb/acre	Apply on water surface
Filamentous algae	Copper sulfate[e]	1 ppm	Postemergent
	Endothall [di(N, *N*-dimethyl-alkylamine)salt] (liquid)	0.3 ppm	Postemergent
	Mariner[f]	15 lb/acre	Spread on water
	Simazine (80-WP)	0.5 ppm	Apply to total water volume

Note: Additional information — *The Chemical Control of Some Aquatic Plants,* Mimemeographed leaflets A5-16, Section of Aquatic Biology, Illinois Natural History Survey, Urbana. *Aquatic Plants and Their Control,* Fishery Bull. No. 4, Illinois Department of Conservation, Springield.

[a] ai — active ingredient; ae — acid equivalent.

[b] This product has been withdrawn from the market pending recertification hearings.

[c] The formulation currently available may contain 10% dichlobenil; therefore the amount of granular formulation used will have to be adjusted.

[d] The preemergent herbicides have not given satisfactory season-long control of leafy pondweed.

[e] Crystalline copper sulfate can be used; however, there are several copper-containing formulations that contain copper-chelating compounds, which prevent the immediate precipitation of copper as copper carbonate. Check the label for instructions concerning their uses and rates of application. A lower rate of application of copper can be used with these latter formulations. Their copper contents may vary.

[f] A copper-containing formulation developed by the 3M Company.

REFERENCES

1. **Balcom, R. B.**, What price weeds? *U.S. Bur. Reclam. Reclam. ERA*, 35, 171, 1949.
2. **Balcom, R.B.**, Weeds — water robbers, *J. Soil Water Conserv.*, 5, 165, 1950.
3. **Bartley, T. R.**, Control of submersed aquatic weeds in irrigation, in *ASCE Annu. Meet. Natl. Meet. Water Resources Engineering*, American Society of Civil Engineers, New York, 1969.
4. **Blackburn, R. D. and Weldon, L. W.**, Results of three years of testing diquat as an aquatic herbicide in Florida, *Proc. South. Weed Conf.*, 16, 365, 1963.
5. **Blackburn, R. D.**, Control of southern naiad in Florida drainage and irrigation channels, *Weeds*, 12, 295, 1964.
6. **Bond, E. E., Lewis, R. H., and Fryer, J. L.**, Toxicity of various herbicides and herbicidal materials to fish, *Robert A. Taft Sanit. Eng. Cent. Tech. Rep.*, W-60-3, 96, 1960.
7. **Bruns, V. F., Hodgson, J. M., Arle, H. F., and Timmons, F. L.**, The use of aromatic solvents for control of submersed aquatic weeds in irrigation channels, *U.S. Dep. Agric. Circ.*, 971, 1955.
8. **Fassett, N. C.**, *A Manual of Aquatic Plants*, The University of Wisconsin Press, Madison, 1957.
9. **Grant, T. C.**, Aquatic weed control program of the central and southern Florida flood control district, *Hyacinth Control J.*, 1, 24, 1962.
10. **Grigsby, B. H.**, Response of certain unicellular green algae to several herbicides, *Proc. North Cent. Weed Control Conf.*, 14, 29, 1958.
11. **Hildbrand, E. M.**, Herbicidal action of 2,4-dichlorophenoxy acetic acid on the water hyacinth *Eichornia crassipes*, *Science*, 103, 447, 1946.
12. **Hiltibran, R. C.**, Aquatic weed control in Illinois farm ponds, *Proc. North Cent. Weed Control Conf.*, 17, 28, 1960.
13. **Hiltibran, R. C.**, Aquatic weed control in central Illinois, *Res. Rep. North Cent. Weed Control Conf.*, 18, 102, 1961.
14. **Hiltibran, R. C.**, Aquatic weed control research in central Illinois, *Res. Rep. North Cent. Weed Control Conf.*, 19, 125, 1962.
15. **Hiltibran, R. C.**, Effect of endothall on aquatic plants, *Weeds*, 11, 256, 1963.
16. **Hiltibran, R. C.**, Effect of herbicides on several aquatic plants in central Illinois, *Res. Rep. North Cent. Weed Control Conf.*, 21, 33, 1964.
17. **Hiltibran, R. C.**, The effect of diquat on aquatic plants in central Illinois, *Weeds*, 13, 71, 1965.
18. **Hiltibran, R. C.**, Eradication of curlyleaf pondweed from Mansion Pond, *Abstr. Weed Soc. Am.*, 30, 1967a.
19. **Hiltibran, R. C.**, Effect of some herbicides on fertilized eggs and fish fry, *Trans. Am. Fish. Soc.*, 96, 414, 1967b.
20. **Hiltibran, R. C.**, Aquatic weed control in Illinois, *Hyacinth Control J.*, 7, 11, 1968.
21. **Hiltibran, R. C.**, Aquatic herbicides and the aquatic environment, *Proc. North Cent. Weed Control Conf.*, 27, 12, 1972.
22. **Hiltibran, R. C.**, Some experiences in aquatic plant management, *Proc. North Cent. Weed Control Conf.*, 29, 65, 1974.
23. **Hiltibran, R. C.**, Is there a relationship between aquatic herbicides and water quality? *Proc. North Cent. Weed Control Conf.*, 31, 103, 1975.
24. **Hiltibran, R. C.**, Miller Pond: a review, *Proc. North Cent. Weed Control Conf.*, 31, 103, 1976.
25. **Hiltibran, R. C.**, Aquatic plant management: long term studies, *Proc. North Cent. Weed Control Conf.*, 32, 118, 1977.
26. **Hiltibran, R. C.**, Aquatic plants, aquatic herbicides, and water quality, *Abstr. Weed Sci. Soc. Am.*, 47, 1978.

26a. **Hiltibran, R. C.**, Unpublished data, 1978.

27. **Hughes, J. S. and Davis, J.T.**, Toxicity of selected herbicides to bluegill sunfish, *Proc. La. Acad. Sci.*, 25, 86, 1962.
28. **Jones, R. O.**, Tolerance of the fry of common warm-water fishes to some chemicals employed in fish culture, *Proc. Southeast. Assoc. Game Fish Comm.*, 16, 436, 1962.
29. **Lawrence, J. M.**, *Aquatic Herbicide Data*, Agricultural Handbook No. 231, U.S. Department of Agriculture, Washington, D.C., 1962.
30. **Lawrence, J. M., Funderburk, H. H., Blackburn, R. D., and Beasley, P. G.**, The status of diquat and paraquat as aquatic herbicides, *Proc. Southeast. Assoc. Game Fish Comm.*, 16, 247, 1962.
31. **Moore, G. T. and Kellerman, M. F.**, A method of destroying or preventing the growth of algae and certain pathogenic bacteria in water supplies, *U.S. Dep. Agric. Bur. Plant Ind. Bull.*, 64, 15, 1904.
32. Subcommittee of Standardization of Common and Botanical Names of Weeds, *Weed Sci. Soc. Am. Weed Sci.*, 19(4), 435, 1973.
33. **Surber, E. W.**, Controlling vegetation in fish ponds with sodium arsenite, *U.S. Dep. Commerce Invest. Rep.*, 11(1), 1, 1932.

34. **Surber, E. W.**, Control of aquatic plants in ponds and lakes, *U.S. Fish and Wildl. Serv. Leafl.*, 344, 20, 1949.
35. **Surber, E. W. and Pickering, O. H.**, Acute toxicity of endothall, diquat, hyamine, dalapon, and silvex to fish, *Prog. Fish. Cult.*, 164, 24, 1962.
36. **Timmons, F. L.**, Weed Control in Western Irrigation and Drainage Systems, Losses Caused by Weeds, Costs and Benefits of Weed Control, Joint Rep. U.S. Department of Agriculture and U.S. Department of the Interior, ARS 34-13, Washington, D.C., 1960.
37. **Timmons, F. L.**, Present and future of aquatic weed control, *Proc. Northeast. Weed Control Conf.*, 16, 1, 1962.
38. **Walker, C. R.**, Control of certain aquatic weeds in Missouri farm ponds, *Weeds*, 7, 310, 1959.
39. **Walker, C. R.**, Endothall derivatives as aquatic herbicides in fishery habitats, *Weeds*, 11(3), 226, 1963.
40. **Walker, C. R.**, Simazine and other s-triazine compounds as aquatic herbicides in fish habitats, *Weeds*, 12(2), 134, 1964.
41. **Walker, C. R.**, Dichlobenil as a herbicide in fish habitats, *Weeds*, 12, 267, 1964.
42. **Wunderlick, W. E.**, History of water hyacinth control in Louisiana, *Hyacinth Control J.*, 1, 14, 1962.

Chapter 3

CONTROL OF MACROPHYTIC GROWTH BY NATURALLY PRODUCED SUBSTANCES*

INTRODUCTION

Complex interactions of factors are involved in the colonization and establishment of aquatic plants. Many of these factors include some already well studied in the aquatic environment: competition for light, space, and nutrient,[1-5] exploitation by herbivorous fish,[6,7] inorganic carbon relationships,[8-10] light quality effects,[11] and attack on plants by pathogenic microorganisms.[12,13] Such studies into the physiologic requirements of aquatic weeds and the relationships of weeds to their environments have led to the development of control measures alternative to chemical means,[14-19] representing a direct translation of research results into weed control techniques which are effective as well as environmentally compatible. These recent developments are particularly fortunate in light of serious problems associated with the continued use of herbicides in aquatic environments (cost, human toxications, governmental restrictions, etc).

Other factors operating in aquatic systems have yet to be fully investigated. This category includes chemical, more specifically allelopathic, interactions. Allelopathic interactions among plants involve substances elaborated by macrophytes, algae, and bacteria whose influence in aquatic environments is largely unknwon. Allelopathy in terrestrial systems has received much attention chiefly in the agricultural contex.[20,21] However, some basic research has suggested a significant role for allelopathy in plant successional patterns.[22,23] Plant growth inhibitors isolated from terrestrial environments have been characterized as being of terpenoid or phenolic natures and have been isolated from whole plants and from soil used to grow inhibitor-producing plants.[22,24,25] There is no reason to presume that compounds of a nature and importance similar to those recovered from the terrestrial environment do not occur in aquatic systems.

The second category of chemical interactions — those between plants and substances produced by the action of physical processes on natural material — has long been a subject of active research by agriculturists. Again, attention has only recently been directed to analogous processes in the aquatic environment. Here, primary consideration has revolved around organic substances observed in highly colored lakes and streams. Originally thought to inhibit the growth of rooted, submersed vegetation simply by light intensity reduction, organic matter imparting color to natural waters is now believed to achieve its effect partially through chemical phytotoxic activity. Such compounds have been shown to be related to humic acids, a class of naturally occurring materials of a highly complex nature whose mode of influence on plant growth is thought to involve physiologic reactions requiring metal ions.

These two classes of chemical interactions are being considered in a volume such as the present one because it is hoped that their identification and definition will lead to the development of new weed control methods associated with reduced environmental disturbance.

Control techniques whose effectiveness is based on an exploitation of natural processes rather than the application of toxic chemicals are more desirable in the long-term view and are consistent with the multiple purposes of water resource management activities.

* By Drs. P. M. Dooris, Southwest Water Management District, Brooksville, Fla., and D. F. Martin, University of South Florida, Tampa.

EVIDENCE SUGGESTIVE FOR "NATURAL" CONTROL OF MACROPHYTES

The search for naturally produced chemical substances as control agents for aquatic weeds has been a very short one. Therefore, in seeking evidence that such materials are active in the aquatic environment, we are led first to indirect evidence based on observations of natural systems. These observations include (1) the recording of the relatively sudden disappearance of a plant species from a lake or stream and (2) the apparent exclusion of a hardy, prolific macrophyte from a particular water body. Using these observations, the results of experimental work describing organic compounds produced by both algae and macrophytes can be evaluated. Finally, direct evidence is available from the sparse, but encouraging, laboratory research which has demonstrated the growth-inhibiting character of substances derived from natural materials. Taken together, this information suggests that, while this field is as yet unexploited, the potential for the development of useful weed-control agents from natural products is real.

Plant Disappearance

The literature contains many references to long-term observations of water bodies documenting the rapid decline and eventual disappearance of a dominant aquatic plant and later replacement of the plant by one of another species. The entire process may occur over a long or short period, but, in either case, it may signal major changes in the environment of the lake or stream in question. Some of these changes may be reflective of the activity of chemicals from natural sources.

Elodea canadensis is an interesting example of a species that spread rapidly then diminished some time after being introduced into England. The full story is given by Salisbury[26] and only the salient features need be reviewed here. The plant exhibited a marked vigor and an astonishing abundance after its appearance in a locality, then a diminution occurred. First reports of a significant appearance in England (1847) at Foxton Reservoir, near market Haybrough (Leicestershire) caused concern because this was the strategic center of the canal system of England. Within 2 years after its initial appearance, elodea was noted as being in greatest abundance in Derbyshire, Strattfordshire, and Warwickshire. An interesting and remarkable experiment was then tried. A plant from Foxton was transplanted to the Botanic Garden at Cambridge, then deliberately introduced into a stream nearby. Within 4 years, the plant was regarded as a "menace" around Cambridge. Water courses were blocked, and, in most areas, the plant became superabundant. Yet in these same areas, about 100 years later, the plant was "a normally common, or even sometimes rather sparse, member of the aquatic community."[26]

Why did the plant disappear? The plant is exclusively vegetive in its propagation, but this is not the explanation for its apparent reduction in vigor, according to Salisbury.[26] When fragments were removed from those areas where the plant had diminished noticeably and were introduced into ponds where the species was absent, the same increase-abundance-decline sequence was observed.

Elodea is not the only plant to show the introduction-abundance-decline sequence. Water soldier *(Stratiotes aloides)* once was quite abundant in the East Anglian Broads and water courses, but became less abundant. Normally only the female plant occurred in Britain, and production occurred by means of sucker-like offsets. The little water fern *(Azolla filiculoides)* also has followed the abundance-decline sequence.

Why did the plants decline following some period of rapid growth? The physiological basis for rapid fluctuations in biomass of elodea or even the gradual decline pattern remain obsure. Brennan[27] made the interesting suggestion that the cause may have been a lack of a sexual reproduction phase and an overall loss of vegetative vigor. While this might be true, the lack of such a phase has not been demonstrated to be detrimental to general vigor over repeated generations in other plants.[28]

Salisbury himself was led to the same view that it was more probable that some mineral nutrient or combinations of nutrients taken up by the plant was unduly depleted.[26]

Striking declines in the amounts of rooted vegetation has been observed by others. Morgan[29] noted declines in the amount of submerged aquatic vegetation in the Norfolk Broads between World War II and 1968 to 1969. He has provided[29] an extensive tabulation of the distribution, both spatially and temporally, of a number of these plants.

The basis for these declines was unavailable at the time of his tabulation,[29] though some interesting possibilities were raised. A great increase in phytoplankton density, owing to enhanced nutrient levels from sewage, occurred in Lake Lingby and reduced light intensities may have been responsible for the complete elimination of the submersed macrophytes from the lake.[30] Perhaps enhanced phytoplankton densities may have been responsible for reduced light penetration and subsequent decline in rooted macrophytes in Loch Leven and in the Norfolk Broads,[29] although Morgan[29] notes that this explanation is not completely satisfactory.

In Florida, the phenomenon of plant disappearance has been recorded frequently. For example, in Lake Apopka (12,268 ha), all submersed plant species disappeared within an 8-year period from 1963 to 1971, and water hyacinth became the dominant species.[31] No explanation of these observations was given, however, a combination of factors was no doubt involved. First the vigor of the water hyacinth itself resulted in the formation of luxuriant mats on the lake surface, excluding light to the water below. Second, Lake Apopka was (is) undergoing extremely accelerated eutrophication. The process has resulted in, among other things, significantly diminished water clarity and a destruction of benthic substrate suitable for the rooting of macrophytes. Third, blooms of cyanobacteria were an almost constant feature of Lake Apopka, and may have been a distinct influence in the plants' disappearance, as water-soluble compounds produced from blue-greens are known to affect macrophytic populations.[32]

Other explanations for declines in submersed macrophytes have been presented. In some lakes, the substrate may change as considerable quantities of semiliquid ooze accumulate, causing anaerobic conditions, and resulting in an unsuitable environment for some macrophytes.[29] Also, increased boat traffic may have contributed to the decline of some plants,[29] through mechanical damage, and pollution by boat-associated sewage and waste, and stirring of muds. Other areas that noted declines in plants did not have boat traffic.[29] Finally, it was suggested that, in the case of the Norfolk Broads, large numbers of coypu [*Myocaster coypus* Molina] may have been responsible for denudation of a number of plants.[33]

No doubt, many observations of macrophyte decline are related to changes brought about by human activities which alter the loads of nutrients and toxic substances entering a lake or stream. The influence of water chemistry has been recognized as a plausible explanation for the decline of a formerly superabundant plant. Salisbury,[26] for example, suggested that "some mineral nutrient or nutrients taken up by the plant was unduly depleted." Unfortunately, few studies seem to have been directed toward an understanding of why declines occurred, and the absence of information may be easily understood. Among the various factors that have been proposed, one that seems to have been neglected is the explanation that a natural product is providing natural control. It would seem that the absence of a plant might provide suggestive evidence of natural control, an observation worthy at least of additional investigation.

Apparent Exclusion of Macrophytes from Some Water Bodies

A few peculiarities have been noted in the distribution of certain macrophytes from water bodies, and these observations suggest that all aquatic systems do not possess equal ability to support the growth of even the hardiest aquatic weed.

For example, in three Florida lakes, the production of hydrilla (as dry weight) and the density of hydrilla tubers varied substantially.[34] While the differences among the lakes were ascribed to geographic, hence climatic influences, no data on other lake characteristics — water chemistry, algae density and composition, sediment type and chemistry — were provided. These factors, in addition to natural regulating substances, may have been at work in some of these lakes.

Studies on lakes in which geographic location was not a factor also indicated that hydrilla does not exhibit equal growth potential in all lakes. For example, hydrilla was present in Lake Carroll near Tampa in Hillsborough County, although it had not been observed in adjoining White Trout Lake.[14] Some studies were made of the water quality, and no obvious differences were discovered. The two lakes were once connected so that fragments of hydrilla might have floated from Lake Carroll to White Trout Lake. Boats were used frequently on both lakes and the lakes were regularly visited by ducks, providing excellent opportunity for the lakes to become infested with hydrilla.

One difference did exist; whether significant or not is a matter of judgment. White Trout Lake is the site of a former sawmill operation, and a number of cypress trees were cut there. In some areas, the cypress sawdust appeared to be about 2 ft thick on the lake bottom. The importance of this fact will be apparent later in this discussion.

Hydrilla was also sparse in Lake Starvation, a lake with public boating facilities some 10 mi north of White Trout Lake. It is difficult to find hydrilla in Lake Starvation, but it was present, and during a 10-year period, the plant has not spread notably. Further, the plants which are in Lake Starvation are thin, stunted, stalky specimens, in contrast to the luxuriant growth that is so typical of hydrilla in Florida lakes and streams.

The case of Lake Starvation is made the more interesting when it is realized that it was heavily supplemented with ground water in an effort to maintain the usefulness of the lake for public recreation. The large-scale addition of ground water to nearby lakes apparently had made these lakes more susceptible to hydrilla infestation by virtue of the changes brought about in water chemistry.[35] Water from Lake Starvation, nevertheless, failed to support lush hydrilla growth either in the field or under laboratory conditions.[35] The "hydrilla-containment" in the lake was still operable despite dilution by ground water.

The apparent exclusion of plants is even more puzzling in the instance of water hyacinth in Polk County, Florida, south of Bartow in the phosphate-mining area. Two different sites have been recognized, a pair of holding ponds and an artificial lake that was divided.

The holding ponds are used in the phosphate mining industry in Florida to hold the "phosphate slimes" from the land-pebble washer plants.[36] The slimes constitute about one half of the plant wastes and roughly one third of the total matrix. The chief characteristic is the slow dewatering of the slimes: from 6% solids at the end of the first day to 15% after 3 months, after which time further dewatering is very slow. After 40 years of operations (in 1972), some 4.5 billion tons of water were associated with the contained slimes.

Water hyacinths have been used in the holding ponds for two purposes: to reduce nutrients through harvesting and to protect the ponds from wind-induced erosion. Although water hyacinths have thrived in one holding pond, they have simply died in another. Truckloads of water hyacinths were added, and the plants survived less than a month in this unusual pond. While differences in the clay distribution of suspended materials are suspected as being a significant control factor, an adequate explanation remains to be presented.

The second site was a pond that was created, then roughly bisected with a causeway in 1972. The water hyacinths thrive in the relative shallow northern one third, and the southern, deeper two thirds of the original pond is devoid of water hyacinths, despite several additions of truckloads of water hyacinths. In 1978, a culvert was built under the causeway, and the water hyacinths appear to be slowly diminishing in the northern section of the pond. Clearly, the pond is an interesting case study that begs a more intensive investigation.

From the above, it would be easy to suspect that hydrilla and water hyacinth are sensitive to their environment, but the documented studies of the rate of spread of these plants would indicate that in Florida and the southeastern U.S. the plants are very hardy, so that the absence of growth or even the failure in a given locale in Florida is noteworthy. For example, Ornes and Sutton[37] reported that water hyacinths grew at the rate of 1.9 daughter plants per parent plant per week under ideal conditions. Also, hydrilla is known to grow rapidly. Sutton[38] has reported that hydrilla planted in 25% of a test lake covered 90% of the Florida lake within a year. Hydrilla spread in one Louisiana reservoir from an area of 5 ha to over 10 ha in just over 6 weeks.[39] Finally, one may note the observation of Haller and Sutton:[40] a lake infested with hydrilla could contain 3 million propagules per hectare; a typical corn (maize) field by comparison would contain about 75,000 plants per hectare.

In the case of two species, observations of nature have been directly followed up by laboratory work. For example, it had been noticed that the growth of *Glyceria maxima* apparently excluded the adjacent growth of *Phragmites communis* and vice versa.[41] The possibility of inhibition as a result of toxin production by the plants was investigated, but rejected. However, the study did not employ extracts of the plants. Therefore, the production of a growth inhibiting material by the two plants cannot be ruled out completely.

Evidence from Mixed-Culture Studies

Russell[42] noted the curious lack of interest on the part of marine ecologists in species competition, particularly when one considers the suitability of available techniques for investigating the problem. Fletcher[43] emphasized the utility of the approach in establishing heteroantagonism as observed in mixed cultures. The studies of heteroantagonism for fresh-water species are limited, but are interesting and indicative of the potential of the approach.

Extracellular inhibitory substances (probably long-chain fatty acids) were described from the green alga *Chlamydomonas reinhardi* [44] and these substances inhibited the growth of *Haematococcus plurialis*. In mixed cultures, *C. reinhardi* eliminated the second fresh-water alga within 3 to 5 days, depending upon the light intensity and nitrogen source (and other factors). Proctor[44] tested eight known fatty acids against some six algae, and concluded that palmitic was the most toxic saturated acid tested, and linoleic was the most unsaturated one. The difference between concentrations causing no inhibition and complete inhibition was surprisingly slight.

Lange[45] also described extracellular inhibitory substances from the blue-green *Aphanizomenon*.

Macrophytes secrete large amount of organic material[46,47] some of which may have inhibitory activity. The presence of *Anacharis canadenis* and *Potamogeton folisus* inhibited biomass production of algae in large artificial systems.[48] While nutrient competition was suggested as the responsible cause for the observation, it now seems possible that the macrophytes may have elaborated some active metabolic inhibitor. Boyd[49] suggested that *Chara* sp. secreted antibiotic substances that inhibited the growth of phytoplankton in catfish ponds. It is interesting, in this connection, that farm-pond restoration could be benefited by pioneer *Chara vulgaris*, early growth of which was "found to be associated with clear water conditions and lessened phytoplankton growth."[50] In addition, workers at the University of Southern Mississippi found that the emergent *Elocharis microcarpa* produced water-soluble algacidal materials that were active against species of *Plectonema*, *Anabaena*, and *Anacystis* at concentrations of 1 ppm or less.[51] The observation is interesting because of the possibility that substances elaborated by littoral macrophytes (*Chara* sp. and *Eleocharis* sp.) may inhibit algal production in Lake Okeechobee, Fla.;[52] algal assays of lake water indicated the presence of an algal inhibitor, whether a metabolic inhibitor or another agent was uncertain.

The effects of sediments should not be overlooked in mixed algal assays or in the bioassays that are used for screening for natural products. First, hydrosoil adsorption may reduce the effective concentration of a herbicide, either applied or being sought as a natural product. Second, the hydrosoil may contain or be the source of the natural product that is capable of natural management of macrophytes. Third, sediments may represent the major source of phosphorus nutrition.

The last two points were illustrated in studies by Barko and Smart.[53] The macrophytes investigated (*Egeria densa*, *Hydrilla verticillata*, and *Myriophyllum spicatum*) were demonstrated to be fully capable of deriving their entire phosphorus nutrition requirements from the sediments, despite the root systems that were minor on a biomass basis. The studies consisted of two compartment studies, in which it was possible to separate the roots and the sediments (compartment one) from the overlying phosphorus free "complete" nutrient system. The studies of the three plants and some five sediments indicated that the mobilization of sediment phosphorus by submergent macrophytes "represents an important aspect of the (phosphorus) cycle, and may affect the overall metabolism of the lacustrine system."[53]

An interesting aspect of the study was the observation of the unfavorable growth of hydrilla on one sediment (from Ashtabula Reservoir, North Dakota) and the unfavorable growth of *Myriophyllum* on sediment from W. Thompson Reservoir, in Connecticut. The unfavorable growth was reflected in the biomass category, and the root:shoot ratio. For some four sediments the mean ratio was 2.0 ± 0.5 (SEM)% and 6.1% for Ashtabula-grown plants; for *Myriophyllum* and four sediments the mean ratio was 11.2 ± 2 (SEM)% vs. 23.8% for W. Thompson sediment.

Direct Evidence for Management of Macrophytes by Natural Products

Strong direct evidence for the control of the composition and density of macrophytic communities comes from two studies spaced widely in time but both involving *Typha latifolia*. Extracts of cattail leaves were found to be highly inhibitory to the germination of seedlings of the same species.[54] The water in the marsh in which the cattails were growing also proved somewhat inhibitory. It is possible that an autotoxic mechanism became operative after the stand of cattails reached a particular density. The mechanism may have developed to reduce the chance of habitat overexploitation by the plant. In this study, the "toxin" was believed to be of a phenolic nature, although the material was not fully chemically characterized.

The second study demonstrated that debris from *T. latifolia* as well as from *Phragmites australis* inhibited the growth of some emergent species while stimulating that of others.[55] No pattern could be discerned with respect to the taxonomic relationships among the plants favored or inhibited by plant debris; however, the results indicated a clear case of macrophyte control by naturally produced substances.

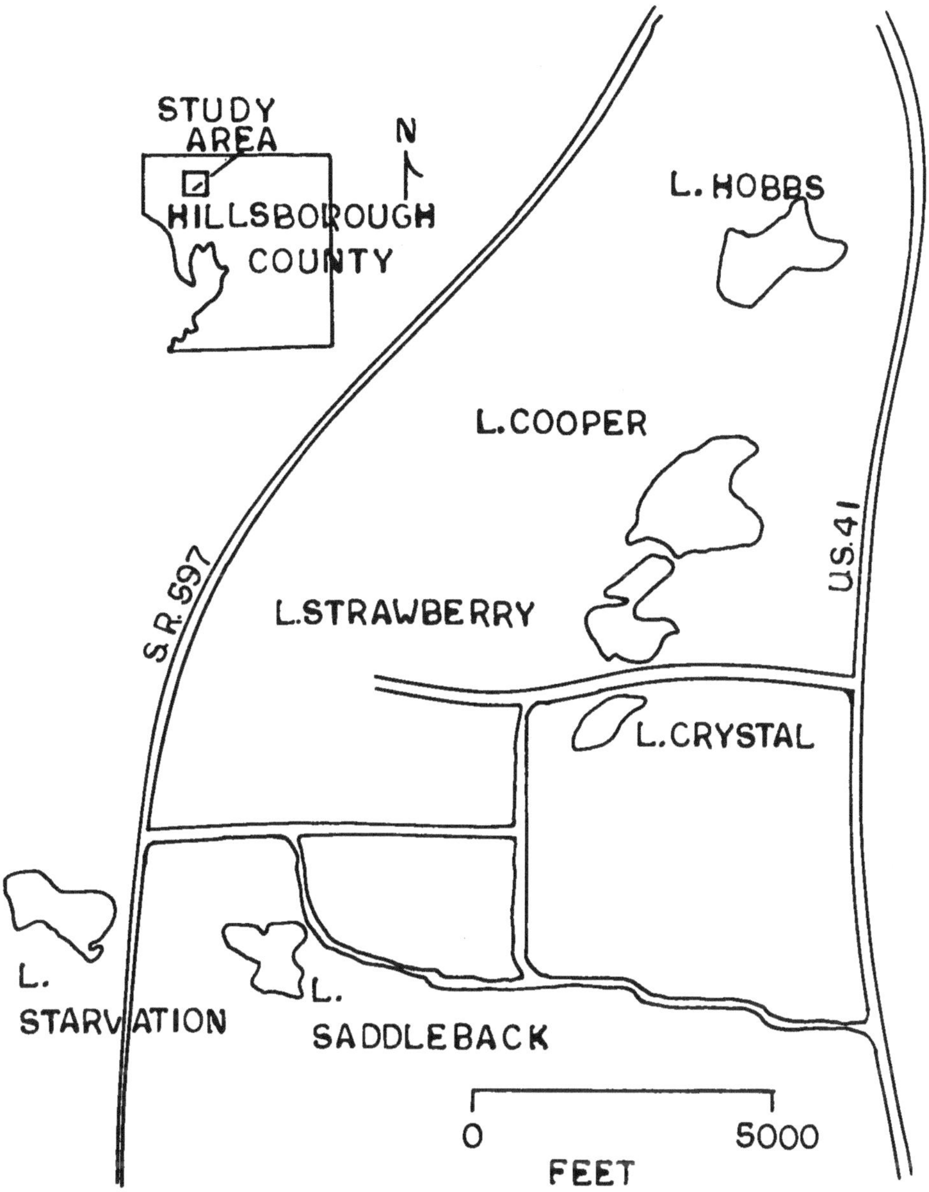

FIGURE 1. Relative location of Starvation, Saddleback, and Hobbs Lakes.

At this point, it would be useful to consider again the case of the two lakes in Florida which did not support hydrilla. The basis for the phenomenon is interesting to the present topic. A number of factors could be eliminated as potential explanations of hydrilla sparsity, but some significant factors remained. Both lakes had been surrounded by adjacent stands of bald cypress *(Taxodium distichum)*, and organic material derived from these trees comprised a considerable portion of the sediment.

In addition, one study[35] had indicated how unique was the water overlying the hydrosoil in the instance of Lake Starvation.

The study is indicated in Figure 1, and the study of the water chemistry is summarized in Table 1, which compares two types of lakes — those that were artificially augmented with water from the Florida aquifer as previously described, and those lakes that had been left alone. Obviously, a number of differences might be considered,

Table 1
COMPARISON OF DATA FOR SELECTED WATER QUALITY PARAMETERS FOR LAKES SUPPLEMENTED WITH GROUND WATER AND THOSE NOT SO SUPPLEMENTED

	Lakes		
Parameter (mg/l)	Hobbs	Saddleback	Starvation
Carbon, inorganic	2.0	16.4	9.3
Carbon, organic	4.0	15.7	24.6
Iron	0.24	0.3	1.6
pH	6.5	6.6	6.4
Hardness	30.7	63.9	86.75

Note: Values are means of several analyses.

but the water quality data were a good starting point. It became evident that the two lakes differed in hardness at the time of the study, and they also differed in the ability to sustain hydrilla. In this bioassay, only the water was studied, and the amount of dissolved oxygen produced during a short-term period was measured. A measure of growth, the growth characteristic, was defined as the percentage increase in dissolved oxygen, relative to the initial value. The data indicate that hydrilla grew better in lakes that had been augmented with ground water (high in inorganic carbon), with one notable exception. Lake Starvation did not show the same relationship between water hardness and growth characteristic. In addition, this lake was highly colored (additional data are presented elsewhere[35]). Moreover, the iron content of this dark water was about 14 times the value observed to be toxic to hydrilla growth in laboratory studies.[56]

Dark-colored water and the iron content appear at the very least to provide useful diagnostic tests for potential hydrilla management by natural substances. The dark water would not be a limiting factor in the laboratory (though it might be in nature), but it does raise the question of what is the cause of the dark color. One way of answering this question is to study the sediment and see whether water extracts or leachates would also be colored and contain inhibitors.

PHYSICAL SEPARATION AND CHARACTERIZATION OF NATURAL MANAGEMENT SUBSTANCES

Separation and Analysis of Sediment Extracts

In the above study, sediment samples were collected from the lakes of interest (White Trout and Starvation) and returned in the cold to the laboratory for extraction. The extraction scheme, the separation procedure, is presented in Figure 2. The properly filtered samples were separated into a parent fraction (A, Figure 2), and subjected to ultrafiltration using a technique recommended by Gjessing[57] for the study of humate substances. An Amicon apparatus was operated at 50 psi nitrogen using filters with the nominal cutoff points indicated in Figure 2.

Other Separations

Other techniques for separation should be considered, including chromatography over appropriate columns. In the study noted above, when samples of parent compound A were passed over a cation-exchange resin, test plants showed slight decreases

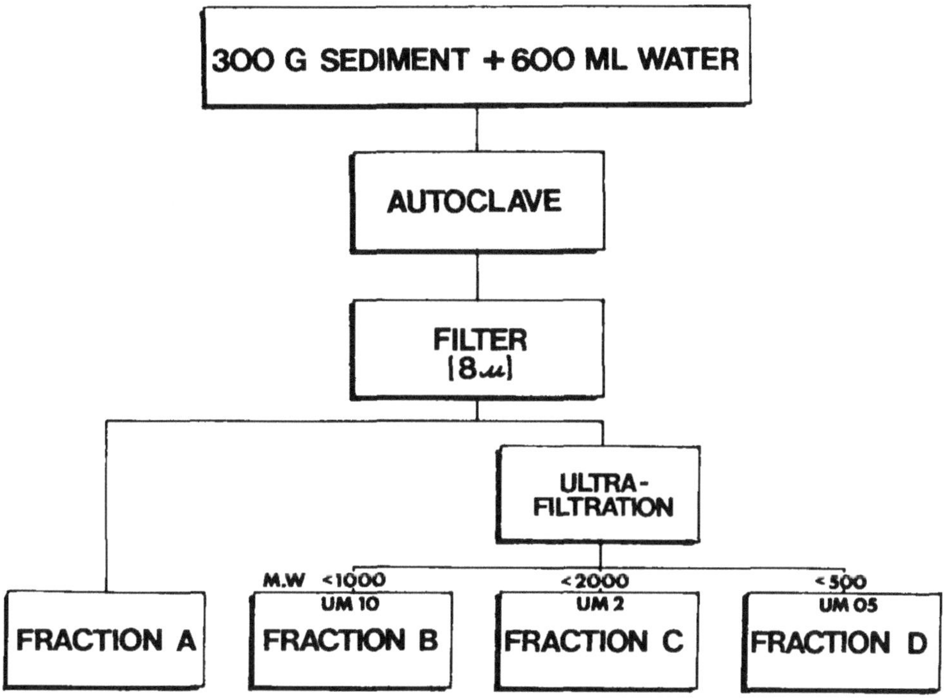

FIGURE 2. Extraction scheme and procedure.

or decreases in weight. Moreover, the control plants were dark green and dense-leaved, while the test plants were yellowish brown and partially defoliated.

Some fractions showed no significant differences between control and test series. Fractions C and D, for example, had no discernible effect on the growth of the test plants (Figure 3). Also, treating the active Fraction A of the Lake Starvation extract with Chelex 100 (a chelating cation-exchange resin) did not alter the activity (Figure 4). In addition, extract or leachates of *Taxodium distichum* were inactive (Figure 4). Finally, passage of Fraction A (Lake Starvation extract) over an anion-exchange column resulted in effluent that was devoid of significant inhibitory activity (Figure 4).

Summary of Observation on Extracts

From the information available, a number of pertinent observations may be made concerning the naturally occurring growth inhibition factor. The inhibitor can be extracted from the sediment, the specific inhibitory activity was enhanced by ultrafiltration, and the active fraction had a nominal molecular weight between 2,000 and 10,000 daltons. The material is likely polyphenolic; activity is associated with some anionic moiety (e.g., A^{n-}), and sorption of this material on an ion-exchange column led to loss of color and activity. Extracts of inhibiting material could be prepared from the sediment of White Trout Lake, and from commercially available humic acid[58,59] but the extracts from cypress leaves were not active.

The latter set of observations may be interpreted as indicating that the active material results from the degradation of cypress material, and that degradation is a critical step. The act of autoclaving may serve two functions: first, any bacteria and fungi that are present should be destroyed and separate fractions eluted over (Bio-Rad® Chelex-100) and over an anion-exchange resin (Bio-Rad® Cellex D). Bioassays were performed on the eluents and on all separated fractions.

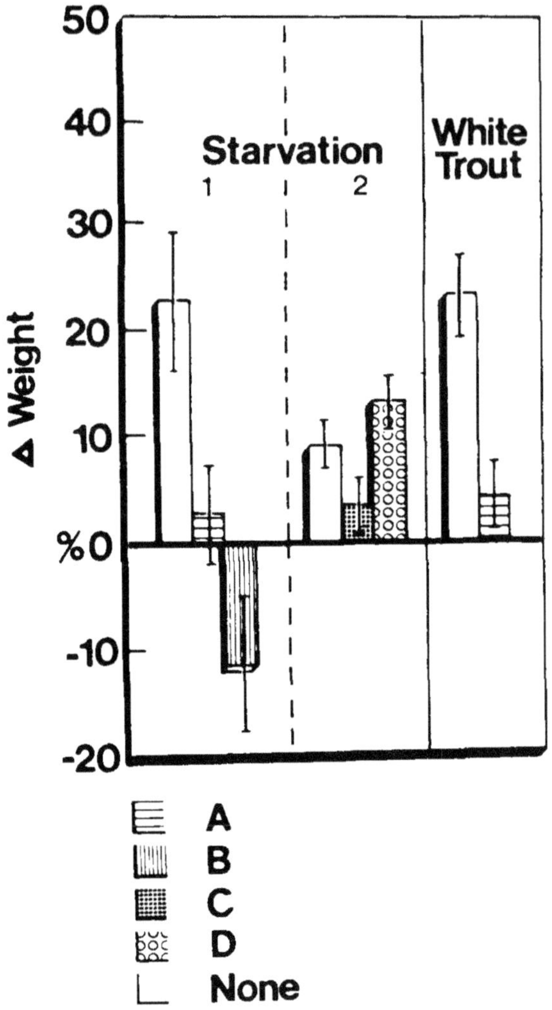

FIGURE 3. Percent growth in weight of test plants for fractions A, B, C, and D.

Pertinent Analyses

It is difficult to determine what analyses may be pertinent for a given problem, but it is useful to know the weight of dissolved material, the amount of carbon present (inorganic and organic), water quality data, etc. Table 1 lists selected chemical and physical characteristics for various fractions of the Lake Starvation extracts. This table indicates that most of the iron was associated with the larger molecules, i.e., those having nominal sizes in excess of 10,000 daltons, and the iron was not removed when the parent fraction A was passed over Chelex-100, though most of the iron was removed when the parent fraction was passed over an anion-exchange resin. The fluorescence data are interesting because fluorescence was also associated with the nominally higher molecular-weight fraction.

Bioassays

Chemical analyses are interesting and important, but bioassays are the point of major interest. The fractions varied considerably in activity which was measured as follows. Some 2 to 3 g of apical plant material was placed in Hoagland's solution[57] in

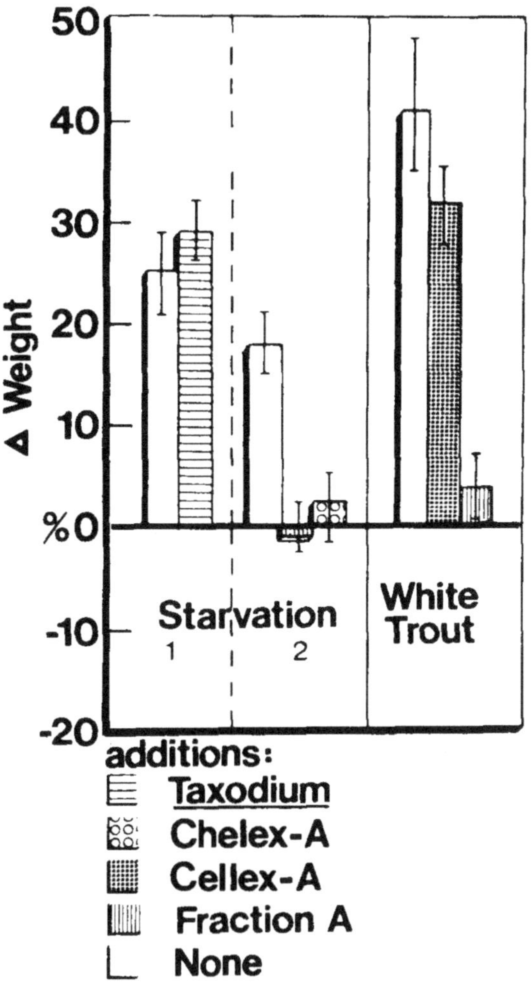

FIGURE 4. Percent growth in weight of test plants for anion-exchange elements.

500-mℓ Erlenmeyer flasks, with 0.8 mℓ of extract (approximately 0.4 ppm organic carbon) or distilled water (for control). Flasks were illuminated for 14 hr/day (800 ueinsteins, $m^{-2}sec^{-1}$) and after 7 days, the change in biomass was determined.

Some fractions inhibited the growth of the test fractions. Fractions A and B from Lake Starvation and Fraction A from White Trout Lake are examples. Control plants had an average increase of 20% or more growth, while destroyed so that the continued production of bacterial secretions would not produce artifactual material. Second, autoclaving may provide further degradation and produce useful inhibitory material. On the other hand, preliminary experiments with autoclaving peat samples did not produce hydrilla growth inhibitors. Finally, if bacterial action on a specific substrate (e.g., cypress material) is necessary, then the activity should vary with depth. In preliminary experiments, this seems to be the case; the extracts of the top 10 cm (Fraction A) of a Lake Starvation sediment column were active, the extract of the next 10 cm was not, and the lower sediment had a greater amount of sand present.

Obviously, additional studies are necessary and are in order, but the observations are presented to suggest that a pattern for control does exist, and that sediments are available that provide suitable leachates perhaps through bacterial mediation. The degradation of the sediments produces polyphenolic material that is related to the complex material called humic acid.[57]

IMPLICATIONS

As explained previously, the search for natural chemical agents which may control the density and composition of aquatic macrophyte communities has not been extensive. This situation has probably come about because it has been customary to attribute changes in plant communities to alterations in nutrient availability, light intensity, or exploitation of plants by wildlife. Further, it is generally held that introduced plant species are very successful in the new environment because of the absence of well-evolved control mechanisms. While the influence of factors such as water chemistry, grazing, and damage due to insects and microorganisms in aquatic plant community dynamics is indisputable, the role of allelochemics and other naturally produced chemicals should be explored on a serious basis. This area of research, after a period of time, should yield two important products: first, an enhancement of basic knowledge of plant community metabolism in aquatic systems, and second, the development of new weed-control methods.

Results of the research along these lines (summarized here) are quite preliminary; nevertheless, it is possible to anticipate some of the ways in which natural products may be used in aquatic weed control.

It is likely that additional compounds will be isolated from plant material and from the sediments of certain types of lakes that will prove inhibitory to undesirable macrophytes and algae. The chemical characterization of such compounds will follow this work, and toxicity studies will be done to define the specificity of the inhibitory action. Then, it will be necessary to determine the most efficient means of producing the required quantities of the compound — by synthetic mechanisms or by recovery from natural sources. The material then may be introduced to the environment for the purpose of weed control. This procedure somewhat resembles that employed to bring any herbicide to work in the field, and it could be argued that problems similar to those related to herbicide use would attend the large-scale introduction of even a "natural" compound to the environment. However, a control substance derived from natural sources would probably have the advantage of being less persistent in the environment as a result of rapid decomposition. In addition, toxicity problems *should* be minimized by using naturally derived materials. This may not be the case should excessively large quantities of the material be injected into a lake. This fact may necessitate a shift in the goal of weed control efforts — from eradication to containment of the pest species.

It is also a distinct possibility that the intentional planting of certain plants may become a viable method of control, particularly in new, man-made aquatic systems. For example, it has been shown that some plants produce water soluble materials which inhibit the germination of other plants. In a newly created lake or pond, it would be a preventive practice to cultivate those plants known to be antagonistic to local nuisance weeds. Under such circumstances, the successional pattern in the lake would be a more favorable one, and the use of herbicides could be delayed.

A future possibility for potential control through natural agents involves the color-producing materials in lakes and streams. As already described, the substances contributing to organic color in water have been implicated as inhibiting certain aquatic weeds, both through light reduction and chemotaxis activity. Whatever the mode of action, it stands to reason that increasing or maintaining the concentration of color-producing substances in a lake should discourage the growth of noxious submersed plant species.[60] Accomplishing this in colored lakes whose elevations are controllable by means of hydraulic structures would not be difficult. Discharges could be timed suitably to take advantage of the sometimes highly seasonal inflow of colored water to a lake in a effort to discourage weed growth.

The knowledge gained about potential weed control through naturally produced chemicals will be of direct benefit to those in aquatic plant management. It will also be of value to those involved in the broad field of water resource management, particularly in lake restoration. The excessive growth of water weeds usually accompanies the overenrichment and eventual deterioration of a lake. Initially the result of eutrophication, aquatic weeds often become a large part of the problem confronting the rehabilitation effort, and the solution to the situation is usually to remove the weeds by chemical and/or mechanical means. Little thought is given to the postrestoration period with respect to the development of desirable plants in the lake. Here "biomanipulation" of the new lake habitat through the use of natural chemicals or by the planting of desirable species would prolong the benefits and increase the cost-effectiveness of restoration.

Research Directions

As this field is very new, a good deal of work remains to be done, and a number of directions could be pursued. It would seem very obvious to investigate further those lakes in which hydrilla does not thrive in an effort to pinpoint precisely why this is so. As a sufficient number of lakes are studied, a pattern should emerge relating lakes within close geographic proximity. Investigations of this nature should include controlled laboratory growth studies using lake water in continuous-culture and circulated systems.

Bioassays should be done employing a nuisance weed as the stable component and desirable plants as the variable component. Systems should be arranged in such a way that water soluble substances of particular molecular weights will contact the growth medium containing the nuisance weed. In this way, it should be possible to identify desirable plants which produce compounds antagonistic to aquatic weeds. Field certification of results obtained from such studies would, of course, be necessary. Performing bioassays, using random combinations of plants, would be a laborious process to say the least. Clues as to which plants may be more likely candidates for the production of weed-inhibiting materials may be found through observations of plant communities in natural systems. Indeed, because of the relatively little activity in this field of natural chemical control of water weeds, useful progress will be achieved by research programs designed to answer basic questions formulated from careful observations of nature.

ACKNOWLEDGMENTS

The authors would like to acknowledge with appreciation the interest and encouragement of Dr. Edward Gangstad and the assistance of Mrs. Miriam Morcom and Mrs. Frances Butler in the preparation of the manuscript. Financial assistance for some of the research described here has been provided by The Florida Department of Natural Resources (Bureau of Aquatic Plant Research and Control), The Southwest Florida Water Management District, and the U.S. Army Corps of Engineers.

REFERENCES

1. **Chadwick, M. J. and Obeid, M.**, A comparative study of the growth of *Eichornia crassipes* Solms and *Pistia stratiotes* L. in water culture, *J. Ecol.*, 54, 563, 1966.
2. **Bowes, G., Van, T. K., Garrard, L. A., and Haller, W. T.**, Adaptation to low light levels by *Hydrilla*, *J. Aq. Pl. Man.*, 15, 32, 1977.
3. **Van, T. K., Haller, W. T., Bowes, G., and Garrard, L. A.**, Effects of light quality on growth and chlorophyll composition in *Hydrilla*, *J. Aq. Pl. Man.*, 15, 29, 1977.
4. **Tag El Seed, M.**, Effect of pH on the nature of competition between *Eichornia crassipes* and *Pistia stratiotes*, *J. Aq. Pl. Man.*, 16, 53, 1978.
5. **Van, T. K., Haller, W. T., and Garrard, L. A.**, The effect of day length and temperature on *Hydrilla* growth and tuber production, *J. Aq. Pl. Man.*, 16, 57, 1978.
6. **Osborne, J. and Sassic, N.**, Biological control of *Hydrilla verticillata* Royle with grass carp *(Ctenopharyngodon idella)*, *J. Aq. Pl. Man.*, 17, 45, 1979.
7. **Miley, W. M., II, Leslie, A. J., Jr., and VanDyke, J. M.**, The Effects of Grass Carp *(Ctenopharyngodon Idella* Val.) on Vegetation and Water Quality in Three Central Florida Lakes, Final Rep., Bureau of Aquatic Plants Research and Control, Florida Department of Natural Resources, Tallahassee, 1979.
8. **Spence, D. N. N.**, Factors controlling the distribution of freshwater macrophytes with particular reference to the lochs of Scotland, *J. Ecol.*, 55, 147, 1967.
9. **Raven, J. A.**, Exogenous inorganic carbon sources in plant photosynthesis, *Biol. Rev.*, 45, 167, 1970.
10. **Uetsch, G. R. and Anthony, D. S.**, The role of aquatic exchange of carbon dioxide in the ecology of the water hyacinth *(Eichornia crassipes)*, *Fla. Sci.*, 36, 16, 1973.
11. **Haller, W. and Sutton, D.**, Community structure and competition between hydrilla and vallisneria, *Hyacinth Control J.*, 13, 48, 1975.
12. **Chardattan, R. and Liu, C. Y.**, Isolates of *Penicillium*, Aspergillus, and Trichoderma to aquatic plants, *Hyacinth Control J.*, 12, 70, 1974.
13. **Rintz, D. E.**, A zonal leaf spot of water hyacinth caused by *Cephalosporium zonatum*, *Hyacinth Control J.*, 11, 41, 1973.
14. **Martin, D. F., Doig, M. T., III, and Millard, D. K.**, Potential control of Florida *Elodea* by nutrient control agents, *Hyacinth Control J.*, 9, 36, 1971.
15. **Michewicz, J., Sutton, D. L., and Blackburn, R.**, The white amur for aquatic weed control, *Weed Sci.*, 20, 106, 1972.
16. **Wilson, C. G.**, Method for controlling the growth of aquatic plants, *Off. Gaz. U.S. Patent Off.*, 961, 1240, 1977.
17. **Bateman, J. M. and Laing, R. L.**, Restoration of water quality in Lake Weston, Orlando, Florida, *J. Aq. Pl. Man.*, 15, 69, 1977.
18. **Laing, R. L.**, The use of multiple inversion and Clean-Flo lake cleanser in controlling aquatic plants, *J. Aq. Pl. Man.*, 17, 33, 1979.
19. **Freeman, T. E., Chardatten, R., and Conway, K. E.**, Biological control of water weeds with plant pathogens, in Fla. Dept. Nat. Res. Res. Rev. Conf., Orlando, Fla., 1977.
20. **Overland, L.**, The role of allelopathic substances in the "smother" crop, barley, *Am. J. Bot.*, 53, 423, 1966.
21. **Putnam, A. R. and Duke, W. B.**, Biological suppression of weeds: evidence for allelopathy in access cans of cucumber, *Science*, 185, 370, 1974.
22. **Brown, R. T.**, Influence of naturally occurring compounds on germination and growth of jack pine, *Ecology*, 48, 542, 1967.
23. **Rice, E. L.**, *Allelopathy*, Academic Press, New York, 1974.
24. **Toussoun, T. A., Weinhold, A. R., Linderman, R. G., and Patrick, L. A.**, Nature of phytotoxic substance produced during plant residue decomposition in soil, *Phytopathology*, 58, 41, 1968.
25. **Blaschke, H.**, Untersuchungen uber das Vorkommen und die Wirkung von biogenen Wachstumpregulatoren in waBrigen Nadelstreuasziigen, *Flora Bd.*, 166, 537, 1977.
26. **Salisbury, E.**, *Weeds and Aliens*, Collins, London, 1961, 78.
27. **Brennan, J. P. M.**, personal communication, April 1, 1979.
28. **Brennan, J. P. M.**, personal communication, April 11, 1979.
29. **Morgan, N. C.**, Problems of conservation of freshwater ecosystems, in *Conservation and Productivity of Natural Waters*, Edwards, R. W. and Garrod, D. J., Eds., Academic Press, London, 1972, 148.
30. **Olsen, S.**, Vegetation saendringer i Lingby So Bidrag til analyse af kulturpavirkninger pa vand-og Sumpplante vegetationen, *Bot. Tidsskr.*, 59, 273, 1964.
31. **Chesnut, T. L. and Barman, E. H., Jr.**, Aquatic vascular plants of Lake Apopka, Florida, *Fla. Sci.*, 37, 60, 1974.

32. **Jupp, B. P. and Spence, D. H. N.**, Limitations on macrophytes in a eutrophic lake, Loch Leven. I. Effects of phytoplankton, *J. Ecol.*, 65, 175, 1977.
33. **Ellis, E. A.**, Some effects of selective feeding by the Coypu *(Myocastor coypus)* on the vegetation of broadland, *Trans. Norwich Nat. Sci. Soc.*, 20, 32, 1963.
34. **Bowes, G., Holaday, A. S., and Haller, W.**, Seasonal variation in the biomass, tuber density and photosynthetic metabolism of *Hydrilla* in three Florida lakes, *J. Aq. Pl. Man.*, 17, 61, 1979.
35. **Martin, D. F., Victor, D. M., and Dooris, P. M.**, Effects of artificially introduced ground water on the chemical and biochemical characteristics of six Hillsborough County (Florida) lakes, *Water Res.*, 10, 65, 1976.
36. Staff, Bureau of Mines, The Florida Phosphate Slimes Problem, A Review and a Bibliography, Information Circular 8668, U.S. Bureau of Mines, Washington, D.C., 1975.
37. **Ornes, W. B. and Sutton, D. L.**, Removal of phosphorus from static sewage effluent by waterhyacinth, *Hyacinth Control J.*, 13, 56, 1975.
38. **Sutton, D. L.**, Fort Lauderdale Agricultural Research Center, Research Report FL72-1, University of Florida, Gainesville, 1972.
39. **Johnson, B. and Manning, J.**, Hydrilla — the most serious threat to Louisiana's water resources, *La. Conserv.*, 20, 26, 1974.
40. **Haller, W. T. and Sutton, D. L.**, Community structure and competition between *Hydrilla* and *Vallisneria, Hyacinth Control J.*, 13, 48, 1974.
41. **Buttery, B. R., Williams, W. T., and Lambert, J. M.**, Competition between *Glyceria maxima* and *Phragmites communis* in the region of Surlingham Broad. II. The fen gradient, *J. Ecol.*, 53, 183, 1965.
42. **Russell, G.**, The Phaeophyta: a synopsis of some recent developments, *Oceanogr. Mar. Biol. Annu. Rev.*, 11, 45, 51973.
43. **Fletcher, R. L.**, Heteroantagonism observed in mixed algae cultures, *Nature*, 253, 534, 1975.
44. **Proctor, V. F.**, Studies of algal antibiosis using *Haematococcus* and *Chlamydomonas, Limnol. Oceanog.*, 2, 125, 1957.
45. **Lange, W.**, Limiting nutrient element infiltrates of Lake Erie water, *Water Res.*, 5, 1031, 1971.
46. **Wetzel, R. G.**, Excretion of dissolved organic compounds by aquatic macrophytes, *Bio. Sci.*, 19, 539, 1969.
47. **Otsuki, A. and Wetzel, R. G.**, Release of dissolved organic matter by autolysis of a submersed macrophyte, *Scirpus subterminalis, Limnol. Oceanog.*, 19, 842, 1974.
48. **Hasler, A. D. and Jones, E.**, Demonstration of the antagonistic action of large aquatic plants on algae and rotifers, *Ecology*, 30, 359, 1949.
49. **Boyd, C. E.**, Summer algal communities and primary productions in fish ponds, *Hydrobiologica*, 41, 357, 1973.
50. **Crawford, S. A.**, Farm pond restoration using *Chara vulgaris* vegetation, *Hydrobiologica*, 62, 1, 1979.
51. **Clock, L. R., Pessoney, G. F., and VanAller, R. T.**, Extraction and purification of algal inhibitors from *Eleochoris macrocarpa, Am. Chem. Soc. — S.E. Reg. Meet.*, Tampa, Fla., 1977, 57.
52. **Dye, C. W., Jones, D. A., Ross, L. T., and Willmore, R. L.**, Limiting Nutrients in the Kissimmee River-Lake Okeechobee Basin Based on Algae Bioassay Techniques, Tech. Ser., 1(2), Florida Department of Environmental Regulations, Tallahassee, Fla., 1975.
53. **Barko, J. W. and Smart, R. M.**, Mobilization of sediment phosphorus by submergent freshwater macrophytes, *Freshwat. Biol.*, 10, 229, 1980.
54. **McNaughton, S. J.**, Autotoxic feedback in relation to germination and seedling growth in *Typha latifolia, Ecology*, 49, 367, 1968.
55. **Szczepanski, A. J.**, Allelopathy as a means of biological control of water weeds, *Aq. Bot.*, 3, 193, 1977.
56. **Reid, G. A., Jr., Martin, D. F., and Kim, Y. S.**, Stimulation of *Hydrilla* growth by Fe(EDTA), *J. Inorg. Nucl. Chem.*, 37, 357, 1975.
57. **Gjessing, E.**, *Physical and Chemical Characterization of Aquatic Humus*, Ann Arbor Science, Ann Arbor, Mich., 1976.
58. **Dooris, P. M.**, *Hydrilla verticillata:* Chemical Factors in Lakes Affecting Growth, Ph.D. thesis, University of South Florida, Tampa, 1978.
59. **Dooris, P. M. and Martin, D. F.**, Growth inhibition of *Hydrilla verticillata* by selected lake sediments, *Water Resour. Bull.*, 6, 112, 1980.
60. **Crisman, T.**, personal communication, October 30, 1979.

Chapter 4

PROBLEMS OF WEED CONTROL IN AQUATIC SITES

INTRODUCTION

Problems concerned with weed control in aquatic sites are specialized because of conditions relating to the environment, i.e., the environment primarily determines which method is effective and safe for controlling a given aquatic weed. Environmental examples are (1) banks above the water line of impoundments, waterways, and drainage and irrigation channels, (2) shallow edges of impoundments, waterways, and drainage and irrigation channels for quiescent or slowly moving water, and (3) ditch banks and channels of rivers, waterways, and canals for flowing or rapidly moving water. Much discussion and many experimental studies have been made relating to the efficacy and herbicide residues under these different conditions and the facts of each case can generally be stated, but there is a wide difference of expert opinion.

Another determinant for selecting a satisfactory method of control is the nature of the plant — whether it is a grass or grass-like plant (e.g., cattail, sedge) a broadleaf (dicot), emersed or floating plant with all or most of the foliage above the water, or a submersed plant growing entirely under water or with foliage extending to the water surface.

These are further limitations to the methods of aquatic weed control, i.e., the effects on fish, livestock, irrigated crops, and potable drinking water, and the environment, i.e., river and ditch banks, ponds, lakes, reservoirs, and irrigation drainage canals.

MECHANICAL MANAGEMENT PRACTICES

As supplements to design and construction of canals, ponds, lakes, and reservoirs, a number of management practices can be utilized to control aquatic weeds or prevent them from becoming a problem. These practices include draining, flooding, fertilization, and livestock grazing.[1-3]

Draining

For control of submersed weeds in irrigation cannals with uniform flow gradients, drain completely and allow the canals to dry for 3 or 4 days. Repeat as necessary during periods when water is not critically needed for irrigating crops. To control cattails and certain other emersed weeds in ponds, marshes, and wetlands, drain water off, plow the ground, and let it dry for a few weeks. This method is limited to areas which can be readily drained and left dry for a period of time.

Flooding

Maintain the water level at a depth of 1 m or more for several months to control cattails and certain other emersed weeds. This method is limited to areas which can be readily flooded and kept flooded for a period of time.

Fertilization

In the southern states, frequent application of nitrogen, phosphorous, and potassium (NPK) mineral fertilizers in ratios of 8-8-2, 10-10-5, or higher, stimulates a dense growth or "bloom" of microscopic algae which shades the pond bottom and prevents or reduces the growth of rooted submersed weeds. To control submersed weeds in ponds with stable water levels, apply 100 to 200 kg/ha of NPK mineral fertilizer (analysis 8-8-2) beginning in late winter or early spring, and at 10-day intervals thereafter, until a white disk placed 25 to 30 cm below the water surface is invisible. Subsequently, apply fertilizer as necessary to maintain the density of the algal "bloom". Fertilization does not control submersed weeds where the outflow or change of water in a month exceeds the water storage capacity of the canal or pond. Fertilization will not control weeds in cooler waters of the central and northern states.

Livestock Grazing

Intensive grazing by livestock, where practicable and safe, provides effective and economic control of most kinds of emersed and marsh species of weeds, including aquatic grasses, sedges, rushes, reeds, cattails, alligator weed, watercress, and willows. Avoid grazing where it causes serious erosion of steep banks or loose soil.

HAND AND MECHANICAL CONTROL

Although the traditional hand and mechanical methods of controlling aquatic weeds have recently been replaced to a large extent by herbicides, the hand and mechanical methods are still advantageous in many situations.[1-3]

Young plants of cattail, buttonbush, willow, and certain other emersed or marsh species can be eliminated by hand pulling. Frequent inspection of the channel, pond, or marsh and pulling when plants are young and few in number, is necessary for effective and economical control.

Control of submersed weeds by mechanical methods often is less costly than chemical methods in canals with flows of water greater than 70 ft^3/sec. In canals that supply water for sprinkler irrigation, mechanical methods are undesirable because fragments of dislodged weeds that remain in the water often clog sprinkler heads, valves, screens, and other irrigation equipment.

Self-propelled and boat-mounted movers and harvesters that cut off submersed or emersed weeds below the water surface at depths of 0.5 to 2 m are available. They are restricted to operation in open water areas unobstructed by stumps, trees, and other physical limitations such as access and topography of the shoreline. To eliminate cattails, cut the stems off below the water surface during the early heading stage and again 1 to 2 months later when all regrowth is emersed. A third cutting is necessary when regrowth occurs a second time. Small patches of weeds such as water lilies and water shield can be eliminated by cutting the leaves off below the water surface at frequent intervals. Five or six cuttings a year may be necessary.

For temporary control of submersed waterweeds in large canals, rivers, around boat docks, and in fishing and swimming areas of ponds and lakes, mow off as deeply as possible and remove dislodged weeds. In canals the dislodged weeds should be trapped downstream and removed by a dragline or other device to prevent them from lodging against structures, clogging the canal, and causing overflows, canal breaks, or washed-out structures. In ponds and lakes, debris accumulating in shallow water along the shoreline can be removed by draglining, cabling, conveying, harvester-type devices, or other mechanical means.

For removal of submersed waterweeds from irrigation or drainage canals, pull a heavy chain, drag, or disk upstream along the bottom of the canal with a tractor on each bank. Several trips are usually necessary to dislodge all weed growth. Trap the floating masses of weeds at strategic places downstream and remove mechanically or by hand.

CHEMICAL CONTROL OF AQUATIC WEEDS

Herbicides often give more effective, longer lasting, and less expensive control of aquatic weeds than do mechanical or hand methods. Few aquatic herbicides are poisonous to human beings or other warm-blooded animals, but nevertheless, must be handled and used with caution according to special procedures. Some herbicides are toxic to fish, but most do not injure fish at concentrations required to kill weeds. Most aquatic herbicides do not injure crops irrigated with water at concentrations necessary for aquatic weed control. In addition to following all general precautions for the safe use of herbicides, the user of an aquatic herbicide must follow carefully all instructions and restrictions on the label regarding aquatic situations in which the herbicide should not be used. He must also know how much time should elapse after herbicide treatment before treated water may be used for drinking, fishing, swimming, and irrigating crops.[1-3]

Chemical control of streambank vegetation has been a standard operating procedure. Experiments conducted on two small headwater streams of the Newark, N.J. watershed and two similar streams on the Stone Valley Experimental Forest in Huntingdon County, Pa. demonstrated that stream flow contamination is not a problem.

A portable mistblower was used to spray watershed vegetation growing within 3.3 m of each streambank for a distance of 3.3 km. The operator walked in the middle of the stream, spraying on one side only. Vegetation on the other side was sprayed on returning in the other direction. No extra precautions were taken and, hence, some spray settled directly on the water in the stream channel.

Two chemicals were tested. One, an ester of 2,4,5-trichlorophenoxy (2,4,5-T), was considered a representative formulation of the commercially available herbicides. The other was 2,4,5-T in the form of an emulsifiable acid.

Herbicides were applied to all streams by the same operator during a low streamflow period. Flow in all streams was less than 0.1 ft^3/sec (45 gal/min). Water samples were taken periodically after treatment at various locations upstream and downstream. The first samples were collected immediately after spraying followed by a second group of samples 4 hr later. Thereafter, samples were collected daily during the first week and twice a week during the next 3 weeks. Streamflow samples collected were tested for contamination by a calibrated three-member odor panel.

Results indicate that during the 3 weeks following treatment contamination of streamflow occurred only immediately after spraying. Contamination was detectable only within the treated reach of stream and no contamination was ever found in a downstream sample. Downstream samples were collected approximately 1 mi away from the treated areas and in both locations below the junction of the two treated streams.

All areas were examined 1 year after treatment to determine the effect of the herbicides on the riparian vegetation. Streamside vegetation was very brushy. Results indicated that the ester formulation had completely killed 70 to 81% of the stems and the emulsifiable acid killed 58 to 78%. In addition, from 17 to 32% of the remaining stems were partially killed. New streambank vegetation consisted primarily of grasses and herbaceous plants.

Microbial Degradation

At least eight species of bacteria are known to feed actively on chlorinated phenoxy herbicides.[30,37] Still another species that feeds actively on phenoxy compounds has been identified as *Pseudomonas cruciviae*, and is one whose source of isolation and habitat is soil. *P. cruciviae* fed equally as well on 2,4-dichlorophenol and pentachlorophenol.

Laboratory apparatus used in these experiments consisted mainly of a source of supply of phenoxy solutions, an aerator (since biological oxidation by previously mentioned species of bacteria occurs generally under aerobic conditions), a settling chamber, and a proportioning pump to provide continuous feed at a set rate. The effluent solutions were collected and chemically analyzed to determine components or degradation products. Seeding of the aerator to obtain phenol digesting bacteria was done in two ways. Activated sludge from an industrial waste treatment plant which treats both sanitary and industrial wastes provided bacterial "seed" for half of the test equipment. However, since bacteria which decompose these compounds are reported to be found generally in nature, a collection of soil samples was gathered from a typical watershed area and used for comparison in the other half of the laboratory setup. These watershed soil samples were taken from areas on which no prior application of any herbicides had been made.[22]

Known amounts (i.e., 55 ppm) of phenoxy herbicides were fed to the influent of the systems. After 30 hours of exposure to the bacterial cultures, effluent samples were removed and exhaustively analyzed. Besides odor and taste threshold tests, a test to determine free phenols with a sensitivity of 10 ppb was used. No free phenols were detected. Microflora from either the general waste seeding or from watershed soil samples had decomposed the phenoxy compounds to carbon dioxide, inorganic chlorides, and water. A quantitative release of the chloride ion had occurred during the bacterial decomposition. In the effluents there was no detectable dichlorophenol or trichlorophenol residue. The oxygen consumption was that which theoretically would be required to effect complete oxidation. Free phenols as such were also introduced into the system for further tests on biological oxidation. The phenols were reduced in a similar manner. Confirmation of this work with regard to 2,4-dichlorophenoxy (2,4-D) has been reported by Hemmett and Faust.[18]

On the basis of this study it appears that 2,4-D can be used to control watershed vegetation, if properly applied with the normal precautions, without constituting a water pollution hazard. Studies of the degradation products of 2,4-D show conclusively that decomposition under natural conditions does *not* result in the formation of the corresponding phenols. These phenoxy herbicides are decomposed into carbon dioxide, hydrochloric acid, and water. Free phenols are decreased by bacterial degradation and are not the breakdown products of phenoxy herbicides. Aromatic oils have a much lower taste and odor threshold than the active ingredients in phenoxy herbicide formulations. These oils, however, are decomposed more rapidly than the phenoxy components. Therefore, commercial formulations properly applied to watershed areas do not constitute a water pollution hazard.[5,12,18,30,37]

Chemical Control in Slow Moving Water

Water hyacinth, *Eichhornia crassibes* Mart. Solms is an important weed which usually grows as a free-floating plant in fresh water and reproduces primarily by a vegetative root-stalk method, but also by seeds. Since they are free floating, water hyacinths are readily dispersed in water. Their growth can be so extensive that they almost completely cover the water surface, impede boat movement, and reduce oxygen levels in the water so that game fish cannot survive.

In 1944 experiments conducted at the University of Southwestern Louisiana showed that almost any 2,4-D derivative would destroy the water hyacinth, and in 1945 experiments were begun there to determine effective dose levels of 2,4-D compounds for removing this aquatic weed. The effectiveness of 2,4-D derivatives in removing water hyacinths is very good; this plant is well controlled by application of 2.2 to 4.4 kg/ha 2,4-D amine salt or its low-volatile esters.[6,10,15,30]

A

B

FIGURE 1. (A) Water hyacinth completely blocking an irrigation and drainage canal near Belle Glade, Fla. (June 1961); (B) stoppage of Little River Bayou in northern Louisiana (September 1962). (Courtesy of the U.S. Army Corps of Engineers.)

The proper amount of chemical concentrate, depending on the acid equivalent per gallon, is mixed with 50 gal (190 ℓ) of water in a 55-gal (209 ℓ) steel drum connected to a pump applied directly as a spray. The concentrate can also be placed in a tank metered to a source of water where it becomes mixed at proper proportions as it is used. The 2,4-D can be applied in as little as 50 gal of water per acre (47 kg/ha), but may be applied in mixtures from 100 to 200 gal of water per acre (93 to 187 kg/ha). The ester salt of 2,4-D mixed with a light fuel oil is particularly adaptable for treatment by planes when drift is held to a minimum. The spray equipment used consists of Bean Spray Pumps of the 10-gal/min (38 ℓ/min) and 15-gal/min (57 ℓ/min) models driven by suitable air-cooled gasoline engines. The concentrate is mixed-in-line with water from overboard. The gun used is made from pipe fittings and copper tube fittings with a suitable quick action shut-off valve. It is swivel mounted on a stanchion to absorb the recoil. Pressures of up to 400 psi (2760 kg/m^2) are used to penetrate dense vegetation. Less pressure is used in proximity to valuable vegetation to reduce the chance of damage from drifting chemicals. Where extreme infestations occur in small impoundments and where there is a hazard of deoxygenation, a special technique is used. The procedure, essentially, includes one or more of the following: (1) obnoxious growths are treated at or near feeder areas; (2) growths are trapped in small quantities and treated; (3) retreatment is made about 8 weeks following the first application of herbicide; (4) treatment is made of only 25 to 50% of infestations at one time; and (5) dead plants are permitted to sink before treating adjacent live plants.[10,15,20]

Alligator weed, *Alternanthera philoxeroides* (Mart. Griseb), is a rooted emergent which grows best in shallow water over a mud bottom and thrives in the Gulf Coastal Area of the U.S. (e.g., in the delta and flood plain areas of the Mississippi River and in other rivers flowing into the Gulf of Mexico). It grows in fresh as well as in brackish water. Free-floating pieces do not grow well, but they remain alive and begin growing when they reach a suitable area for rooting. The plant is highly aggressive and adaptable, and its unobstructed growth blocks navigation channels, municipal water supplies, and drainage and irrigation ditches. Because of its dense growth it provides optimum conditions for mosquito breeding and few fish or other aquatic animals can live beneath its dense floating mats which create a low-oxygen environment. It competes with food and cover plants and is therefore harmful to the native wildlife. It can even invade cane fields. South Carolina reservoir lakes have been highly infested. In 1949 the lakes contained 20,000 acres of alligator weed, which was reduced to 7000 acres in 1959 and to about 2500 acres in the 1960s. Various herbicidal treatments and the advent of biological control have reduced the alligator weed to a minimum.[10,15-17,20,22-25]

Eurasian watermilfoil, *Myriophyllum spicatum* L., is a rooted perennial which can produce thick surface mats of vegetation (sometimes over 7 ft thick) which seriously interfere with the harvesting of fish, crabs, and shellfish. Overgrowth with this plant also damages waterfowl habitats by crowding out duck food and other native plants, and it can seriously hurt recreational uses and real estate values of the affected areas. It can provide excellent breeding habitats for mosquitoes, a problem which is especially significant in areas such as the Tennessee Valley Authority (TVA) reservoirs where malaria mosquitoes pose a serious problem. This milfoil has invaded fresh and brackish waters along the Atlantic coast and is becoming established in inland habitats throughout the U.S. At present this plant is established in the Northeast in Massachusetts, Vermont, New York, Pennsylvania, New Jersey, and Delaware. In the Midwest it is reported in Ohio, Indiana, Illinois, and Wisconsin. It is now invading lakes or estuarine areas in Maryland, Virginia, North Carolina, Georgia, Florida, Alabama, Louisiana, Texas, and also reported in California. New invasions in the southeastern states have recently occurred at Lake Seminole, Ga.; Chassahowitzha Lake, Fla.;

A

B

FIGURE 2. St. Johns River Basin, Deep Creek, Florida. Large-scale field tests using Silvex® for chemical control of alligator weed. (A) Upstream of State Road No. 207, before treatment, May 1963 and (B) after treatment, October 1963. (Courtesy of the U.S. Army Corps of Engineers.)

A

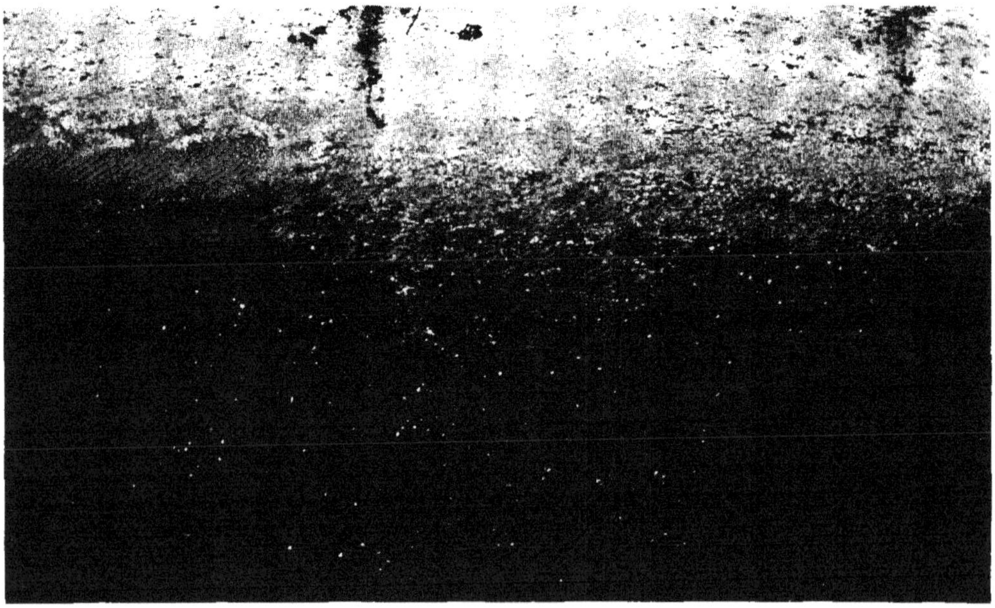

B

FIGURE 3. Eurasian watermilfoil (June 1954) is a rooted perennial that grows on the bottom of many lakes which seriously interferes with the harvesting of fish, crabs, and shellfish. The plant as shown in (A) has a very prolific seed head and can reproduce vegetatively from each node and (B) an infestation growing on the bottom of Lake Bistineau, near Shreveport, Louisiana. (Courtesy of the U.S. Army Corps of Engineers.)

Caddo Lake, La.; and several interior lakes in Texas. Eurasian watermilfoil also became more widely spread in other aquatic habitats in the states, particularly in the Pacific Northwest.

Once this plant has invaded an area it can increase rapidly. For example, it has rapidly spread over 80,000 ha throughout the Chesapeake Bay area and its tributaries during approximately the past 10 years. Another example of Eurasian watermilfoil's impressive growth potential can be provided by citing its recent increase in the Currituck Sound region of North Carolina where in 1965 only 40 ha was classified as heavily infested with milfoil and the plant was established in another 27,000 ha (there are about 40,000 ha of open water in Currituck Sound plus an additional 9300 ha of connecting waters of Back Bay).

Noteworthy economic losses are incurred by overgrowth with Eurasian watermilfoil. In the Chesapeake Bay area under proper management, oystermen, every 3 years, can harvest oyster crops valued at $1000/ha. In the Currituck Sound-Back Bay area in recent years approximately 85,000 large mouth bass and 30,000 waterfowl have been harvested annually (and the majority of this harvest occurred in the areas most endangered by the watermilfoil). With overgrowth of watermilfoil this production is lost.

Herbicides have been successfully used for Eurasian watermilfoil reduction or removal beginning with early tests employing phenoxy formulations. A 2,4-D treatment has been developed which is not harmful to food-chain organisms or to commercial fish growing in the Chesapeake Bay area. Also, oysters and clams are not adversely affected by this treatment of milfoil, and analytical studies have shown that these shellfish tend to cleanse themselves of 2,4-D residues. Experiments of Lowe cited by Rawls[26] on oysters exposed to 2,4-D butyl ethyl ester (BEE), a concentration of 0.1 ppm in flowing sea water for 7 days showed that the oysters accumulated an average of 18 ppm of this herbicide, but after "flushing" for 7 days no 2,4-D BEE residues could be found in their tissues. Thus, even as oysters required as long as several months to free themselves of residual herbicide, treatment of the oyster beds during the vulnerable milfoil period would find the oysters essentially free of residue by the September harvest period.

For control of Eurasian watermilfoil available evidence indicates that 2,4-D is the herbicide of choice. Time and habitat are factors which influence the application of this compound. Differences in habitat also determine specifications for control of Eurasian watermilfoil. A standard procedure for control of this plant is dispersal of 2,4-D ester on attaclay granules. In tidal waters and reservoirs, best results have occurred with butoxyethanol ester of 2,4-D because the balance of residual and release of this formulation seems best adapted in these situations.[21,26,31,32,38,39]

Water chestnut, *Trapa natans* L., is another aquatic weed that can effectively prevent use of water areas by man and destroy animal wildlife habitats. This plant is of Asian origin and was introduced into New York in about 1884, presumably to decorate a pond. In 1900 it was a common decorative plant on the Washington Mall. The water chestnut is an annual aquatic plant which forms a dense cover of overlapping leafy rosettes. Each rosette can produce 15 to 20 hard, thorn-covered seeds, each approximately the size of a hickory nut. The seeds are buoyant and can be transported by water currents. Moreover, they remain viable for several years and therefore complicate control operations.

Water chestnut infestations occur primarily in New York and Maryland. In New York these plants are mainly in the Hudson-Mohawk Valley area where it is well established (as well as more vigorous and more resistant to control than in Maryland). In Maryland infestations occurred in the upper Chesapeake Bay area. Although this plant does not infest saline water, there is a large area of fresh water in the upper Bay which can be invaded. By 1923 water chestnuts were found in Oxon Run opposite Alexandria and 10 years later 10,000 acres were occupied by this plant in the Potomac.

Removal of water chestnuts from the Potomac area was instituted because these plants were seriously interfering with navigation. The U.S. Army Corps of Engineers successfully utilized underwater mowing techniques for its removal, and during recent years water chestnuts have seldom been seen in the Potomac.

In the New York area good progress has been made in controlling water chestnuts with 2,4-D, although this compound has not been uniformly successful. 2,4-D was applied with variable success in the fresh and slightly brackish waters of the upper Chesapeake Bay. Extensive cutting and pulling operations are also conducted in the Bay area. The herbicide 2,4-D has been used successfully for controlling blanket growths of water chestnut where there is little or no water movement.[21,31]

Water lettuce, *Pistia stratiotes* L., is a floating water plant which develops an extensive root system (18 to 36 in. long); the plants interlock to form thick masses of vegetation which are strong enough to stop a boat, interfere with fishing and lower the dissolved oxygen in the water below them. It is a problem primarily in areas that remain warm all year. Cold weather, not necessarily freezing, will kill it. Mats of water lettuce serve as breeding grounds for a number of species of disease-carrying insects including those insect vectors for encephalomyelitis and filariasis.[36]

Chemical Control in Rapidly Flowing Water

Irrigation Canals and Ditches

Aquatic vegetation in irrigation conveyance systems is difficult to control. The water is constantly flowing (usually at a fairly high velocity) which results in the transport of herbicide chemicals from the site of application. In order to provide adequate contact time for lethal dosages to be absorbed by the vegetation, herbicides must be applied continuously to the water for periods known to be effective. Consequently, large volumes of water must be treated at considerable expense. In addition, herbicides must usually be applied to water that will, in a short time, be diverted onto cropland. In irrigation areas, water is a scarce commodity and not to be wasted. During the time of season when weeds are most troublesome, the canals must operate continuously at full capacity to deliver the volumes of water required to satisfy the needs of the farmers. Thus, it is impractical to interrupt the flow of water to cropland by shutting down the flow in the canals or by denying diversion of water to crops.[3,7-9,11]

Acrolein

Acrolein is usually used as a contact herbicide in large irrigation canals where aromatic solvents do not perform well. Application rates vary from about 0.1 to 0.6 ppmw. Treatment times (contact periods) range from 8 to 48 hr. In general, the larger the canal the lower the treatment rate and the longer the contact period. Infrequently, smaller canals may be treated with rates up to 15 ppmw for periods of 30 to 60 min. Irrigation canals and ditches with little grade and very slow moving water may be treated in the same manner as impounded water by injecting 4 to 7 ppmw of acrolein beneath the water surface along the entire length of the canal. Acrolein is very toxic to fish. Use of acrolein must be restricted to waterways where fish are of minor importance, and treated water should not be permitted to enter other waters until the acrolein has been dissipated, unless the dilution is great enough to provide a suitable safety factor.

Xylene®

Xylene® of not less than grade b has been used in irrigation canals of up to 100 ft^3 sec of flow for more than 25 years. Application rates vary from 6 to 10 gal/ft^3 sec of water flow applied over a period of 30 to 60 min. An emulsifier is added to the Xylene® (1 to 2%) before injection into the water. Xylene® is volatile and is lost quite rapidly making booster applications necessary at intervals of 5 to 10 km. Xylene®, like acrolein, is a contact herbicide and is effective on algae as well as vascular aquatic weeds. It is very toxic to fish, crayfish, and other aquatic animals, but has an exceedingly good record of safety to crops and warm-blooded animals. It is common practice for farmers to continue diverting water onto crops during periods when Xylene® is used for weed control. Because Xylene® is used primarily in the smaller canals of water distribution systems, there is usually very little waste or return water. Because of its volatility, and breakdown of the emulsion during the irrigating process, practically no solvent remains in drainage water. Livestock and other animals refuse to drink water containing more than trace amounts of Xylene®.

Copper Sulfate

The continuous-feed method of using copper sulfate for control of submersed aquatic plants has been used only rarely, and mainly on experimental bases. Field trials have shown that 1 ppmw used early in the season, followed by about 0.5 ppmw later in the season when the water is warmer, will control algae and some species of vascular aquatic plants. No acute toxicity to fish has been observed.

Diquat

This herbicide has had limited use in irrigation systems, being restricted to situations where there is virtually no movement of the water. Diquat is ineffective in turbulent water and in water containing appreciable quantities of suspended solids. It is very readily absorbed to plant or mineral matter, and water residues are usually present in only trace amounts 2 to 4 days after treatment. Present restrictions require a waiting period of 10 days. It is not toxic to fish when used at the recommended rates of 0.25 to 1.0 ppmw. Effects on other organisms appear to be largely indirect and caused by habitat modification.

Endothall

The restrictions concerning the use of the sodium and alkylamine salts of endothall, 7-oxabicyclo-(2.2.1) heptane, 2,3, dicarboxylic acid, are the same as for diquat. Treatment rates vary from 1 to 5 ppmw, depending on the levels of weed infestation. The water should not be used for any purpose for 14 days or for 25 days following treatment with rates up to 3 and 5 ppmw, respectively. The alkylamine salts of endothall are very toxic to fish at levels above 0.3 ppmw. Fish may be used for food 3 days following treatment.

Monuron

This herbicide, 3-(*p*-chlorpheny 1)-1, 1-dimethylurea, is not useful for control of aquatic weeds in the main channels of irrigation systems. Its principal use is for control of vegetation in lateral ditches that carry water intermittently and directly to irrigated fields. Since herbicides used in these ditches are not linked directly to other water systems, they are only of secondary importance. The use of diuron, 3-(3,4-dichlorophenyl)-1, 1-dimethylurea, is the same as for monuron.

Drainage and Dry Flood Control Ditches

Most of the factors and conditions governing the uses of herbicides in irrigation canals and ditches apply equally to drainage and flood control ditches. In some drainage ditches fish may be of considerable importance, in others they may be nonexistent or of little importance. Also, some drainage ditches may be nonexistent or of little importance. Also, some drainage ditches may have sizable water flows, while others have essentially no flow during much of the year. Drainage ditches commonly empty into water impoundments or into streams. Water from drainage ditches is also used occasionally for irrigation. In treating drainage ditches for control of vegetation, the precautions that must be observed are dependent upon the circumstances that apply to the time of the year; the ditch may be treated as noncropland without the limitations of aquatic sites.

Problems of Chemical Control

The use of herbicides for the control of aquatic weeds poses some unique problems due in part to differences in plant morphology and in part to the differences in the plant environment. When trees, brush, and other terrestrial weeds are sprayed with herbicide, relatively large quantities of the herbicide-spray actually contact the foliage, stem, and in some cases the roots (i.e., following precipitation). A similar direct application of herbicide spray to the exposed portion of aquatic vegetation is possible and for some weeds this is enough since the translocation process carries the herbicide throughout the plant. On the other hand, for the submersed weeds and for the parts of the other weeds under the surface of the water, the physical problem of carrying the herbicide directly to the plant without great dilution in a large mass of water is much greater than for the land weeds. Another problem is caused by differences in the plants themselves; a number of aquatic weeds such as the water hyacinth, pond lily, etc., are covered with a waxy coat which is only slightly permeable to aqueous solutions. Some of the other weeds such as elodea are highly permeable and nearly anything soluble in water can get into the plant.

The application methods most used for the control of aquatic and ditchbank weeds involve the spraying of a herbicide (1) as an aqueous solution and (2) a nonaqueous solution (e.g., fuel oil solution of a 2,4-D emulsion which inverts when placed in large quantities of water). An alternative method of dissemination often used consists of the broadcast distribution of a granular formulation (such as 2,4-D compound on attaclay granules). The latter is used particularly as a means of getting release of the herbicide close to the roots in the mud or silt bottom.[4,10,19,20,27-29]

Ecological Consequences of Chemical Control

A toxicological study on the effects of 2,4-D BEE impregnated attaclay pellets was carried out in experimental 1-acre tidewater plots in Potomac tributaries in the Chesapeake Bay area. For this study caged oysters, crabs, clams, and fish were held near the center and outside of plots of Eurasian watermilfoil, *Myriophyllum spicatum* L., during and after treatment with the 2,4-D preparation. When normal aerobic conditions were maintained, treatments neither caused mortality to the fauna nor killed the native plants. However, a serious threat to bottom organisms can occur when large mats of decomposing milfoil remain on the bottom.[3,13,14,38,39]

In another study in the Chesapeake Bay area on the toxicity to caged estuarine animals of herbicides used to control Eurasian watermilfoil, of the compounds tested only 2,4-D acetamide appeared to be dangerously toxic to test animals at levels needed for milfoil control. Blue crabs, *Callinectes sapidus,* eastern oysters, *Crassostrea virginica,* softshell clams, *Niya arenaria,* and various species of fish were studied using principally 2,4-D formulations (PBBEE, BEE, IOC, and acetamide 2,4-D compounds and Silvex® were among the agents tested).

A field study was conducted to determine the uptake and dissipation of the dimethylamine salt of 2,4-D in water, hydrosoil, and fish in ponds located in Florida and

Georgia. The residue in the treated water was dissipated to less than the negligible level of 100 ppb in 2 weeks, and did not attain this level at any time in the hydrosoil or fish flesh from a treatment of 2.24 to 8.96 kg/ha acid equivalent (ae) of 2,4-D.

The highest residues found in any fish were samples from the 1-day harvest from ponds in Florida. These fish had been kept in live cages for up to 3 days after treatment to facilitate sampling. Hence, they were unable to escape from the applied herbicide. No residues were detected in fish from Florida ponds at the 3- or 7-day samples; however, negligible residues were detected in fish at the 14-day sampling. Only one fish from the Florida ponds contained a detectable residue after 14 days and this was less than the negligible level.

The highest 2,4-D residue found in fish from the Georgia ponds was 0.075 mg/kg found in one of three bluegills harvested at 14 days from the pond treated at 8.96 kg/ha. No detectable residues were found in any fish from the Georgia ponds at the 3- or 7-day harvests, paralleling the results found in the Florida ponds.

None of the control fish contained detectable residues of 2,4-D and the 2,4-D residues found in the exposed fish were well below the toxic levels for the dimethylamine salt formulations.[27-29]

Parameters Affecting Residue Levels

Under the present requirements of the Environmental Protection Agency (EPA) for herbicide registration, it is necessary to establish that residue levels in treated water are those environmentally acceptable for human and animal health. The concentration of detectable amounts of the herbicide 2,4-D in the water after weed control treatment is known to be determined by such abiotic (nonliving) factors as the rate of treatment and the extent of dilution as affected by depth, flow, or addition of water into the system. Photodegradation and absorption of the hydrosoil are *not* found to be major routes for reduction of 2,4-D residues but biodegradation by microorganisms and metabolism by plants are found to be major routes of herbicide decomposition. Temperature and time are important variables affecting the rate of dissipation because they are known to influence biological processes to a large degree.[5,7,8,12,13]

Biodegradation in Aquatic Systems

The persistence of 2,4-D in aquatic environments has been studied. Aly and Faust[5] used bottom mud samples collected from two lakes. The first lake had been treated the previous year with 2,4-D while the second lake had never been treated with 2,4-D. It was observed that when 2,4-D was applied to the mud from the lake previously treated with 2,4-D, it disappeared in 35 days, whereas 2,4-D persisted in muds from the untreated lake for 65 days. Schwartz[30] found that only a small fraction of 2,4-D was degraded by a mixed microbial population in a dilute medium of mineral salts. The presence of large amounts of nutrient broth as a supplemental source of organic carbon had no appreciable effect on the rate of decomposition.

Biodegradation Kinetics of 2,4-D

In a study of the biodegradation kinetics of 2,4-D[18] three variables were considered: (1) microorganism concentration, (2) substrate concentration, and (3) ratio of microorganism concentration to substrate. These organisms converted 2,4-D into carbon dioxide, water, and hydrochloric acid. The biodegradation appeared to follow zero-order kinetics, with the oxidation rate independent of the substrate concentration. When the ratio of microorganism concentration to 2,4-D concentration was held constant, constant rates of oxidation were observed which suggested the enzyme systems were saturated with substrate. Acclimatization time appeared to be an important factor. The longer the period of acclimatization, the shorter was the period of time to effect "ultimate" biodegradation of 2,4-D. The rates at which the biodegradation is performed are dependent on several variables: (1) the time necessary for the enzyme systems to become acclimatized to the herbicide, (2) the age of the acclimatized microorganisms, and (3) the natural condition of the aquatic environment.

Program of Herbicide Treatment

Direct effects of repeated use of herbicides such as 2,4-D on the aquatic environment are relatively minor if the herbicides are used properly. Indirect effects may conceivably result in particular changes in the environmental system over a period of time. These changes, however, are not necessarily detrimental for single organisms are not indispensable to an ecological community and many different organisms compete for the same niche. The most obvious effect of herbicides is the reduction of the total population to an earlier point in the succession. This again is not a serious or lasting problem because many different organisms are known to be able to adapt their metabolic processes to utilize the herbicide as a sole source of carbon and reduce it to a nontoxic entity.[3,18]

Registration of 2,4-D

There are special data requirements necessary for the registration of herbicides with the EPA.[33,34] For a number of years the Bureau of Reclamation and the Bureau of Sport Fisheries, U.S. Department of the Interior, the Agricultural Research Service, U.S. Department of Agriculture; and the Corps of Engineers, U.S. Department of Defense, have been studying the aquatic-use pattern of 2,4-D.

Tolerance of 2,4-D

Tolerances are established for residues of the herbicide 2,4-D as follows:

1. 2 ppm in the milled fractions (except flour) derived from barley, oats, rye, and wheat to be ingested as food or to be converted to food. Such residues may be present therein only as a result of application to the growing crop of the herbicides identified in 40 CFR 180.142.
2. 0.1 ppm (negligible residue) in potable water. Such residues may be present therein only: (a) as a result of the application of the dimethylamine salt of 2,4-D to irrigation ditch banks in the western U.S. in programs of the Bureau of Reclamation; cooperating water user organizations; the Bureau of Sport Fisheries, U.S. Department of the Interior; Agricultural Research Service, U.S. Department of Agriculture; and the Corps of Engineers; U.S. Department of Defense or (b) as a result of the application of the dimethylamine salt of 2,4-D for water hyacinth control in ponds, lakes, reservoirs, marshes, bayous, drainage ditches, canals, rivers, and streams that are quiescent or slow moving in programs of the Corps of Engineers or other federal, state, or local public agencies.

BIOLOGICAL CONTROL OF ALLIGATOR WEED

Insects have been used successfully as a form of biological control to suppress alligator weed, *Alternanthera philoxeroides* (Mart.) Griseb, in Florida and other states of the Southeast under the Corps' aquatic plant control program in cooperation with the Division of Entomology Research of the U.S. Department of Agriculture. The alligator weed flea beetle, *Agasicles hydrophila* (Selman and Vogt) was the first host-specific insect to be introduced. Other insects, alligator weed thrips, *Amynothrips andersoni* (O'Neill), and a stem-boring moth, *Vogtia malloi* (Pastrana), are also host-specific and have been introduced for alligator weed control. Infestations of alligator weed are reduced to a maintenance status in most situations in the southeastern states where these insect controls have been released.[23-25,35]

Exploration for Natural Enemies

A research agreement was initiated by the Corps of Engineers in 1959 with the U.S. Department of Agriculture, Division of Entomology Research, to conduct explorations in South America for natural enemies of alligator weed. A research laboratory was set up in Argentina for further research.[18]

Host Specificity

Laboratory feeding studies on the alligator weed flea beetle were conducted on *Polygonum, Fagophysrum, Rheum, Chenopodium, Atriplex, Amaranthus,* and *Alternanthera.* These feeding tests also included *Oryza, Nasturtium,* and *Nymphaea. Atriplex hastata* L. was the only species other than alligator weed that indicated a response, and this species did not permit completion of the life cycle. In the feeding experiments with *A. hastata* L. both larval and adult flea beetles fed on the test plant. As a result of this feeding, the adults lived an average of 21.7 days but showed abnormal behavior and died without producing eggs. In fact, dissection of the females subsequent to death demonstrated no ovarian development. Furthermore, *A. hastata* does not have a hollow stem required in nature as a site for pupation. Observations of the plant under growing conditions have failed to demonstrate feeding by either larvae or adults of the alligator weed flea beetle.

Insect Introduction

The decision was made that this flea beetle was the most promising biocontrol for alligator weed, and the insect was brought, under quarantine, to the research laboratory in Albany, Calif. for further study in 1963. The first release was made in South Carolina in 1964, at the Savannah National Wildlife Refuge under conditions which appeared to be similar to those in South America. The first successful biological control was observed in the Ortega River, near Jacksonville, Fla. in 1965. Most of the alligator weed flea beetles distributed in the U.S. have come from this area. Conditions in Florida were apparently more favorable than elsewhere. Successful control, however, may have been due in part, to the effects of 2,4-D on the alligator weed mat from incidental treatment of water hyacinth, shortly after release of the beetle.[23-25]

Distribution in Southeastern U.S.

During May 1965, two trips were made into southeastern U.S. to distribute the alligator weed flea beetle and to initiate evaluation studies. The first trip was made by Dr. W. H. Anderson, U.S. Department of Agriculture, from May 7 to 12. His primary objective was to release beetles at selected localities in Georgia, South Carolina, and North Carolina. On May 9th, with Messrs. Charles Zeiger and James McGeehee of the Corps of Engineers, Jacksonville, he visited the original release site on the Ortega River. The alligator weed infestation at this site showed no evidence of recovery. At another site on Black Creek some 15 mi south of Jacksonville, where there was a considerable amount of alligator weed in small as well as extensive patches along the banks, all plants were found to be under heavy attack. Additional releases were made at the Jim Woodruff Reservoir in Georgia and Florida, at three sites near Mobile, Ala., at Gulfport and Yazoo City, Miss., at two sites on the Dam B Reservoir near Jasper, Tex., and at three sites in the J.D. Murphree Wildlife Area near Port Arthur, Tex.[23-25]

Vegetative Control

Throughout its lifestages, the flea beetle attacks the alligator weed in different ways. The adults feed on surface leaves, females lay their eggs — 1000 or more — on the undersides of leaves, young larvae then feed on the undersurface of the leaf and, as mature larvae, chew their way into the stems. Larvae develop into adults within the stems, eat their way out, and return to the leaves to start the cycle again. Damage by the beetle and larvae either kills the alligator weed outright, or weakens it, making it vulnerable to disease, competition from other aquatic plants, and wind and wave action.[23-25]

BIOLOGICAL CONTROL OF WATER HYACINTH

As pointed out by Zettler and Freeman[40,41] plant pathogens have many characteristics that make them ideal candidates as biocontrols for aquatic weeds. They are (1) numerous and diverse, (2) frequently host specific, (3) easily disseminated and self-perpetuating, (4) will not completely eliminate a host species, and (5) do not normally affect man or other animals. With these points in mind, a modest program was begun at the University of Florida in 1970, with the object of the evaluation and subsequent use of plant pathogens as biocontrol agents for aquatic weeds.

The Florida Department of Natural Resources estimates over $15 million are expended annually in Florida for aquatic weed control. These control efforts are concentrated primarily on the estimated 100,000 ha of water hyacinth, *Eichhornia crassipes*, and 40,000 ha of hydrilla, *Hydrilla verticillata*, that occur in the state. Lesser attention is given to the approximately 20,000 ha of other aquatic weeds, such as Eurasian watermilfoil, *Myriophyllum spicatum*, and alligator weed, *Alternanthera philoxeroides*. Despite these efforts, aquatic weed infestations have increased steadily in the years since these plants were introduced into Florida. This is alarming when one considers that hydrilla was introduced only 15 years ago.

Florida is by no means unique in having a tremendous aquatic weed problem. Proliferating waterweed populations are of concern in the rest of the U.S., Middle Europe, Africa, Asia, and South and Central Americas. Indeed the problem is worldwide, but is more acute in the warmer latitudes where water hyacinth, hydrilla, watermilfoil, alligator weed salvinia, *Salvinia* spp., and water lettuce, *Pistia stratioties*, are the major offenders.

In recent years, biological control methods have received considerable attention. Various species of herbivorous insects, fish, snails, and even the manatee have been, or are being, investigated for their ability to exert some control pressure on noxious aquatic plants. Some of them, such as the alligator weed flea beetle, have been reasonably effective, especially in an integrated control program. Surprisingly, until our program was initiated, plant pathogens had been rarely considered as biocontrol agents. They have all the prerequisites of a biocontrol agent and thus offer an untapped reservoir of potential usefulness, either alone or in an integrated control program with insects and, perhaps, chemicals.

Research efforts have been directed toward those pathogens with definite biocontrol potential. These are the endemic pathogens of water hyacinth — *Acremonium zonatum, Cercospora piaropi, C. rodmanii, Rhizoctonia solani* — and three exotic ones — *Bipolaris stenospila, Uredo eichhorniae,* and *Rhizoctonia* sp. Extensive cultural studies in the laboratory, and greenhouse, including cross-inoculation onto other plant species, host-range studies and, in the case of the endemic pathogens, small-scale field tests have been carried out. These latter studies have shown *A. zonatum* and *C. rodmanii* to have considerable potential as biocontrols at locations in Florida and Lake Concordia in Louisiana.[40,41]

REFERENCES

1. **Anon.**, Herbicide Manual for Noncropland Weeds, Department of the Army, the Navy and the Air Force, U.S. Government Printing Office, Washington, D.C., 1970, 201.
2. **Anon.**, Guidelines for Weed Control, Agricultural Research Service, U.S. Department of Agriculture, Agricultural Handbook No. 447, Washington, D.C., 1972, 121.
3. **Anon.**, Herbicide Report, Chemistry and Analysis of Agricultural and Other Applied Uses, EPA-SBA-74-001, U.S. Environmental Protection Agency, Washington, D.C., 1974.
4. **Allison, R. V. and Seale, C. C.**, Waterhyacinth control through the use of 2,4-dichlorophenoxyacetic acid in different chemical forms and at various rates of application, *Soil Crop Sci. Soc. Fla. Proc.*, 8, 97, 1946.
5. **Aly, O. M. and Faust, S. D.**, Studies on the fate of 2,4-D and ester derivatives in natural surface water, *J. Agric. Chem.*, 12, 541, 1964.
6. **Averitt, W. K. and Gangstad, E. O.**, Dissipation of 2,4-D residues in static water, *J. Environ. Qual.*, 5, 145, 1976.
7. **Bartley, T. R. and Hattrup, A. R.**, 2,4-D contamination and persistence in irrigation water, *Proc. West. Soc. Weed Sci.*, 23, 10, 1970.
8. **Bartley, T. R. and Gangstad, E. O.**, Environmental aspects of aquatic plant control, *J. Irrig. Drain. Div. ASCE*, 100, 231, 1974.
9. **Blackburn, R. D., Seaman, D. E., and Weldon, L. W.**, A preliminary note on the control of southern naiad *(Najas guadalupensis)* in south Florida drainage and irrigation channels, *Proc. 14th South. Weed Conf.*, 302, 1961.
10. **Brown, C. A. and Carter, W. H.**, Weed investigations, *La. Agric. Exp. Stn. Bull.*, 402, 24, 1946.
11. **Bruns, V. F., Hodgson, J. M., Arle, H. F., and Timmons, F. L.**, The use of aromatic solvents for control of submersed aquatic weeds in irrigation channels, *USDA Circ.*, 971, 1955.
12. Council for Agricultural Sciences and Technology, The phenoxy herbicides, *Weed Sci.*, 23, 259, 1975.
13. **Crafts, A. A.**, Herbicide behavior in the plant, in *The Physiology and Biochemistry of Herbicides*, Audus, L. J., Ed., Academic Press, New York, 1964, 75.
14. **Frank, P. A. and Comes, R. D.**, Herbicide residue in pond water and hydrosoil, *Weeds*, 15, 210, 1967.
15. **Gangstad, E. O.**, Herbicidal control of aquatic plants, *J. Sanit. Eng. Div., Am. Soc. Civ. Eng.*, 98, 397, 1972.
16. **Gangstad, E. O., Spencer, N. R., and Foret, J. A.**, Towards integrated control of alligatorweed, *Hyacinth Control J.*, 13, 30, 1974.
17. **Hawkes, R. B., Andres, L. A., and Anderson, W. H.**, Release and progress of an introduced flea beetle, *Agasicles* n. sp., to control alligatorweed, *J. Econ. Entomol.*, 60, 1476, 1967.
18. **Hemmett, R. B. and Faust, S. F.**, Biodegradation kinetics of 2,4-dichlorophenoxyacetic acid by microorganisms, *Herbicide Rev.*, 29, 191, 1969.
19. **Hilderbrand, E. M.**, Herbicidal action of 2,4-dichlorophenoxy acetic acid on the waterhyacinth, *Eichhornia crassipes, Science*, 103, 477, 1946.
20. **Hitchcock, A. E., Zimmerman, P. W., Kirkpatrick, H., Jr., and Earle, T. T.**, Growth and reproduction of waterhyacinth and alligatorweed and their control by means of 2,4-D, *Contr. Boyce Thompson Inst.*, 16, 91, 1950.
21. **Holm, L. F., Weldon, L. W., and Blackburn, R. D.**, Aquatic weeds, *Science*, 166, 699, 1969.
22. **Lawrence, J. M. and Funderburk, H. H., Jr.**, Effectiveness of herbicidal combinations on alligatorweed growing in flowing water, *South. Weed Conf., Proc.*, 18, 558, 1965.
23. **Maddox, D. M., Andres, L. A., Hennessey, R. D., Blackburn, R. D., and Spencer, N. R.**, Insects to control alligatorweed, an invader of aquatic ecosystems in the United States, *Bioscience*, 21, 985, 1971.
24. **McGilvrey, F. B. and Steenis, J. H.**, Control of alligatorweed in South Carolina and granular silvex, *Weeds*, 13(1), 66, 1966.
25. **Power, A. E. and Walsh, W. C.**, Operational control of alligatorweed on Santee-Cooper Lakes, South Carolina, *South. Weed Conf., Proc.*, 18, 555, 1965.
26. **Rawls, C. K.**, Field test of herbicide toxicity to certain estuarine animals, *Chesapeake Sci.*, 6(3), 150, 1965.
27. **Schultz, D. P.**, Dynamics of a salt of (2,4-Dichlorophenoxy) acetic acid in fish, water and hydrosol, *J. Agric. Food Chem.*, 21, 186, 1973.
28. **Schultz, D. P. and Harman, P. D.**, Residues of 2,4-D in pond waters, mud and fish, *Pestic. Monit. J.*, 8, 173, 1974.
29. **Schultz, D. P. and Gangstad, E. O.**, Dissipation of residues of 2,4-D in water, hydrosoil and fish, *J. Aquatic Plant Manage.*, 14, 43, 1976.
30. **Schwartz, H. J.**, Microbial degradation of pesticides, *J. Water Pollut. Control Fed.*, 39, 1701, 1967.

31. **Steenis, J. H.,** Specifications for control of aquatic plants for the northeast, *Northeastern Weed Control Conf., Proc.,* 20, 18, 1966.
32. **Timmons, F. L.,** Herbicides in aquatic weed control, *South. Weed Conf., Proc.,* 16, 5, 1963.
33. **Walker, C. R.,** Toxicological effects of herbicides on fish environment, *Water Sewage Works,* 3, 113, 1969.
34. **Walker, C. R.,** Problems in clearance and registration of herbicide for aquatic areas, *Hyacinth Control J.,* 9, 5, 1971.
35. **Weldon, L. W. and Blackburn, R. D.,** Recent studies on the control of alligatorweed, *South. Weed Conf., Proc.,* 18, 567, 1965.
36. **Weldon, L. W. and Blackburn, R. D.,** Water lettuce — nature, problem, and control, *Weeds,* 15(1), 5, 1967.
37. **Winston, H. W. and Ritty, P. M.,** What happens to phenoxy herbicides when applied to a watershed area, *Northeastern Weed Control Conf., Proc.,* 14, 396, 1960.
38. **Whitney, E. W., Montgomery, A. B., Martin, E. C., and Gangstad, E. O.,** The effects of a 2,4-D application on the biota and water quality in Currituck Sound, North Carolina, *Hyacinth Control J.,* 11, 13, 1973.
39. **Wojtalik, T. A., Hall, T. F., and Hill, L. O.,** Monitoring ecological conditions associated with wide-scale applications of DMA 2,4-D to the aquatic environment, *Pestic. Monit. J.,* 4, 189, 1971.
40. **Zettler, F. W. and Freeman, T. E.,** Plant pathogens as biocontrols of aquatic weeds, *Annu. Rev. Phytopathol.,* 10, 455, 1972.
41. **Zettler, F. W. and Freeman, T. E.,** Potential for the use of plant pathogens as biocontrol agents of weeds, Proc. 2nd Int. Congr. Plant Pathology, St. Paul, Minn., 1973.

Chapter 5

THE ROLE OF HERBICIDE RESIDUES OF 2,4-D IN THE AQUATIC ECOSYSTEM

INTRODUCTION

Aquatic macrophyte problems have become common in the reservoirs and inland waterways of the U.S. This is attributed to: (1) greater light penetration resulting from reduced sediment loads, (2) increased human habitation of areas adjoining shorelines with accompanying increases in water usage for domestic, industrial, and recreational purposes, (3) increased eutrophication from natural and sewage nutritive enrichment of the water, and (4) the presence of native and introduced species that possess high colonization potential.

To understand both the desirable and undesirable aspects of herbicide control in the water, it is necessary to know something about the residues, their disappearance or degradation, and their effects on target and nontarget organisms. This paper discusses selected portions of the published literature dealing with the use of phenoxy herbicides in aquatic sites.[1-3,12-14,28]

SOURCE OF HERBICIDE RESIDUES

Before discussing the degradation of herbicides, some consideration should be given to the levels of residues which occur in water following the use of chemicals for weed control. The usual means by which herbicides enter water are (1) from surface runoff during irrigation and/or heavy rainfall, (2) by application of herbicides to soil or water for control of submersed weeds in canals, ponds, or lakes, (3) by herbicide treatment of floating and emersed weeds, and (4) from treatment of banks of streams and canals for control of bank and marginal weeds.

Surface runoff — Where residues are shown to occur, the concentrations are low and persist for only short periods. Columbia Basin irrigation supply and waste waters were monitored for several pesticides during the 1961 irrigation season. Only on rare occasions were small quantities of 2,4-dichlorophenoxy acetic acid (2,4-D) found in the supply water. Esters of 2,4-D were found more in waste drains; however, the highest level measured was 18 ppb, which was much higher than the average of all levels of 2,4-D.[9] Schulze et al.,[58] while monitoring many streams in the western U.S. for presence of pesticides, occasionally found low levels (10 ppb) of 2,4-D in the water.

Pond and lakes — Granular formulations may prevent occurrence of high concentrations of certain herbicides in water by confining portions of the chemicals at the soil (hydrosoil) surface. Granular formulations of phenoxy herbicides are notable in this respect. In a pond treated with 1.33 ppm of granular butoxyethanol ester of 2,4-D, the maximum residue level observed was 0.067 ppm.[21,25,31]

Floating weeds — In similar studies involving applications of the dimethylamine salt of 2,4-D and Silvex® applied at 4.5 kilogram per acid equivalent per hectare (kg/ae/ha) on water hyacinth *(Eichhornia crassipes)* and alligator weed *(Alternanthera philoxeroides)* almost all of the highest residue levels were between 1 and 0.65 ppm. The highest concentrations of herbicide did not appear in the water until 1 to 2 weeks after the treatments. These results suggested that the herbicides are absorbed by the floating plants and later released into the water through roots and other submersed plant tissues.[7-9,33,35,54-57]

Marginal and ditchbank weeds — The proximity of marginal and ditchbank weeds to water almost invariably results in some herbicide entering the water following herbicide application. Principal factors affecting the amount of herbicide found in the water are (1) treatment rate, (2) volume-flow of water, (3) nature of the weed growth, and (4) the spray overlap at the water's edge. Considerable data with respect to residues in irrigation water following use of herbicides on canal banks were recently reported. In all but a few instances, concentrations of the herbicides in irrigation water were quite low, being under 50 ppb in most cases, despite the small size of canals involved.[9,25,32,35]

DISSIPATION OF HERBICIDE RESIDUES

Dissipation in Static Water

Most of the herbicides registered for use in aquatic situations have water-use restrictions which require at least partial dissipation of the herbicides before normal use of the water resource is resumed. The pathways leading to dissipation are dependent largely upon the nature of the herbicides. Biological degradation accounts for much of the loss of 2,4-D.[32]

Dissipation in static water frequently requires 2 to 3 weeks. Averitt and Gangstad[8] found that the rate of dissipation of residues of 2,4-D at the point of application, applied for control of aquatic vegetation, was affected by the (1) rate of treatment, (2) depth of water, (3) mean temperature, and (4) time after treatment. For the standard rate of treatment, i.e., 4.48 kg/ae/ha 2,4-D declined by 58 ppb in residue for each 0.61 m increase in depth of water treated; also, a decrease of 115 ppb in residue for each 7°C increase in mean temperature, and a decrease of 53 ppb in residue for each 7-day interval of time after treatment was observed.

A pond study was conducted to determine the uptake and dissipation of the dimethylamine salt of 2,4-D in water, hydrosoil, and fish in ponds located in Florida and Georgia. The residue in the treated water declined to less than the negligible 100 ppb in 2 weeks, and did not attain this level at any time in the hydrosoil or fish flesh, from a treatment of 2.24 to 8.96 kg/ae/ha of 2,4-D.[55-57]

Pond treatment with the propylene glycol butyl ether (PGBE) ester of 2,4-D in liquid formulation resulted in lower residue levels. Dissipation from water followed the usual pattern of 4 to 6 weeks; however, at treatment rates as high as 5 and 10 ppm, no 2,4-D was detected in the soil 2 weeks after application of the herbicide. The dissipation rate of the dimethylamine salt of 2,4-D is similar to those reported for the ester derivatives. Residues of 2,4-D in water following treatment of water hyacinth at 4.48 kg/ha and 5.60 kg/ha of the amine salt were dissipated from 0.53 and 0.97 ppm to 4 and 15 ppb, respectively, after 29 days.[4-9]

Dissipation in Slow Moving Waters

Residue studies in Louisiana and Florida indicate that the residue of 2,4-D from spraying operations to control water hyacinth at 2.24 to 4.48 kg/ae/ha is related to: (1) volume dilution as a function of the surface area to the total area, (2) temperature in degrees Celsius, and (3) time as a function of the rate of flow or displacement.[33]

Dissipation in Large Reservoirs

The use of phenoxy-type herbicides in aquatic situations is almost entirely limited to derivatives of 2,4-D. While amines of 2,4-D are commonly used for control of floating and emergent species, esters of 2,4-D in granular form are frequently applied to water infested with submersed weeds.

In Tennessee Valley Authority (TVA) milfoil studies, over 18,000 surface acres of reservoirs in Alabama and Tennessee were treated with about 170,000 gal of dimethylamine salt of 2,4-D during April to June 1969 to control invading Eurasian watermilfoil. The DMA 2,4-D was applied at the rates of 22 and 45 kg/ae/ha of 2,4-D.

Representative habitat types were selected and monitored for 2,4-D content in water, plankton, and sediment and for plankton species composition, distribution, abundance, change, and response. The treated water had no apparent effects on most marginal plants. No harmful response to the herbicide was observed in zooplankton, phytoplankton, benthic macroinvertebrates, or fish. Water continued to be used for domestic purposes without user complaints. Dimethylamine salt of 2,4-D appears to be noncumulative.[10,35,66]

Residues from Granules

Residues in water, whether applied as the ester or the acid, are generally reduced to a nondetectable level within 3 to 4 weeks. To establish this pattern under field conditions a series of relatively small ponds were treated with the granular formulation of the 2,4-D butoxyethanol ester. Seven locations in the U.S. and one in Canada with a total of 15 ponds were used. Of those analyzed, 5 at 1 ppm and 2 at 5 ppm, 6 had reached a nondetectable level by the 30th day. These ponds all behaved in a similar manner with detectable residue levels peaking between 8 and 15 days posttreatment, followed by a rapid decline to the nondetectable stage.[67]

Early work done by Wilkinson[64] indicates the action of granular 2,4-D formulations in the water. Wilkinson studied several factors affecting the subaqueous rate of release from clay granules. His finding showed the following:

1. Temperature had an effect only during short periods of exposure.
2. The derivative of 2,4-D employed was the most significant factor affecting the rate of release. The rank in order of rate of release was 2,4-D acid; 2,4-D butyl ethylester (BEE); 2,4-D propyl glycol ester (PGE); and 2,4-D isooctyl ester.
3. The type bottom was shown to affect the total quantity of herbicide recovered. Bentonite absorbed significant quantities of herbicides.
4. The quantity of the herbicide recovered was shown to depend on concentration.

Data from 256 recoveries showed almost complete release from the granule by the 10th day. This substantiates the peaking factor noted in most field tests. Averitt and Gangstad[8] analyzed waters receiving 2,4-D amine treatments for water hyacinth control and forced a peak at 1 week posttreatment. Small enclosed ponds within the city of New Orleans showed almost 100% removal at the end of 31 days. Frank and Comes,[32] reported on a pond treated with granular 2,4-D butoxyethanol ester at a rate of 1.3 ppmw, found a peak of 0.067 ppmw which occurred between 12 and 18 days. This was approximately one third of the initial calculated concentration. The hydrosoil showed a much higher initial concentration with residues lasting to 55 days before a nondetectable level was reached as opposed to the 24-day residual in the water.

More recent work by Steward and Nelson[62] supports the commonly observed pattern of initial peaking and rapid decline. They compared polyvinyl chloride controlled release pellets with the attaclay formulation. The concentrations from the two formulations at 1 and 2 ppm, peak out between 7 and 14 days and the nondetectable level was reached between the 2nd and 4th week.

Residues in Fish and Shellfish

Rodgers and Stallings,[51] using the standardized techniques, noted that maximum whole body and tissue residues of 2,4-D were related to the concentration of the ester and that they declined as the ester was hydrolyzed. Rapid uptake may result from increased solubility of 2,4-D ester and transport across the gill membrane. Absorption of the ester by the lipid and lipoprotein in the blood may enhance the distribution of 2,4-D to various organs. In summary, they note that the maximum residue levels were observed in most organs of fed fish within 1 to 2 hr exposure and within 1 to 8 hr for the starved fish. After maximum residue levels (19.5 μg/g for catfish and 54.6 μg/g for bluegills were reached), the herbicide or its metabolite was eliminated rapidly.

Whitney et al.[63] in North Carolina, carrying out biological studies in conjunction with a 2,4-D butoxyethanol ester treatment for milfoil control in Currituck Sound, sampled fish over a 3-week period. Of 39 samples, the highest concentration detected was 0.24 ppm found in a large-mouth bass 8 days following application. Of the six fish species, pumpkinseed sunfish, white and yellow perch, catfish, large-mouth bass, and bluegill sunfish, no one species appeared to accumulate more residue than the others. Twenty-six of the 39 samples showed no residue, while the remaining samples had less than 0.2 ppm, except for one bass which contained 0.24 ppm.

Rawls,[44-50] working with the problem of residues in oysters in the Chesapeake Bay area, comments in his final report "crabs, fish, clams, and oysters will accumulate and lose 2,4-D butoxyethanol ester residues in varying amounts depending on the bioassay animals, the distance they are held from the treated area, water movement through the study plots, and the degree of vegetation within and around the sites of herbicidal application." He further notes that clams and oysters acquire greater residues than do fish and crabs. Maximum residue level is reached from 1 to 9 days following application and rapidly declines thereafter. The real question of residues found in aquatic animals lies in the probability of continuous exposure and the animals' ability to take certain actions to avoid residue buildups. Crabs and fish can readily move out of treatment areas while oysters and other bivalves have the ability to flush out toxic compounds as the dosage levels decline.

Wilkinson[64] included 30 posttreatment and one pretreatment sample of benthic organisms among those analyzed for 2,4-D butoxyethanol ester. The highest concentration, 0.23 ppm, was recorded in a sample collected 1 day after treatment. Twelve samples showed very slight residues, and the remainder no residues.

WATER USE

The use of herbicides in water assumes that in the normal process of degradation certain amounts may deposit on the bottom muds. The use of rapidly sinking granular formulations thus increases the probability that residues can be found in the bottom muds. Frank and Comes[32] showed that greater residues of 2,4-D butoxyethanol ester were found in the hydrosoil than in the water.

Residues in Irrigation Water

Bruns[15] made the following statement after irrigating red Mexican beans with 2,4-D-treated water:

> The application of 2.24 kg/ae/ha 2,4-D in irrigation water during the seedling stage caused pronounced injury to the top growth and roots of the bean plants. However, within 16 days after the application, plants began to recover and no significant reduction in yield resulted.
>
> The rate of 6.6 kg/ae/ha of 2,4-D during the seedling stage caused severe damage to the top growth and destroyed nearly all of the root systems of the plants. Amazingly, within 3 weeks, a number of plants began to recover with the development of club-like adventitious roots.
>
> Application of 2.24 and 6.72 kg/ae/ha of 2,4-D in irrigation water during the bloom stage caused some leaf chlorosis and damage to the root systems and hastened somewhat the maturity of the bean plants. The

2.24 kg/ae/ha treatments caused no significant reduction in yield, but the 6.72 kg/ae/ha applications reduced the yield 29%.

Laboratory and greenhouse tests exhibited no carry-over effects of 2,4-D in bean seeds harvested from treated plots.

A second phase of Bruns' work indicated that:

1. The rates of 2.24, 4.48, 8.96 and 17.92 kg/ae/ha were lethal to 13, 22, 65, and 54% of the plants, respectively, whereas lesser rates caused no death of plants.
2. Root systems of injured plants were partially or completely destroyed. Remaining portions of fragments showed varying degrees of discoloration and of tip enlargement. Depth, distribution, and development of the root systems at the time of treatment apparently was an important factor in the experiment.

Residues in Potable Water

Potable water supplies frequently become infested with algal blooms which produce objectionable flavors and odors. These may become offensive enough to make the water unfit for consumption. Another cause of flavors and odors in water is the death and decay of large masses of vascular aquatic weeds. The ensuing anaerobic conditions produce hydrogen sulfide and other substances typical of stagnant water.

There have been relatively few instances of undesirable flavor where herbicides were used to control aquatic weeds in drinking water. During 1966, the TVA used large-scale applications of granular butoxyethanol ester of 2,4-D at rates of 44 to 112 kg/ae/ha for control of Eurasian watermilfoil. The highest concentration of 2,4-D recorded at any of the water-treatment plants downstream where water was monitored was 2 ppb.[66] In the U.S. the combined permissible level of phenoxy-type herbicides in potable water sources is 0.1 ppm.

Residue in Fish and Wildlife Management

The use of herbicides as tools in fish and wildlife management is becoming more important each year. Whitney et al.[3] have pointed out that weed control is a major phase of fish management in over a million acres of ponds in the southern U.S. In recent years, Florida alone has seen the construction of more than 4000 farm ponds, and during the 20 years preceding 1965, over 1.25 million ponds plus numerous other water retaining structures, were built by the Soil Conservation Districts throughout the U.S. Fishing and other recreational activities in many of these waters, especially those of the southern states, would be impossible without some form of chemical weed control. In some regions of the U.S., notably the eastern and southeastern states, large areas of weed growth have eliminated almost entirely the harvest of fish and waterfowl. The harvest of oysters *(Crassotrea virginica)*, clams *(Mya arenaria)*, and crabs *(Callinectes sapidus)* in certain estuarine water long noted for production of seafood, is being restricted by weed species such as watermilfoil and eelgrass *(Zostera marina L.)*. The use of 2,4-D for control of undesirable weeds in tidal areas opens the water to commercial and sport fishing and releases from competition the native plants which attract waterfowl. Maintenance of the extensive and valuable areas of water and marsh would be impossible without 2,4-D herbicides.[7,10,16-29]

Residues in Mammals

Chronic or acute toxicity of aquatic herbicides in water to humans, livestock, and wildlife are obvious but perhaps overemphasized dangers. Cases of acute poisoning have been accidental and due to drinking by livestock from left over herbicide solutions. Where herbicides are used in water with due regard to the welfare of fish, there is little danger of livestock poisoning. Because of the lower tolerance levels for fish than for livestock, the presence of live fish is ample evidence that the water is safe for livestock. Numerous feeding studies show that herbicides, unlike certain insecticides, are rapidly excreted in body wastes with little or no accumulation in tissue. Where residues were found in body tissues, or in milk, they persisted for very short periods, usually not more than several days. Persuasive evidence was presented to show the wide margin of safety involved in the use of herbicides at concentrations normally applied for weed control.[25-27,38,39,43]

Residue studies and observations are reported on the effects of herbicides on phytoplankton and invertebrate fauna.[10,17,19,25,26,43,58] Where the aquatic biota have been affected, the common pattern observed has been (1) the death of macrophytes and a reduction in number of phytoplankton, (2) a loss in the number of insects associated with the rooted plants, (3) a decrease in the microfauna population, (4) an increase in density of phytoplankton, and finally (5) an increase in population of various forms of aquatic fauna. The time required for this sequence of events has varied, but most studies report periods of 2 to 3 weeks. Death and decay of the rooted plants created a nutrient-enriched environment ideal for the growth of phytoplankton that serve as the basic link in the aquatic food chain. Only in rare cases were overall effects detrimental to fish-food organisms or fish production. Rawls[45] has suggested that where fish-food organisms and bottom-dwelling invertebrates were adversely affected, the problem was linked directly to anaerobiosis in the immediate environment.

DEGRADATION OF HERBICIDES

Pathways of herbicide degradation in the aquatic environment (photo, chemical, and biological) presumably are similar to the pathways involved in degradation of herbicides in the terrestrial environment. Specific modes of degradation, however, may be significantly different. Whereas terrestrial degradation occurs predominantly in an aerobic medium, the aquatic environment, particularly beneath the water-soil interface, is largely deficient in oxygen. The microbe populations, their metabolic processes, and the end products of metabolism should be expected to vary widely from the aerobic forms. All degradation processes may occur at rates which differ greatly from those which take place in the familiar terrestrial environment.[30,31-42,52,53]

Photodegradation

Solutions of 2,4-dichlorophenol (DCP), sodium salt of 2,4-D, and butyl and isopropyl esters of 2,4-D were irradiated with ultraviolet light from a 660-w mercury discharge lamp. Degradation of all compounds was relatively rapid and pH dependent. DCP was decomposed most rapidly, with 50% decomposition of 0.26 m*M* solutions occurring in 5 min at pH 7. Similar solutions of the sodium salt, butyl, and isopropyl derivatives of 2,4-D were 50% degraded in 71, 54, and 48 min, respectively. Despite these results, it was concluded that ultraviolet energy from sunlight would not effectively decompose 2,4-D in natural waters.[5,6,11,25]

Chemical Degradation

Most organic herbicides used for control of aquatic weeds are resistant to chemical attack. While herbicides, such as the phenoxyalkanoic acids, are rapidly converted to salts and ester derivatives are hydrolyzed to acids, the basic herbicide molecules are quite stable. During studies on the kinetics of degradation of the PGBE ester of Silvex®, most organisms were unable to degrade this herbicide to 2,4,5-trichlorophenol. Later studies have shown that the herbicide is degraded by co-metabolism. Neither chlorine nor $KMnO_4$ were effective in removing 2,4-D or its ester derivative from surface waters; however, neither treatment successfully removed the closely related DCP.[4-6,21,27,36]

Biological Degradation

Among the agents of biological degradation are the various forms of aquatic fauna, flora, and the microbial population. All major groups of fauna present in water containing a herbicide are found to have some residue of the chemical. Whether the eventual loss of herbicides from these organisms is due to metabolism or to excretion of the unaltered compound to the surrounding medium is a basic question. The large volume of water, compared to the total mass of animal life in a natural body of water, makes it unlikely that degradation by aquatic fauna plays a major role in reduction of herbicide levels.[35,36,66]

Laboratory studies show that microbial activity is the major factor involved in degradation of herbicides in natural water. Aly and Faust[5,6] observed that 2,4-D in lake water alone was not degraded during a period of 120 days. However, by "seeding" 20 ppm solutions of 2,4-D with lake mud, total degradation of 2,4-D occurred in 65 days. Successive additions of 2,4-D to the same media resulted in shorter periods required for complete degradation of the herbicide. Adapted populations of microorganisms were capable of degrading 81 to 85% of the 2,4-D in 24 hr. Using mud from a lake treated with 2,4-D the previous year, the initial addition of 2,4-D was decomposed in about 35 days. The adapted organisms were capable of persisting in the lake mud for an extended period. Hemmett and Faust[36] in a study of the biodegradation kinetics of 2,4-D observed that three variables must be considered: (1) microorganism concentration, (2) substrate concentration, and (3) ratio of microorganism concentration to substrate concentration. These three variables affect the rate at which 2,4-D is converted to carbon dioxide, water, and hydrochloric acid.

The biodegradation appeared to follow zero-order kinetics, with the oxidation rate independent of the substrate concentration. When the ratio of microorganism concentration to 2,4-D concentration was held constant, constant rates of oxidation were observed which suggested the enzyme systems were saturated with substrate. Thus, substrate concentration was not a limiting factor in these kinetic studies, further substantiating the zero-order kinetic model. Acclimatization time also appeared as a factor. The longer the period of acclimatization, the shorter was the period of time to effect "ultimate" biodegradation of 2,4-D. In a mixed culture, chlorophenoxy intermediates were not detected.

UPTAKE AND METABOLISM

The uptake and metabolism of phenoxy herbicides has been studied for the past three decades.[40-41] The herbicide 2,4-D is excreted very nearly 100% by the animal body,[39,43] as the parent acid or a detoxified chemical and there are no apparent detrimental effects at the levels used in commercial agriculture. In plants, there are specific metabolites related to a particular plant species.[40-42]

Rodger and Stallings,[51] studying the dynamics of ^{14}C labeled 2,4-D butoxyethanol ester in organs of rainbow trout, channel catfish, and bluegills, noted difference in the hydrolysis rate in aquariums with and without fish. Without fish, 40 hr were required to degrade 50% of a 1.0 mg/ℓ solution of the 2,4-D butoxyethanol ester. With 1.5 g/ℓ of fish present, the same amount of chemical was hydrolyzed in 3 hr. At the end of 24 hr, 0.01 mg/ℓ remained as opposed to 0.5 mg/ℓ after 90 hr in the water without fish. Although the ratio of fish to total water would be considerably higher than the maximum standing crop, it can be assumed that the presence of fish will aid in a more rapid hydrolysis and decline of residues in the field.

Schultz[55] studied the uptake and dissipation of 2,4-D amine salt in fish in growth pools. He found that the herbicide was rapidly dissipated but residues of ^{14}C carbon were subsequently determined in the muscle tissues of the fish flesh. It was postulated that the herbicides excreted by the fish were degraded by microflora of the growth pool and were resorbed as radio active carbon dioxide fragments. Sikka et al.[59] demonstrated that this was indeed the case.

The growth pool studies by Schultz were repeated by Stallings and Huckins.[61] They demonstrated that the radioactive carbon dioxide fragments were incorporated into the fatty acids of the fish flesh and that these residues are not the parent herbicide per se.

TOXICOLOGY

Chronic toxicity studies on this herbicide coupled with a multigeneration reproduction study using the technical grade 2,4-D, free of 2.7-dicholor- and 2,3,7,8-tetrachlorodibenzo-*p*-dioxin, have been reported by Hansen et al.[34]

Feeding of dietary levels of 5,25,125,625, or 1250 ppm 2,4-D for 2 years to rats had no adverse effect on growth, hematology, survival, organ weights, pathology, or on the pattern of tumor development. The tendency for the proportion of tumor-bearing animals to rise with increasing dietary levels of 2,4-D was not found to reflect important pathological differences. Dogs fed 10,50,100, or 500 ppm 2,4-D for 2 years showed no adverse effects on survival, organ weights, or degeneration of tissue.

In a three-generation reproduction study in rats, dose levels of 100,500, and 1500 ppm had no effect on the fertility of males or females or on litter size, but the 1500-ppm diet reduced the survival and weight of the pups at 21 days. Assay of aliesterase and acylamidase activity in liver homogenates, microsomes, and mitochondria from 90-day-old second-generation rats fed 0,100, or 500 ppm 2,4-D revealed no differences between test and control rats.

In an assessment of the safety of 2,4-D residues in food, it is calculated that the maximum possible intake based on U.S. legal tolerances would be 0.3 ppm of the total diet. Thus a wide margin of safety exists between this level and levels employed in the above 2-year rat and dog studies.

The toxicology of 2,4-D to avian species has also been studied in great detail, and no adverse effects have been noted on normal levels of applications. Somers and co-workers[60] sprayed hen and pheasant eggs with DDT, 2,4,5-T, with mixtures of 2,4-D and picloram, as well as with mixtures of 2,4-D and 2,4,5-T using up to 20 times the usual amount. These experiments showed no adverse effects with regard to hatching rate, malformations, or chick mortality. Hilbig and co-workers,[37] sprayed incubated eggs of pheasants, quails and hens with 2,4-D, 2,4,5-T, as well as with a mixture of 2,4-D and 2,4,5-T in amounts 10 times the recommended dosage and produced no effects on the hatching rate, body weight, or number of malformed chicks compared to the controls.

Chicks hatched from (Japanese) quail eggs sprayed with 2,4-D-containing agents were raised and paired, in order to investigate the influence of 2,4-D on sexual differentiation in bird embryos. No significant differences were found between the experimental and control groups either in the laying yield, or in the fertilization and hatching rates. Therefore, it was concluded that 2,4-D does not cause any reproductive disturbances in quails.

ENVIRONMENTAL ASPECTS

Biological transformation of chemical residues involves a variety of intermediate substances until one or more final end products are obtained. In the case of 2,4-D these end products are carbon dioxide, water, and a chloride ion or a terminal residue of a more complex configuration.

The process of co-metabolism is recognized as an important mechanism of metabolism for certain phenoxy herbicides.[3,4,35,36] The principal reactions associated with the metabolism of 2,4-D include dealkylation, dehalogenation, amide or ester hydrolysis, oxidation, reduction, aromatic ring hydroxylation, ring cleavage, and condensate or conjugate formation. Detoxication and excretion are the ultimate results of 2,4-D ingestion by animals and fish. The formation of glycosides and glucuronides by many organisms constitutes a mechanism for detoxication. The glycosides and glucuronides are generally more water soluble than the parent herbicide molecule and are therefore more readily excretable by living organisms.

Many organisms obtain energy for other life processes in the metabolism of 2,4-D. The initial metabolic reactions require an expenditure of energy by the organism. Only when the herbicide is fragmented into compounds that can be channeled into oxidative cycles (e.g., the Krebs cycle) does the organism derive useful energy. The complete oxidation of a chemical to carbon dioxide and water, although perhaps desirable from a "complete degradation" point of view, may be a wasteful process. Actually by a series of intermediate processes the energy originally incorporated into chemical bonds in herbicides may be transferred into ecologically acceptable products that can be used beneficially by many organisms when needed. Thus the overall trend in herbicide metabolism is toward less complex molecules. It follows that fragmentation of the molecule through metabolism generally results in detoxication and the formation of utilizable products. This is exemplified by the ultimate degradation of 2,4-D to alanine and succinic acid. This represents the ultimate conversion to an ecologically acceptable metabolite.

Only in a few instances are the toxicological properties of one or more of the metabolites fully known. Although the general trend of herbicide metabolism is toward increasingly less toxic molecules, in some instances the toxicity of the metabolite may be equal to or greater than that of the parent pesticide molecule. Some herbicides, such as sesone and 2,4 DEP, MCPES, and 2,4-DEB, are not biologically active themselves but actually require metabolic hydrolysis for activation to a phytotoxicant. In all cases studied, however, even the more toxic metabolite is ultimately degradable to an ecologically acceptable entity.

SUMMARY AND CONCLUSIONS

The herbicide 2,4-D is found to provide an ecologically acceptable means of control for susceptible aquatic macrophytes such as Eurasian watermilfoil.

REFERENCES

1. **Adams, M. S. and McCracken, M. D.**, Seasonal production of the *Myriophyllum* component of the littoral of Lake Wingra, Wisconsin, *J. Ecol.*, 62(2), 457, 1974.
2. **Adams, M. S., Titus, J., and McCracken, M.**, Depth distribution of photosynthetic activity in a *Myriophyllum spicatum* community in Lake Wingra, *Limnol. Oceanogr.*, 19(3), 377, 1974.
3. **Anderson, R. A.**, Ecology and Mineral Nutrition of *Myriophyllum spicatum* L., M.S. thesis, University of Maryland, College Park, 1964, 42.
4. **Altom, J. D., and Stritzke, J. F.**, Degradation of dicamba, picloram, and four phenoxy herbicides in soils, *Weed Sci.*, 21(6), 556, 1973.
5. **Aly, O. M. and Faust, S. D.**, Studies on the fate of 2,4-D and ester derivatives in natural surface waters, *J. Agric. Food Chem.*, 12, 541, 1964.
6. **Aly, O. M. and Faust, S. D.**, Removal of 2,4-Dichlorophenoxyacetic acid derivatives from natural waters, *J. Am. Water Works Assoc.*, 57, 221, 1965.
7. **Ashton, F. M. and Crafts, A. A.**, *Mode of Action of Herbicides*, John Wiley & Sons, New York, 1973.
8. **Averitt, W. K. and Gangstad, E. O.**, Dissipation of Residues of 2,4-D in static water, *J. Environ. Qual.*, 5, 145, 1976.
9. **Bartley, T. R. and Hatrrup, A. R.**, 2,4-D Contamination and persistence in irrigation water, *Proc. West. Soc. Weed Sci.*, 23, 10, 1970.
10. **Bates, L. A., Hall, T. F., and Stanley, R. A.**, Aquatic weed problems in the Tennessee River watershed, in South. Weed Sci. Soc. Proc. 25th Annu. Meet., New Orleans, La., 1972.
11. **Baur, J. R. and Bovey, R. W.**, Ultraviolet and volatility loss of herbicides, *Arch. Environ. Contamin. Toxicol.*, 2, 275, 1974.
12. **Bevenue, A., Zweig, G., and Nash, N. L.**, Residue determination of 2,4-dichlorophenoxyacetic acid in dry crops and walnuts, *J. Assoc. Off. Anal. Chem.*, 45(4), 990, 1962.
13. **Bristol, D., Cook, L., Koterba, M., and Nelson, D. C.**, Determination of trace residues of 2,4-dichlorophenol in potato tubers, in Am. Chem. Soc. Abstr. Pap. Natl. Meet. 168th, New York, 1974.
14. **Bristol, D. W., Cook, L. W., and Nelson, D. C.**, Studies of 2,4-D and 2,4-DCP residues in potato tubers, *Am. Potato J.*, 51(9), 310, 1974.
15. **Bruns, V. F., Carlile, B. L., and Kelley, A. D.**, Responses and Residues in Sugar Beets, Soybeans, and Corn Irrigated with 2,4-D or Silvex-Treated Water, Tech. Bull. No. 1476, Agricultural Research Service, U.S. Department of Agriculture, Washington, D.C., 1974.
16. **Cope, O. B.**, Effects of pesticides on fish and wildlife, *Fish Wildl. Ser. Circ.*, 226, 51, 1964.
17. **Cope, O.B.**, Some response of fresh-water fish to herbicides, *Proc. South Weed Conf.*, 18, 439, 1965.
18. **Cope, O. B., Wood, E. M., and Wallen, G. H.**, Some chronic effects of 2,4-D on the bluegill *(Lepomis machrochirus)*, *Trans. Am. Fish Soc.*, 99, 1, 1970.
19. **Crosby, D. C. and Tucker, R. K.**, Toxicity of aquatic herbicides to Daphnia magna, *Science*, 154, 289, 1966.
20. **Crowell, T. E., Steenis, J. H., and Sincock, J. L.**, Recent observations of Eurasian watermilfoil in Currituck Sound, North Carolina and other coastal southeastern states, *Proc. South. Weed Conf.*, 18, 485, 1967.
21. **Daly, R. W., Jr.**, Degradation of 2,4-D BEE in Aquatic Environment, Ph.D. dissertation, Auburn University, Alabama, 1971.
22. **Davis, J. T. and Hardcastle, W. S.**, Biological assay of herbicides for fish toxicity, *Weeds*, 7, 397, 1959.
23. **Davis, J. T. and Hughes, J. S.**, Herbicide residue studies, Louisiana Statewide Fisheries Investigations, Fish and Game Proj. Annu. Rep., Louisiana Wildlife and Fisheries Commission Publication, New Orleans, 1965, 21.
24. **Dewitt, J. B.**, Wildlife studies, Patuxent Wildlife Research Center USDI Fish and Wildlife, *Ser. Circ.*, 167, 86, 1962.
25. **Davis, F. S.**, Review of Persistence and Mobility of Phenoxy Herbicides Environment, Tech. Rep. 7, Office of the Chief of Engineers, Department of the Army, Washington, D.C., 1974, E3.
26. **Elliston, R. A. and Steward, K. K.**, The response of Eurasian water milfoil to various concentrations and exposure periods of 2,4-D, *Hyacinth Control*, 10, 38, 1972.
27. **End, C. S. and McDade, J. J.**, Biodegradation of 2,6-dichlorophenoxy acetic acid and related chlorophenols, Am. Soc. Microbiol. Abstr. 72nd Annu. Meet., New York, 1972.
28. **Farewell, S. O., Robinson, E., Powell, W. J., and Adams, D. F.**, Survey of airborne 2,4-D in south-central Washington, *J. Air Pollut. Control Assoc.*, 26(3), 224, 1976.
29. **Faurot, C. W.**, Experience with Eurasian watermilfoil at Edgewood Arsenal, Maryland, *Hyacinth Control J.*, 11, 19, 1973.

30. **Feung, C. S., Hamilton, R. H., and Witham, F. H.**, Metabolism of 2,4-dichlorophenoxy acetic acid by soybean cotyledon callus tissue cultures, *J. Agric. Food Chem.*, 19, 475, 1971.
31. **Fleeker, J. and Steen, R.**, Hydroxylation of 2,4-D in several weed species, *Weed Sci.*, 19(5), 507, 1971.
32. **Frank, P. A. and Comes, R. D.**, Herbicidal residues in pond water and hydrosoil, *Weeds*, 15, 210, 1966.
33. **Gangstad, E. O.**, *Weed Control Methods for River Basin Management*, CRC Press, Boca Raton, Fla., 1978.
34. **Hansen, W. H., Quaife, M. L., Habermann, R. T., and Fitzhugh, O. G.**, Chronic toxicity of 2,4-Dichlorophenoxyacetic acid in rats and dogs, *Toxicol. Appl. Pharmacol.*, 20, 122, 1971.
35. Hazardous Materials Committee, Herbicide Reports, EPA-SAB-74-001, U.S. Environmental Protection Agency, Washington, D.C., 1975.
36. **Hemmett, R. B., Jr. and Faust, S. D.**, Biodegradation kinetics of 2,4-dichlorophenoxyacetic acid by aquatic microorganisms, *Residues Rev.*, 29, 191, 1969.
37. **Hilbig, V. V., Lucas, K., Sebek, V., and Munchow, H.**, Studies on the development of the toxic effects of 2,4-dichloro and 2,4,5-trichlorophenoxy acid (2,4-D and 2,4,5-T) derivatives on the young of pheasants, quail and chickens. II (Eng. transl.), *Anz. Schaedlingskd*, 49(5), 65, 1976.
38. **Leng, M. L.**, Residues in milk and meat and safety to livestock from the use of phenoxy herbicides in pasture and rangeland, *Down Earth*, 28(1), 12, 1972.
39. **Pimentel, D.**, Ecological Effect of Pesticides on Non-Target Species, Office of Science and Technology, Washington, D.C., 1971.
40. **Menzie, C. M.**, Metabolism of Pesticides, Special Sci. Rep.-Wildl. No. 127, Fish and Wildlife Service, U.S. Department of the Interior, Washington, D.C., 1969.
41. **Menzie, C. M.**, Metabolism of Pesticides, and Update, Wildl. No. 184, Fish and Wildlife Service, U.S. Department of the Interior, Washington, D.C., 1974.
42. **Menzie, C. M.**, Metabolism of Pesticides, Update II, USDI, BSF and W, Wildl. No. 212, Fish and Wildlife Service, U.S. Department of the Interior, Washington, D.C., 1978.
43. **Mullison, W. R.**, Effects of herbicides on water and its inhabitants, *Weed Sci.*, 18, 738, 1970.
44. **Rawls, C. K.**, Field tests of herbicide toxicity to certain estuarine animals, *Chesapeake Sci.*, 6(3), 150, 1965a.
45. **Rawls, C. K.**, Toxicity to certain estuarine animals in field tests of selected herbicides, *Univ. Md. Nat. Resour. Inst.* Ref. No. 65-10; *Proc. 18th Annu. South. Weed Conf.*, 1965b.
46. **Rawls, C. K.**, Providing estuarine fauna for herbicide residue analyses, *Univ. Md. Nat. Resour. Inst.*, Ref. No. 65-22; *Proc. 19th Annu. Meet. South. Weed Conf.*, 1966, 397.
47. **Rawls, C. K.**, Opinions on the status of certain marine vegetation species in the Chesapeake Bay and its tributaries, Ref. No. 66-96, *Univ. Md. Nat. Resour. Inst.*, 1966.
48. **Rawls, C. K.**, The accumulation and loss of field-applied butoxyethanol ester of 2,4-D dichlorophenoxyacetic acid in oysters *(Crassostrea virginica)* and soft-shelled clams *(Mya arenaria)*, *Hyacinth Control J.*, 9(1), 62, 1971.
49. **Rawls, C. K.**, Mechanical control of Eurasian watermilfoil in Maryland with and without 2,4-D application, *Chesapeake Sci.*, 16(4), 266, 1975.
50. **Rawls, C. K.**, Field Studies of shell regrowth as a bioindicator of eastern oyster *(Crassostrea virginica)* response to 2,4-D BEE in Maryland tidewaters, *Chesapeake Sci.*, 18(3), 266, 1977.
51. **Rodgers, C. A. and Stallings, D. L.**, Dynamics of 14-C-labeled 2,4-D butoxyethanol ester in organs of rainbow trout, channel catfish, and bluegills, *Weed Sci.*, 20, 101, 1972.
52. **Sears, H. S. and Meehan, W. R.**, Residues in fish, wildlife and estuaries, *Pest. Monitoring J.*, 5(2), 213, 1971.
53. **Sears, H. S. and William, R. M.**, Short-term effects of 2,4-D on aquatic organisms in the Nakwasina River watershed, southeastern Alaska, *Pest. Monitoring J.*, 5(2), 213, 1971.
54. **Schultz, D. P. and Whitney, E. W.**, Monitoring 2,4-D residues at Loxahatchee National Wildlife Refuge, *Pest. Monitoring J.*, 7, 146, 1974.
55. **Schultz, D. P.**, Dynamics of a salt of (2,4-Dichlorophenoxy) acetic acid in fish, water, and hydrosol, *J. Agric. Food Chem.*, 21, 186, 1973.
56. **Schultz, D. P. and Gangstad, E. O.**, Dissipation of residues of 2,4-D in water, hydrosoil and fish, *Aquatic Plant Manage. J.*, 14, 43, 1976.
57. **Schultz, D. P. and Harman, P. D.**, Residues of 2,4-D in pondwaters, mud, and fish, 1971, *Pest. Monitoring J.*, 8(3), 173, 1974.
58. **Schultze, J. A., Manigold, D. P., and Andrews, F. L.**, Pesticides in selected western streams — 1968-1971, *Pest. Monitoring J.*, 7(1), 73, 1973.
59. **Sikka, H. C., Appleton, H. T., and Gangstad, E. O.**, Uptake and metabolism of dimethylaminesalt of 2,4-dichlorophenoxyacitic acid by fish, *J. Agric. Food Chem.*, 25, 1030, 1977.

60. **Somers, J., Moran, E. T., Jr., Reinhart, B. S., and Stephenson, G.**, Effect of external application of pesticides to the fertile egg on hatching success and early chick performance. 1. Pre-incubation spraying with DDT and commercial mixtures of 2,4-D: picloram and 2,4-D: 2,4,5-T, *Bull. Environ. Contamin. Toxicol.*, 11, 33, 1974a.
61. **Stallings, L. D. and Huckins, J. N.**, Metabolism of 2,4-dichlorophenoxyacitic acid (2,4-D) in bluegills and water, *J. Agric. Food Chem.*, 26, 447, 1978.
62. **Steward, K. K. and Nelson, L. L.**, Evaluations of controlled release PVC and attaclay formulations of 2,4-D on Eurasian watermilfoil, *Hyacinth Control J.*, 10, 35, 1972.
63. **Whitney, E. W., Montgomery, A. B., Martin, E. C., and Gangstad, E. O.**, The effects of a 2,4-D application on the biota and water quality in Currituck Sound, North Carolina, *Hyacinth Control J.*, 11, 13, 1973.
64. **Wilkinson, R. E.**, Subaqueous release of herbicides from granules, *Weeds*, 12, 69, 1964.
65. **Wilson, R. G., Jr. and Cheng, H. H.**, Breakdown and movement of 2,4-D in the soil under field conditions, *Weed Sci.*, 24(5), 461, 1976.
66. **Wojtalik, T. A., Hall, T. F., and Hill, L. O.**, Monitoring ecological conditions associated with wide-scale applications of DMA 2,4-D to aquatic environment, *Pest. Monitoring J.*, 4, 184, 1971.
67. **Amchem Products, Inc.**, unpublished data, Ambler, Pa., 1970.

Chapter 6

THE ROLE OF ALGAE IN AQUATIC ECOSYSTEMS

INTRODUCTION

Algae and other aquatic plants play a major role in the functioning of aquatic ecosystems. They may act as shelters or substrates for many different types of invertebrates and fish. They reoxygenate the water by the process of photosynthesis and they form an important food base for the whole ecosystem. In the process of performing their function within the aquatic ecosystem they assimilate during the growing season large amounts of nutrients and also they may assimilate considerable amounts of trace metals which may be concentrated within the cells. It is also known that they excrete various soluble organic compounds which return elements such as nitrogen and phosphorous to the aquatic ecosystem. These then become food for various kinds of organisms.

The immediate effect on water quality is the removal of considerable amounts of nutrients and, in some cases, trace metals which are not known to be nutrients. These nutrients and trace metals are then locked up within the protoplasm of these plants. In the case of emergent vegetation, these nutrients are dispersed to the terrestrial ecosystem as well as being returned to the aquatic ecosystem in the form of detritus and directly as food.

Algae are found everywhere: in the air, in the soil, on trees, in the fur of some animals, in the intestines of most animals, and in the water. This paper will be concerned with algae found in the water and particularly those algae found in reservoirs and lakes.

In water, algae may be found drifting on or within the water (phytoplankton), attached to a substrate (periphyton), or growing on other plants (epiphyton). Most phytoplankton are microscopic in size, while many periphyton can be seen with the naked eye (macroscopic). There are approximately 1500 genera and 17,400 species of algae.[1]

ECOSYSTEM ENERGY FLOW

The algae occupy a unique position among the organisms of the aquatic world because they are able to utilize light energy in the process of reducing CO_2 to the oxidation state of cellular carbon. As a result, the algae are an important link in the food chain and are often called the primary producers in aquatic systems. Because the algae are the ultimate source of both cellular carbon and chemical energy for other organisms in aquatic systems, the biological activity of an aquatic ecosystem is very much dependent upon the rate of primary production. The biological activity within the ecosystem is in turn affected by the physical environment.

The energy found as organic matter in the primary producers reaches the later stages of the food chain in several ways. Some of the organic matter excreted in soluble form or from decaying algal cells serves as nutrient material for the growth of heterotrophic bacteria. The algae as well as the bacteria may also be consumed directly and in this respect are a major source of food for the zooplankton and young fish. The principle importance of algae is, therefore, their ability to give rise to large quantities of organic matter in aquatic systems.

In the process of producing cellular carbon with the energy derived from photosynthetic processes, the algae acquire reducing power by using water as an electron donor. Oxygen is also produced from water as a by-product of the photosynthetic light reactions during daylight hours. As respiration is carried out by the nonphotosynthetic organisms, carbon dioxide is released and oxygen is consumed from the environment. For this reason, the levels of oxygen and carbon dioxide in aquatic environments depend to a large degree upon the relative rates of photosynthesis and respiration being carried on collectively by the algae, bacteria, and other organisms in the immediate area.

The algae make possible important chemical changes through the release of oxygen and the consumption of carbon dioxide during daylight hours. Oxygen is made available for respiration carried on by all types of organisms from fish to the smallest bacteria. The algae constitute the primary source for the continuous daytime renewal of essential oxygen in lakes and reservoirs. Oxygen release by algae and oxygen uptake by aeration are the two primary sources of oxygen in flowing surface waters. As algae remove carbon dioxide from their surroundings they cause an increase of the pH in surface waters. At night when photosynthetic activity is at a minimum, the reaction is reversed and the pH decreases.

Obviously, lakes and rivers normally contain many genera of planktonic and benthic algae. Limited numbers of algae are not troublesome, but rather a necessary link in the aquatic food chain. Algae frequently do become a problem in surface waters because of their capacity for rather prolific growth under certain conditions. Algal abundance varies with the degree of enrichment with algal nutrients, the presence of toxic substances, temperature, turbidity, and other parameters. Under conditions where the proper nutrients are available, cell counts as high as 171,000/ℓ have been recorded. Such large quantities of algal material are the usual cause of difficulties (eutrophication) in surface waters. The simple solution to the problems caused by excessive growths of algae would seem to be to have the proper kind of algae present in the appropriate amounts.

Growth and Nutrition

Algae need various elements as macro- or micronutrients to live, photosynthesize, reproduce, etc. The macronutrients needed by algae are carbon, hydrogen, oxygen, nitrogen, phosphorous, sulfur, potassium, magnesium and sodium. The micronutrients required by most algae are iron, manganese, copper, calcium, zinc, molybdenum, vanadium, and zinc; these are needed during the photosynthetic process. Iron, calcium, boron, and molybdenum are associated with nitrogen utilization in most algae, and in the heterocyst producing blue-green algae they are also needed for nitrogen fixation. Water contains most of the elements needed by algae in ample supply from chemical actions occurring between the air-water and water-substrate interfaces.

The major sources of nitrogen and phosphorous in water are domestic sewage effluents, animal and plant processing wastes, fertilizer and chemical manufacturing spillage, and agricultural runoff.[2] Nitrogen can also get into the water via the atmosphere, runoff from animal manure, and the aforementioned nitrogen fixation by blue-green algae and bacteria. Additional phosphorous can be released into the water by the grubbing and burning of plant growth surrounding the shore line of a new impoundment.[3]

Carbon can generally be obtained in adequate concentration for algal growth from the atmosphere.[4] Additional carbon may be found in waters within calcareous basins since limestone yields considerable quantities of carbonate alkalinity.[5] Other theorized sources of carbon are the respiration of pelagial bacteria and the extra cellular dissolved compounds of higher aquatic vascular plan communities;[6] in a recent bioassy study of Norris Reservoir water[7] there was a significant increase in phytoplankton productivity when the concentrations of nitrogen and phosphorous, and even iron were limiting phytoplankton growth in Lake Travis during different times of the year.[8] Laboratory studies indicated that the addition of phosphorous to the lake would enable nitrogen-fixing blue-green algae to develop and produce additional nitrogen. Similar results were obtained in Lake Livingston, an eutrophic lake, simultaneously studied by the same researchers. Algal bioassy studies of Oregon lakes indicated algal growth rate to be directly proportional to total dissolved phosphorous concentrations but not to carbon and nitrogen concentrations.[9] However, in general, no simple relationship exists between phytoplankton activity and nitrogen and phosphorous concentrations in the photic zone (i.e., the area in which there is enough light to supply the energy needed for photosynthesis). Carbon dioxide depletion during photosynthesis may be an algal-production regulating factor at very high levels of primary productivity.[10]

Effects of Reservoirs and Lakes on Algae

Nutrients are not the only factors affecting algal growth in water impoundments. Algae, as members of a biotic community in which they are primary producers, require optimum conditions of light, temperature, tubulence, grazing, predation rate, etc. to produce a bloom. Therefore, before we look at the effect of algae on water impoundments (which generally occurs only when they are in bloom condition) we should look at the effect of water impoundments of algae which lead to bloom conditions.

The location of a water impoundment can greatly affect algal conditions. Brylinski and Mann[11] state that algal productivity data from lakes within restricted latitudes are more easily compared by nutrient concentrations. However, on a global scale energy-related variables appear to have a greater effect on algal productivity than those related to nutrient availability. Except for total nitrogen, nutrient concentration explains less than 15% of production variance whereas latitude alone explains about 57%. Corroborative results were obtained during a study of Lake Erken, a moderately eutrophic lake in central Sweden. Like many other lakes, the spring outburst of logarithmic algal growth was not caused by temperature but by solar radiation.[12] Shallow lakes are more productive than deep lakes when they are compared on a production per unit volume basis.

Studies on Norris and Beech Reservoir[7,13] indicated that solar radiation either directly or indirectly controlled seasonal nutrient distribution and therefore governed primary productivity. Thus, even though more light is available for phytoplankton growth, productivity is limited because the nutrient-depleted epilimnion (i.e., the layer of water above the thermocline) cannot be restored by nutrients from the hypolimnion (i.e., the layer of water below the thermocline).

Additional research in the Tennessee Valley Authority (TVA) system indicated that long water-detention periods in lakes or reservoirs would increase phytoplankton growth because more of the nutrients in the "older" water would be in dissolved or particulate form.[14] Furthermore, lake morphology and selective withdrawal of the water from storage impoundments could contribute to the separate aging of the hypolimnion and the epilimnion. Stated another way, the operation of a reservoir can affect phytoplankton growth. For example, the Volta River system in Ghana, West Africa, was dammed; an initial decrease in phytoplankton was followed by an increase in density and diversity. The phytoplankton then stratified by the dam, for a while, until the generating plant began operation. The continual release of water reduced phytoplankton stability but increased phytoplankton concentration, which was dominated by *Synedra acus*, a diatom.[15]

In general, hard water lakes have a greater abundance of algae than soft water lakes. This phenomenon is probably due to the increase in bicarbonate content. British Columbia lake studies indicated that total dissolved solids was the most important factor related to productivity, yet plankton concentrations could not be predicted by fluctuation in this parameter.[16]

The Petenwell Reservoir[17] receives high nutrient loadings from pulp and paper industries upstream of the reservoir. Previous nutrient studies indicated nutrient levels were adequate to support algal growth but most of the inorganic nutrients were probably used by the high level of bacterial activity. In addition, most of the light in the blue and red spectrum was absorbed in the top 40 cm of the reservoir. Thus, at this time there is no algal problem in Petenwell Reservoir but as the effluents from the upstream industries are reduced it is likely that algal growth in this over-fertilized reservoir will cause nuisance growths.

Algae get into lakes and reservoirs from the feeder waters, "seeding" from the gut or body surfaces of birds and insects, and from the air. Thus, there are many avenues for algal entry into a body of water. The two general classifications of problem-causing algae are phytoplankton and periphyton. For example, *Cladophora* spp. a green, filamentous, periphytic algae has caused problems in some of the Great Lakes because the abundant growths became detached during wave and wind action and washed up on the beaches. Once these growths were on the beach their decomposition produced a smelly general nuisance.[1] Filamentous algae in general (i.e., phytoplankton or periphytic) can remove the nutrients in a water body and thus prevent unicellular algae from developing. Unicellular algae are more desirable than filamentous because fish feed upon them, they do not interfere with various harvesting methods, and they do not provide a habitat for aquatic insect growths.[16] Additionally, abundant growths of filamentous algae can block intake structures and trash racks. However, unicellular algae are not blameless since many of them can cause problems such as taste and odor in the water, decreasing filter runs in water treatment plants, and forming unaesthetic green scums on the water.

Methods and Analysis

Many methods have been devised to determine if the algae present are causing actual or potential problems in the lake or reservoir. All of these methods have their advantages or disadvantages. One of the first methods devised was the enumeration of phytoplankton. This method is good for determining what species are present in the water but it tends to overemphasize large concentrations of small-celled organisms. Chlorophyll *a* is a measurement which provides as easy and relatively quick estimate of algal concentration but it overemphasizes the presence of large-celled phytoplankters. Carbon-14 uptake is a measurement of the vitality of the organism rather than the population size, but smaller algae generally photosynthesize more actively than larger forms and therefore incorporate Carbon-14 faster and more efficiently than larger forms.[18] Bioassy procedures provide an indication of the effect of different variables on the lake or reservoir water; however, indigenous algae, bacteria, etc., can produce results that are quite different from those indicated in the bioassy test. Remote sensing is the newest form of monitoring a water body. Cameras containing films sensitive to different wavelengths are mounted in airplanes or satellites and are used to take synoptic views of the water body. At this time remote sensing, in regard to algae, is primarily used for determining trouble areas such as algal bloom location or movement of bloom in relation to time. For remote sensing to be meaningful, ground crews should be in communication with the aerial crew and be able to sample those features noted by them.

While all of the above methods have their drawbacks, they can all provide meaningful data if applied judiciously. For example, Mathis[19] studied a lake using chlorophyll *a* measurement and the ratio of "yellow/green" pigments. The researcher had to use judgment since one method indicated the lake was oligotrophic and contained a young immature ecosystem, whereas the ratio method indicated a mature ecosystem tending toward eutrophication.

Not only the method of algal sampling, but the location and season of sampling can reduce the errors and variance in sampling. Temperature, and thus the time of year samples are taken, can provide a different estimate of the species composition of a water impoundment. Some researchers indicate that optimum temperature for blue-greens is 15 to 27°C, green algae 10 to 21°C, and yellow-green algae 4 to 10°C. Other researchers indicated 30 to 40°C to be the optimum temperature range for blue-green and 25 to 35°C for green algae. We believe that temperature is important for algal growth rate and is directly involved in algae acclimatization,[16] but more research is required for a satisfactory answer.

Another variable that adds to the difficulty of obtaining a meaningful algal sample is the lack of spatial homogeneity in algal distribution within a lake or reservoir. Various species of algae have different rates of flotation and sinking becase of their shapes, oil contents, and air vacuoles. Planktonic and benthic blue-green algae[20] develop in the hypolimnion on or near the bottom of a water impoundment, and as they mature and develop gas vacuoles, they become buoyant and rise to the surface. A turbidimeter may be used for profiling the phytoplankton[21] to obtain a precise and rapid estimate of the phytoplankton population.

Thermal stratification and its breakdown to the isothermal condition are shock periods to the algae and various species may not survive these multiple shocks. Thus during these periods algal distribution and species may change. The application of fertilizer to a lake or reservoir can induce high concentrations of algae, as can the intensive use of herbicides which causes large concentrations of organic matter from the dead aquatic weeds resulting from treatment.[22]

As a lake or reservoir becomes nutrient enriched there is an increase in algal concentrations, an increasing duration of high algal concentrations, and a decrease in diversity. In an eutrophic lake the primary productivity increases in early spring and contin-ues at a high level thoughout the summer. Blue-green algae are usually the dominant form in an eutrophic lake.[23]

Nitrogen fixing, toxicity, and taste and odor production are three reasons that blue-green algae are looked upon by some with disdain. Toxicity, and taste and odor production may be produced by some species in other algal groups, but studies indicate that only heterocyst-producing blue-green algae fix nitrogen.[24] Heterocysts are produced when there is no combined nitrogen in the water but phosphates are present. The nitrogen fixing allows other algae to utilize the nutrients in the water thus thwarting attempts to produce oligotrophication of a water body by removal of nitrates from influent waters. Algal poisoning in birds, unlike botulinus poisoning, does not affect their nictating membranes.[25] Most toxic algae are blue-greens, and the toxins they produce are generally not detoxified by the laboratory equivalent of water treatment.[26] However, hepatic necrosis was produced in mice by feeding them *Scenedesmus* sp., a green algae. This last statement and many more of the "medical aspects of phycology" are mentioned in the excellent compendium by Schwimmer and Schwimmer.[27]

As indicated by Palmer[28] every algal group found in fresh water contains species that produce taste and odor problems at certain levels of population density. Research is being performed on some of these organisms to isolate the taste- and odor-producing chemical. With this type of specific information, treatment methods can be devised to remove the specific nuisance when it occurs. One of the first taste- and odor-producing chemicals to be isolated was geosmin. This musty/earthy smelling metabolite of *Oscillatoria* spp. and *Symploca* spp. is a tertiary alcohol with the formula $C_{12}H_{22}O$.[29] Incidentally, geosmin is said to inhibit heterocyst formation in *Anabaena circinalis* early in its life cycle and enhance it during the later stages. Another taste- and odor-producing chemical that has been isolated is the sulfur compound, isopropyl mercaptan, produced by actively growing *Microcystis flos-aquae (=Anacystic cyanea).*[30] In general, odor problems caused by blue-green algae can be controlled by activated carbon, whereas chlorination tends to intensify their odor. Addition of the bacteria *Bacillus cereus* from cultures has also proved effective. Some other methods of taste and odor control for blue-green algae that are being researched are cell lysis by a weak electrolyte, viral and bacterial agents, and slow-release floating and sinking algicides.[20]

ALGAL MANAGEMENT METHODS

There are many methods of killing algae, but good management requires a tool that will affect the target organism and not affect the rest of the biota or ecosystem. One of the original tools used in water management, and which is still being used today, is copper sulfate. At present, this chemical is still the algicide most used by managers of reservoirs, canals, and waterways in general. Some managers, however, are now combining this compound with citric acid, or buying the premixed cuprose citrate complex from various companies, to use in place of copper sulfate. Cuprose citrate is less sensitive to pH than copper sulfate and therefore is supposed to require a lower copper ion application for the same amount of algal control. A lower copper ion content in the water is advantageous since many fish are highly sensitive to this method. However, there are instances when cuprose citrate appears to provide no more algal control than copper sulfate.

To determine the correct dosage of copper sulfate one should know the algal species present, the density of the population, and the condition of the receiving water. Some managers have been able to attain algal control by trickling in low concentrations of copper ion, less than 0.1 ppm during certain seasons of the year. To do this a good history of the waterbody is needed as is some experimentation with the predominant nuisance algal species. The waters of the California State Water Project have a relatively high pH (i.e., usually in excess of 8.4) which tends to precipitate low dosages of copper sulfate to copper carbonate before there is any measurable uptake by the algae. We have found that 1 ppm of copper ion, applied as copper sulfate pentahydrate, produces the needed control. Experiments are being performed to determine if less copper, applied in the form of cuprose citrate, would produce similar results. In application rate calculations for this chemical we include a factor for depth. For example, if the algal organism in bloom is found to a depth of 10 ft, the amount of water to this depth is included in the calculation to obtain a 1 ppm dose.

There are two reasons that an entire lake is seldom treated with copper. If a heavy bloom of algae is treated with copper or any algicide, the decomposition products will use the oxygen needed by fish, thus leading to asphyxiation of the fish. Additionally, as mentioned previously, copper is toxic to most fish. However, most healthy fish have an avoidance reaction to this chemical. Therefore, treatment methods should always provide for a chemical-free access to untreated open water. It is because of these reasons that most lakes undergoing a heavy algae bloom are generally treated in sections rather than in their entirety.

Copper sulfate is useful in killing most algal growths; however, since it is not specific and tends to accumulate in the benthos, the entire ecosystem is affected by this chemical. Many chemicals have been formulated and researched for algicidal properties[32] and any of these when used correctly and at the right time (i.e., as a preventive application rather than to remove existing blooms), will provide control. However, other methods that are less damaging to nontarget species, and the ecosystem in general, are desirable.

One method used for preventing algal blooms is the removal of nutrients from impounded waters. Eutrophication has become an increasing problem in lakes and reservoirs particularly since wastewater has undergone secondary treatment.[33] This treatment produces a water with a higher degree of mineralization than does primary treatment. The nutrients in the mineralized state are more readily used by the algae. One method of nutrient removal researched was the addition of low-nutrient dilution water to lake water to reduce the blue-green algae population.[34] This study indicates that the blue-green algae population was decreased without any significant effect on either green algae or diatoms. Another method for preventing algal blooms while depositing sewage into a lake or reservoir is to discharge the treated wastewater into the hypolimnion. This method is useful in deep lakes having a small surface area.[35]

Biological Methods

There is a growing awareness on the part of those who are charged with the responsibility for any phase of environmental protection and management that aquatic macrophytes and associated organisms are important to the maintenance and enhancement of water quality. The control and management of these aquatic plants therefore becomes an important responsibility.

Aquatic macrophytes and even the tiny algae which occupy the same environment can be considered as weeds when they grow in profusion where they are not wanted. Although in general there was agreement throughout the conference relative to the types of aquatic plants which can be considered harmful and unwanted, it was often difficult to say that a species should be completely eliminated rather than reduced in numbers and managed. However, it is pointed out that weed removal and control can be very expensive and there were differences relating to the economics of the problem. When biological control is considered there are disagreements relative to the agents which should be employed, and the technology has not yet developed for practical application.[36]

Effect of Herbicide Mixtures

Endothall is used in combination with other herbicides like diquat, Silvex®, and copper sulfate to control certain aquatic plants which are not controlled by endothall alone. Present knowledge of the effects of combinations of endothall with diquat, Silvex®, and copper sulfate is limited to their effects on target organisms, namely, that they result in an increased toxicity to certain aquatic plants.[37-40] However, little is known about the effects of these herbicide combinations on nontarget aquatic microorganisms such as algae and protozoa. These two classes of microorganisms comprise important links in the aquatic food chain and any adverse effects on these organisms may lead to significant changes in the aquatic ecosystem.

Six species representing two algal divisions were used in a study of nontarget organisms: *Chlamydomonas reinhardi, Chlorella pyrenoidosa,* and *Scenedesmus obliquus* (Chlorophyta) and *Nostoc muscorum, Anacystic nidulans,* and *Anabaena variabilis* (Cyanophyta). The Chlorophyta were obtained from the Culture Collection of Algae, Indiana University, and they were grown on Bristol's medium.[41] The Cyanophyta were obtained from the Laboratory of Algal Physiology, University of Texas, and they were grown on Stainer's medium.[42]

The species of protozoa used in this study was *Tetrahymena pyriformis,* obtained from the Gulf Breeze Environmental Research Laboratory, Gulf Breeze, Fla. The protozoan was grown axenically in a proteose peptone (2%, w/v) medium, supplemental with 0.1% (w/v) dehydrated yeast extract and 0.5% (w/v) glucose.[43] The cultures were grown on a reciprocating shaker (300 rpm) at 26 ± 1°C. For toxicity studies, 50 mℓ of cell suspension (1×10^4 cells/mℓ) were incubated in 300-mℓ Erlenmeyer flasks containing an appropriate concentration of herbicide combination. After 1, 2, 3, 4, and 7 days, growth was determined by measuring the absorbance of 540 nm.

Endothall and Diquat

The only significant effect of endothall was a stimulation in the growth of *Nostoc* (0.00447 to 0.447 ppm) and a small inhibition in the growth of *Chlamydomonas* (4.47 $\times 10^{-5}$ to 0.447 ppm), both after 14 days. At 0.552 and 0.276 ppm, diquat completely inhibited growth of all the algae except *Scenedesmus*. Growth of *Scenedesmus* was reduced 21 and 11%, respectively, after 14 days. The blue-green algae were more sensitive to diquat than the green algae. Growth was completely inhibited by diquat at 0.552 and 0.276 ppm and 0.055 ppm inhibited growth by 5 to 15%. Inhibition of growth by combinations of endothall and diquat (1 $\times 10^{-4}$ to 1.0 ppm) was similar to that for diquat alone. However, diquat (0.055 ppm) alone and in combination with endothall (0.0447 ppm) caused an increase in the growth of *Chlamydomonas*.

It seems that the inhibition of growth by a mixture of diquat and endothall resulted from the effect of diquat since diquat alone (0.276 and 0.552 ppm) was as toxic as a combination of diquat (0.276 and 0.552 ppm) and endothall (0.223 and 0.447 ppm).

Endothall and Silvex®

Endothall (0.47 to 4.65 ppm), Silvex® (0.053 to 5.33 ppm), and combinations (0.1 to 10.0 ppm) of endothall and Silvex® had no significant effect on growth of *Chlorella, Scenedesmus,* or the blue-green algae. Endothall (4.65 ppm) or Silvex® (5.33 ppm) alone reduce the growth of *Chlamydomonas* 5 and 9%, respectively, after 14 days. All of the concentrations (0.1 to 10 ppm) of endothall and Silvex® were more toxic to *Chlamydomonas* than the individual herbicides which may indicate a synergistic effect of the herbicides.

Endothall and Copper Sulfate

Growth of *Scenedesmus* was stimulated by endothall alone (0.33 to 1.66 ppm) while growth of *Chlorella* and *Chlamydomonas* was not affected. Copper sulfate alone (0.66 to 3.32 ppm) had no effect on the growth of the green algae. The only effect of the combinations of endothall and copper sulfate (1 to 5 ppm) was a decrease in growth of *Chlamydomonas* after 7 days. However, the cultures had recovered when the experiment was terminated (14 days).

The blue-green algae were more sensitive to copper sulfate than the green algae. Algal growth was totally eliminated by 0.66, 1.66, and 3.32 ppm copper sulfate, while endothall alone (0.033 to 1.66 ppm) had no effect. *Anacystis* was more sensitive to copper sulfate than either *Nostoc or Anabaena*. Growth of *Anacystic* was completely inhibited at 0.33 ppm, while any inhibition in the growth of *Nostoc* and *Anabaena* was less than 8%. Treatment of *Anacystis* with 0.5 ppm endothall and copper sulfate also caused a complete inhibition of growth of this algae.

Effect on the Growth of Protozoa

Endothall in combination with diquat, Silvex® and copper sulfate had little effect on the growth of *Tetrahymena*. Exposure of *Tetrahymena* to mixtures of endothall and Silvex® caused some growth inhibition 3 days after treatment. However, this inhibition generally disappeared within 24 hr and the organism resumed normal growth.

SUMMARY AND CONCLUSIONS

It is evident that the control of aquatic plants for the purpose of water-quality enhancement is a very complex matter. A legion of methods of control are available. However, no single method seems to be outstandingly good. Instead, it appears that the control or management problem will differ in each geographical area and perhaps for each body of water. The solution then would appear to be an integrated type of control administered by well-informed, intelligent people who can tailor the procedure to the needs of the locality or body of water. At present the "state-of-the-art" is such that we need to devote more time to pertinent experimentation and to the type of research which will provide the answer needed to design a comprehensive and fully integrated management plan.

Treatment of aquatic microorganisms with endothall in combination with other herbicides like diquat, Silvex®, and copper sulfate, at the recommended rate of active ingredients for each mixture (0.25 to 0.5 ppm, endothall and diquat; 2 to 6 ppm, endothall and Silvex®; 1.5 to 4 ppm, endothall and copper sulfate), did not significantly affect growth of the protozoan *Tetrahymena*. However, these herbicide combinations are more toxic to the phytoplankton.

Endothall and Silvex® combinations did not significantly affect the green or blue-green algae, except *Chlamydomonas* which was slightly inhibited. Endothall and diquat combinations were highly toxic to all the algae. Treatment of the algae with a concentration within the range of recommended rates substantially reduced the growth of the organisms tested. Combinations of endothall and copper sulfate did not significantly alter growth of the green algae, but were toxic to the blue-green algae. Treatment with endothall and copper sulfate mixtures might be used as a method for selectively reducing blue-green populations without significantly harming the green algae.

REFERENCES

1. **Machenthun, K. M.,** The Practice of Water Pollution Biology, U.S. Federal Water Pollution Control Administration, Washington, D.C., 1964.
2. Report of the National Technical Advisory Committee to the Secretary of the Interior, Plant nutrients and nuisance organisms, in Water Quality Criteria, U.S. Federal Water Pollution Control Adminsitration, Washington, D.C., 1968, 51.
3. State Water Quality Control Board, Eutrophication — A Review, Publ. No. 34, The Resources Agency, State of California, San Francisco, 1967.
4. **Morton, S. D., Sernau, R., and Derse, P. H.,** Natural carbon sources, rates of replenishment, and algal growth, in *Nutrients and Eutrophication — Special Symposia,* Vol. 1, Likens, G. E., Ed., American Society of Limnology and Oceanography, 1972, 197.
5. **King, D. L.,** The role of carbon in eutrophication, *J. Water Pollut. Control Fed.,* 42(12), 2035, 1970.
6. **Allen, H. L.,** Phytoplankton photosynthesis, micronutrient interactions and inorganic carbon availability in a soft-water Vermont lake, in *Nutrients and Eutrophication — Special Symposia,* Vol. 1, Likens, F. E., Ed., American Society of Limnology and Oceanography, 1972, 63.
7. **Taylor, M. P. and Welch, E. B.,** Norris reservoir fertilizer study II — effects of thermal stratification and nutrient availability on the productivity of reservoir phytoplankton, *J. Tenn. Acad. Sci.,* 46(3), 90, 1971.
8. **Huang, V. H., Mase, J. R., and Fruh, E. G.,** Nutrient studies in Texas impoundments, *J. Water Pollut. Control Fed.,* 45(1), 105, 1973.
9. **Maloney, T. E., Miller, W. E., and Shiroyama, T.,** Algal response to nutrient additions in natural waters I. laboratory assays, in *Nutrients and Eutrophication — Special Symposia,* Vol. 1, Likens, F. E., Ed., American Society of Limnology and Oceanography, 1972, 134.

10. **Mathiesen, H.,** Summer Maxima of Algae and Eutrophication, Internationale Vereingung fur Theoretische and Augewandte Limnologic, Mitteilung 19, Stuttgart, November 1971, 161.
11. **Brylinski, M. and Mann, K. H.,** An analysis of factors governing productivity in lakes and reservoirs, *J. Am. Soc. Limnol. Oceanogr.*, 18(1), 1, 1973.
12. **Pechlaner, R.,** The phytoplankton spring outburst and its conditions in Lake Erken (Sweden), *J. Am. Soc. Limnol. Oceanogr.*, 15(1), 113, 1970.
13. **Taylor, M. P.,** Seasonal plankton changes and primary productivity in beech reservoir, *J. Tenn. Acad. Sci.*, 47(3), 103, 1972.
14. **Taylor, M. P.,** Phytoplankton Productivity Response to Nutrients Correlated with Certain Environmental Factors in Six T.V.A. Reservoirs, Special Publ. No. 8, Reservoir Fisheries and Limnology, American Fisheries Society, Washington, D.C., 1971, 209.
15. **Biswas, S.,** Distribution of phytoplankton during the early development of Volta Lake (1964-1968), *Hydrobiologia*, 40(2), 201, 1972.
16. State Water Resources Control Board, Algae, in Water Quality Criteria, McKee, J. E. and Wolf, H. W., Eds., The Resources Agency, State of California, 1963, 300.
17. **Kluesner, J. W. and Lee, G. F.,** Organic loading of petenwell reservoir, Wisconsin, *J. Water Pollut. Control Fed.*, 45(2), 269, 1973.
18. **Round, F. E.,** The Growth and Succession of Algal Populations in Freshwaters, Internationale Vereingung fur Theoretische and Angewandte Limnologic, Mitteilung 19, Stuttgart, November 1971, 70.
19. **Mathis, B. J.,** Chlorophyll and the margalef pigment ratio in a mountain lake, *Am. Midland Nat.*, 88(1), 232, 1972.
20. **Silvex, J. K., Henley, D. E., and Wyatt, J. T.,** Planktonic blue-green algae growth and odor production studies, *J. Am. Water Works Assoc.*, 64(1), 35, 1972.
21. **Brook, A. J., Baker, A. L., and Klemer, A. R.,** The Use of Turbidimetry in Studies of the Population Dynamics of Phytoplankton Populations with Special Reference to *Oscillatoria agardhii* var. isothrix, Internationale Vereingung fur Theoretische and Angewandte Limnologic, Mitteilung 19, Stuttgart, November 1971, 244.
22. **Palmer, C. M.,** Environmental needs of nuisance algal forms, in Proc. 4th Annu. Water Qual. Res. Symp., New York State Department of Health, Albany, April 19, 1967, 8.
23. **Olson, T. A.,** Substances Added to Water by Algae, Department of Public Health Biology, University of Minnesota, Minneapolis, April 3, 1961.
24. **Ogawa, R. E. and Carr, J. F.,** The influence of nitrogen on heterocyst production in blue-green algae, *J. Am. Soc. Limnol. Oceanogr.*, 14(3), 342, 1969.
25. **Olson, T. A.,** Toxic plankton, in *Proceedings from Waterworks Problems Manual*, University of Michigan, Ann Arbor, 1951, 86.
26. **Ingram, W. M.,** Toxic Freshwater Algae, *Am. Midland Nat.*, 52(1), 75, 1954.
27. **Schwimmer, M. and Schwimmer, D.,** Medical aspects of phycology, in *Algae, Man and the Environment*, Jackson, D. F., Ed., Syracuse University Press, Syracuse, New York, 1968, 279.
28. **Palmer, C. M.,** Algae in Water Supplies, Publ. No. 657, Public Health Service, U.S. Department of Health, Education, and Welfare, Atlanta, 1959, 18.
29. **Rosen, A. A., Mashni, C. I., and Safferman, R. S.,** Recent developments in the chemistry of odour in water: the cause of earthy/musty odour, *Water Treat. Exam.*, 19, 106, 1970.
30. Committee Report, Research on tastes and odors, *J. Am. Water Works Assoc.*, 62(1), 59, 1970.
31. **Prober, R., Haimes, Y. Y., Teraguchi, M., and Moss, W. H.,** An ecosystem model of lake algae blooms, *AICE Symp. Ser.*, 68(124), 402, 1972.
32. **Sladeckova, A. and Sladecek, V.,** Algicides — friends and foes? in *Algae, Man, and the Environment*, Jackson, D. F., Ed., Syracuse University Press, New York, 1968, 441.
33. **Nemerown, N. L. and Rand, M. C.,** Algal growth affected by degree and type of wastewater treatment, in *Algae, Man, and the Environment*, Jackson, D. F., Ed., Syracuse University Press, New York, 1968, 391.
34. **Welch, E. B., Buckley, J. A., and Bush, R. M.,** Dilution as an algal flood control, *J. Water Pollut. Control Fed.*, 44(12), 2245, 1972.
35. **Liebmann, H.,** The regeneration of some large lakes in Bavaria, in *Algae, Man, and the Environment*, Jackson, D. F., Ed., Syracuse University Press, New York, 1968, 403.
36. Department of Environmental Engineering Sciences, Water Quality Management through Biological Control, *Proc. Symp.*, Brezonik, P. L. and Fox, J. L., Eds., University of Florida, Gainesville, 1975.
37. **Blackburn, R. D. and Barlowe, W. C.,** Copper combination treatment of aquatic weeds, *Proc. South. Weed Sci. Soc.*, 23, 311, 1970.
38. **Haller, W. T. and Sutton, D. L.,** Factors affecting the uptake of endothall $-^{14}C$ by *Hydrilla*, *Weed Sci.*, 21, 446, 1973.
39. **Sutton, D. L., Blackburn, R. D., and Steward, K. K.,** Influence of herbicides on the uptake of copper in *Hydrilla*, *Weed Res.*, 11, 99, 1971.

40. **Sutton, D. L., Haller, W. T., Steward, K. K., and Blackburn, R. D.**, Effect of copper on uptake of diquat-^{14}C by *Hydrilla, Weed Sci.*, 20, 581, 1972.
41. **Deason, T. R. and Bold, H. C.**, Phycological Studies I. Exploratory Studies of Texas Soil Algae, University of Texas Publ. No. 6022, Austin, Texas, 1960.
42. **Stanier, R. Y., Kunisawa, R., Mandel, M., and Cohen-Bazire, G.**, Purification and properties of unicellular blue-green algae (order Chlorococcales), *Bact. Rev.*, 25, 171, 1971.
43. **Cooley, N. R., Keltner, J. M., Jr., and Forester, J.**, The poly-chlorinated biphenyls, aroclors 1248 and 1260: effect on and accumulation by *Tetrahymena pyriformis, J. Protozool.*, 20, 443, 1973.

Part II: Special Studies on Small Scale Recreation Water Areas

Chapter 7

STONE VALLEY RECREATIONAL LAKE STUDY*

INTRODUCTION

The Pennsylvania State University owns and operates Stone Valley Lake as a recreational, educational, and research facility. The lake covers 70 acres and is located on a 7000-acre tract of forested land which is also owned by the University. Aquatic weeds have posed a problem at the lake for some time. The first attempt at controlling the plants took place during the winter of 1970 to 1971. The water level in the lake basin was drawn down approximately 6 ft, exposing about 25% of the lake's surface area. It was hoped that the cold winter temperatures would kill the weeds. Unfortunately, just the opposite was the case. Not only did the extent of the weed growth increase, the number of species of weeds inhabiting the lake increased. It was thought that unseasonably warm temperatures, snowmelt, and nutrients from decaying plants combined to help increase the scope of the weed problem.

PLAN OF STUDY

In the summer of 1973, a comprehensive aquatic plant control program and research study was put into action on Stone Valley Lake. This aquatic plant control project was a complex, multidisciplined one designed not only to control weed growth, but also to answer many of the questions that arise where aquatic plant control programs are needed. The study covers interactions of chemical and mechanical controls with respect to water quality parameters, ecosystem stability and productivity, and public usage and response. The difference with the Stone Valley Project is that they are approaching the problem from an operational management standpoint, rather than working with closed, controlled systems. The objective of this project was to eventually provide resource managers with practical management guidelines for aquatic weed control programs, with optimum efficiency and minimal environmental impact.

Seven study sections were established to do detailed studies on all aspects and impacts of the project. Each section was organized under the supervision of a faculty member of the University, with graduate students assisting in gathering and analyzing the data. The control methods and procedures are described in the control-method evaluation section. The effects on recreation usage are described in the recreation user part of the study. The effects of both plant control methods on water quality and aquatic flora and fauna are discussed in other sections of the study.[3-5]

In the control-method evaluation section treatment and control areas were defined, sampling stations were designated, and treatment methods were outlined. Plant biomass sampling included pre- and posttreatment periods to establish baseline data and determine the effectiveness of the control methods used. The sampling was conducted at random within designated areas by scuba divers, using quadrats with an area of one square meter.[10-13]

A 29% active formulation of 2,4-D granules was applied at a rate of 40 lb/surface acre on 4 acres of water in C-E cove (Figure 1).

Mechanical harvesting was done in four areas: MI 1, MI 2, MI 3, and Tranquillity

* Abridged and updated from a research report by Jean Kinnear, Recreation Department, Pennsylvania State University, University Park. Report presented at the Corps of Engineers Aquatic Plant Control Meeting, Fort Belvoir, Va., October 1974.

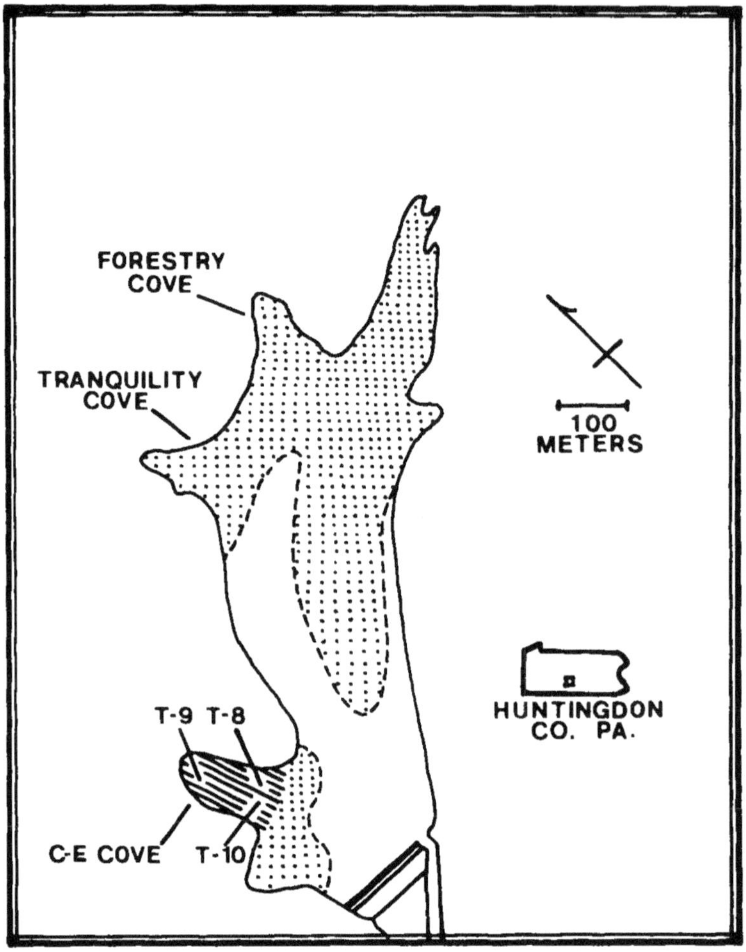

FIGURE 1. Stone Valley Lake, Huntingdon County, Pa., showing the area of the lake infested with weeds (dotted, outlined with broken line) and the area treated with 2,4-D (cross-lined). The insert map of Pennsylvania locates Huntingdon County, Pa. T-8, T-9, and T-10 are sampling transects. (From Marshall, C. D. and Rutschky, C. W., *Proc. Pa. Acad. Sci.*, 48, 127, 1974. With permission.)

Cove. Cutting was only done to within 4 ft of the bottom, to avoid flooded tree stumps (Figure 1).

Effectiveness Evaluation

The overall effectiveness of the chemical and mechanical harvesting was evaluated. Within a month of the application of 2,4-dichlorophenoxy (2,4-D), there was a 100% kill of *Myriophyllum spicatum* in C-E cove. A decrease in the total biomass of *M. spicatum* was also observed all over the lake. The rate of decrease was proportional to the distance from C-E cove. These plants exhibited symptoms similar to 2,4-D poisoning. A year later, in June 1974, there was still a significant decrease in plant biomass when compared to June 1973. It appears evident that the rate of 40 lb of a 29% granular formation of 2,4-D per acre was greater than necessary to control the weeds exclusively within C-E cove. Another good characteristic of 2,4-D is its high degree of specificity toward *M. spicatum*. While *M. spicatum* decreased on all transects in the lake, other plants usually remained unaffected or even colonized areas that *M. spicatum* was removed from.

There were some problems in evaluating the effects of mechanical harvesting on the *M. spicatum* population.

In MI 1, a significant increase in biomass occurred during the 1973 growing season. This was attributed to the fact that the dominant plant species in this area is *Elodea canadensis,* which grows close to the bottom and is therefore unaffected by the harvesting operations.

In the MI 2 and 3 areas a significant decrease in plant biomass occurred during the 1973 growing season. However, in MI 3, a significant decline in biomass occurred prior to harvesting. In Tranquillity Cove a significant decline in biomass also occurred prior to treatment. In all areas, the biomass in June 1974 was significantly less than in June 1973.

The single application of 2,4-D was sufficient to cause a significant decline in *M. spicatum* biomass and maintain control for over a year, thus opening up areas of the lake for their intended recreational usage. The effectiveness of mechanical harvesting could not be clearly determined, because of the unexpected decline of the total biomass prior to treatment. This decline made it impossible to determine whether mechanical harvesting was the true reason for the control gained over *M. spicatum.*

Water Quality

Many quality parameters were monitored to determine effects of both the chemical treatment and the mechanical harvesting on the overall water quality of Stone Valley Lake.

Sampling stations were set up on all tributaries to the lake, discharges from the lake, and the lake itself. Samples were taken from the water surface, mid-depth, and just above the bottom. Bottom sediments and aquatic vegetation were also analyzed. Sampling was done during the period from May 1, 1973, to November 1, 1973.

The temperature gradually increased over the monitoring period. However, this was most likely due to the seasonal solar heating increase, not the weed removal. The dissolved oxygen concentration dropped shortly after 2,4-D treatment in C-E cove, due to the decaying vegetation, but this concentration quickly recovered to acceptable levels. The organic Kjeldahl nitrogen plus ammonia values increased slightly after treatment, but values of tributary organic Kjeldahl nitrogen plus ammonia increased by a similar amount over the same period. The same phenomenon occurred with the ammonia values. The nitrate/nitrate values before and after treatment were below the minimum detectable limit of 0.04 mg/ℓ. A decrease was noted in the bottom water phosphorus concentration. The probable reason for this decrease was the noted increase of the concentration of calcium, since calcium readily precipitates phosphorus. The total carbon value did not show any significant change during the monitoring period. As stated above, the calcium concentration increased during the sampling period. The magnesium concentration also increased, while the sodium concentration decreased. The tributary values of magnesium, calcium, and sodium paralleled those in the lake. No change was noted in the levels of the other metals monitored, potassium, iron, and manganese.

These findings indicate that neither chemical or mechanical control of the aquatic weeds had any significant effect on the water quality.

Recreation Use

A section of the study was devoted to the recreation usage aspect of the project. The objective of this section was to develop a profile of the average recreational user of the Stone Valley Lake, and determine his opinion on the aquatic plant control program. The data was obtained by the use of an opinionnaire. This opinionnaire was distributed to both individuals and groups using the area.

The average user of Stone Valley Lake was somehow associated with Pennsylvania State University and lived in the local area. The most popular activities were canoeing, picnicking, fishing, and hiking, in that order. Most users felt that the aquatic weeds detracted from the quality of the lake, and therefore should be controlled. A slight majority favored mechanical control over chemical control. Most of the users felt that removal of the weeds would improve fishing and boating conditions. Overall, most of the users viewed Stone Valley Lake as an area of high environmental quality.

Another section of the study dealt with the effect of 2,4-D on the diversity of aquatic insects in Stone Valley Lake. Transects were set up in C-E cove, and four samplings were taken, one prior to, and three after the 2,4-D treatment.

It was found that while the total number of organisms dropped markedly, the diversity remained stable. An increase was noted in the relative number of midges and aquatic worms, while the relative number of danselflies and mayflies decreased. This may have been due to the loss of *M. spicatum* for habitat, or it may have been a normal seasonal variation.

The limnology and fisheries of Stone Valley Lake were the subject of another section of the study. Sampling of the fish population was done in the spring and fall, using electrofishing apparatus and trap nets. A creek census was begun in April and continued through early July. This census only applied to the trout fishery. Through the survey, it was determined that the return of trout stocked in 1974 per fisherman was approximately 48%, with a catch of 0.26 trout per fisherman-hour. Stocked trout comprised slightly more than half of the fish population of the lake. Bass, sunfish, catfish, and pickerel were the primary species. In general, the fish populations were large, healthy, and somewhat underexploited, especially the sunfish. Many members of the sunfish population exhibited signs of stunting, due to overcrowding. There was sufficient dissolved oxygen throughout the year, with or without the aquatic weeds. It is not known how the removal of the weeds affected the predator-prey relationships in the lake.

Overall, it was determined that some form of aquatic plant control was needed for Stone Valley Lake. Either the chemical or mechanical control programs were acceptable. The chemical program using the herbicide 2,4-D proved to be very effective in controlling the weed growth without any major effect on water quality, or most of the aquatic biota. The effectiveness of mechanical harvesting was difficult to evaluate because of an unrelated decline in plant biomass prior to harvesting. The specificity of 2,4-D toward *M. spicatum* and the lower cost were major points in its favor. It is believed that mechanical harvesting will provide an adequate amount of control, but not as completely effective as the chemical means.

SUMMARY AND CONCLUSIONS

In mid-June of 1973, a cove approximating 4.5 surface acres of water area was treated at the rate of 40-lb active per surface acre with a 29% granular formulation of the butoxyethanol ester of 2,4-D. This herbicide was chosen for use in the study because of its high order of biological specificity toward *M. spicatum,* the primary problem plant species. Within 2 weeks after treatment the cove area was essentially free of *M. spicatum* and remained in this status throughout the summer and fall of 1973.

Evidence suggests that chemical contamination of other areas of the lake may have occurred following application of 2,4-D as the plant biomass of the nontreated areas were observed to decline during the mid to late summer growing period. It is not known at this time whether this is due to chemical dispersion, a pathologic condition, or whether a subtle disruption of the nutrient balance of the lake occurred to induce this decline.

This distribution of higher aquatic vascular plants was most noticeably affected by herbicide treatment with changes in species composition noted in the area receiving treatment. Some changes in phytoplankton distribution were observed in the treated area. Distribution analysis of bacterial population counts showed no major shift following herbicide treatment and no putrefactive problems were noted. Conversely, damselfly naiads appeared to be adversely affected by chemical application. Preliminary results of the water-quality studies suggest that levels of the primary nutrients, nitrogen and phosphorus, are extremely low and were not significantly affected by herbicide treatment.

REFERENCES

1. **Fitzgerald, G. P.,** Detection of limiting or surplus nitrogen in algae and aquatic weeds, *J. Phycol.*, 4, 221, 1968.
2. **Fitzgerald, G. P. and Nelson, T. G.,** Extractive and enzymatic analysis for limiting or surplus phosphorus in algae, *J. Phycol.*, 2(1), 32, 1956.
3. **Gerloff, G. C.,** Plant Analysis for Nutrient Assay of Natural Wastes, Report No. EPA-R1-73-001, Environmental Protection Agency, Washington, D.C., 1973.
4. **McKee, J. E. and Wolf, H. W.,** Water Quality Criteria, 2nd ed., Publication No. 3-A, State Water Quality Control Board, Sacramento, California, 1963.
5. **Anon.,** Methods for Chemical Analysis Water and Wastes, Water Quality Office, Environmental Protection Agency, Cincinnati, Ohio, 1971.
6. **O'Shaughnessy, J. C. and McDonnell, A. J.,** Criteria for Determining Limiting Nutrients in Surface Waters, Ph.D. thesis, Pennsylvania State University, University Park, 1973.
7. **O'Shaughnessy, J. C. and McDonnell, A. J.,** Criteria for Estimating Limiting Nutrients in Natural Streams, Res. Publ. No. 75, Institute for Research on Land and Water Resources, Pennsylvania State University, University Park, 1973.
8. **Reid, G. K.,** *Ecology of Inland Waters and Estuaries,* D. Van Nostrand, New York, 1961.
9. **Sawyer, C. N. and McCarty, P. L.,** *Chemistry for Sanitary Engineers,* McGraw-Hill, New York, 1967.
10. *Standard Methods for the Examination of Water and Wastewater,* 13th ed., American Public Health Association, Washington, D.C., 1971.
11. Technicon, Autoanalyzer II Technical Publication, No. JTL-0268-1972, New York, 1972.
12. **Weirich, C. B.,** Direct Thermal and Chemical Stratification in Shaver Creek Dam, M.S. thesis, Pennsylvania State University, University Park, 1973.
13. **Marshall, C. D. and Rutschky, C. W.,** Single herbicide treatment, effect on diversity of aquatic insects in Stone Valley Lake, Huntington Co. Pa., *Proc. Pa. Acad. Sci.*, 48, 127, 1974.

Chapter 8

AQUATIC PLANTS IN STONER CREEK IN THE VICINITY OF PARIS, KENTUCKY

INTRODUCTION

Paris, Ky. lies in the central portion of the state (Blue Grass Region) and is the county seat of Bourbon County. The city provides urban services and serves as a regional shopping center for an agricultural region consisting of large horse, cattle, and tobacco farms; however, many residents of the city commute to employment in Lexington, 17 mi to the southwest. The current source of water supply for the city is Stoner Creek, which flows through Paris and into Hinkston Creek, a tributary of Licking River. The Paris, Ky. area, with features pertinent to this investigation, is shown in Figure 1. Other potential sources of water supply are extremely limited due to the existing surface-water drainage pattern and the geology of the area. Ground water is practically nonexisting since the entire area is underlain by limestone.

Local interests have reported taste and odor problems originating within the supply system occurring concurrently with growth of an excessive amount of duckweed in the water supply impoundment near the water intake. This condition was first noted about 1967. The odors have been described as varying from that of cucumbers to very earthy. Various methods of treatment including copper sulfate, carbon filtration, chlorine, and potassium permanganate, have failed to completely alleviate the problem.

Mr. Raymond Bowling, chief operator of the water plant at Paris, contacted Representative John C. Watts to inform him that Paris was having considerable trouble in their domestic water supply with a plant growth identified collectively as duckweed. Several chemicals to control the plant had been recommended but did not meet the approval of the Kentucky Board of Health. He requested that this information be passed on to the proper agency.

Representative Watts contacted the U.S. Department of Agriculture, Agricultural Research Service, concerning plants for control of aquatic plants in potable water supplies. The agency indicated that no research had been carried out directly in potable waters, although control measures had been developed for duckweed in other situations. Two herbicides, 2,4-dichlorophenoxy (2,4-D) and diquat, were effective against duckweed; however, restrictions in their federally registered uses imposed some limitation of use of waters after treatment with these chemicals. The diquat treatment is a satisfactory solution to duckweed problems in lakes for potable water, if an alternate water supply source sufficient to meet demands for a 10-day period is available. No such alternate source is available in the Paris area.

Representative Watts contacted the district engineer, Louisville district, to determine if the Corps of Engineers could be of any assistance to the city of Paris in providing a remedy to this situation. On October 2, 1969, representatives of the Louisville district and the city of Paris met in Paris to discuss the aquatic plants reportedly causing a problem and the procedure to be followed in initiating an investigation.

Mr. Emil Rains, Mayor of Paris, formally requested the Corps of Engineers to undertake a study to determine the practicability and current economic feasibility of controlling or eradicating duckweed. Subsequent actions of the Ohio River Division Engineer and the Office of the Chief of Engineers provide authority for this study. They state that while the legislation does not prohibit participation by the Corps of Engineers in the control of aquatic weeds in subject lake area, it is not current policy to assume costs for aquatic weed control in nonfederal lakes, or in waterways not desig-

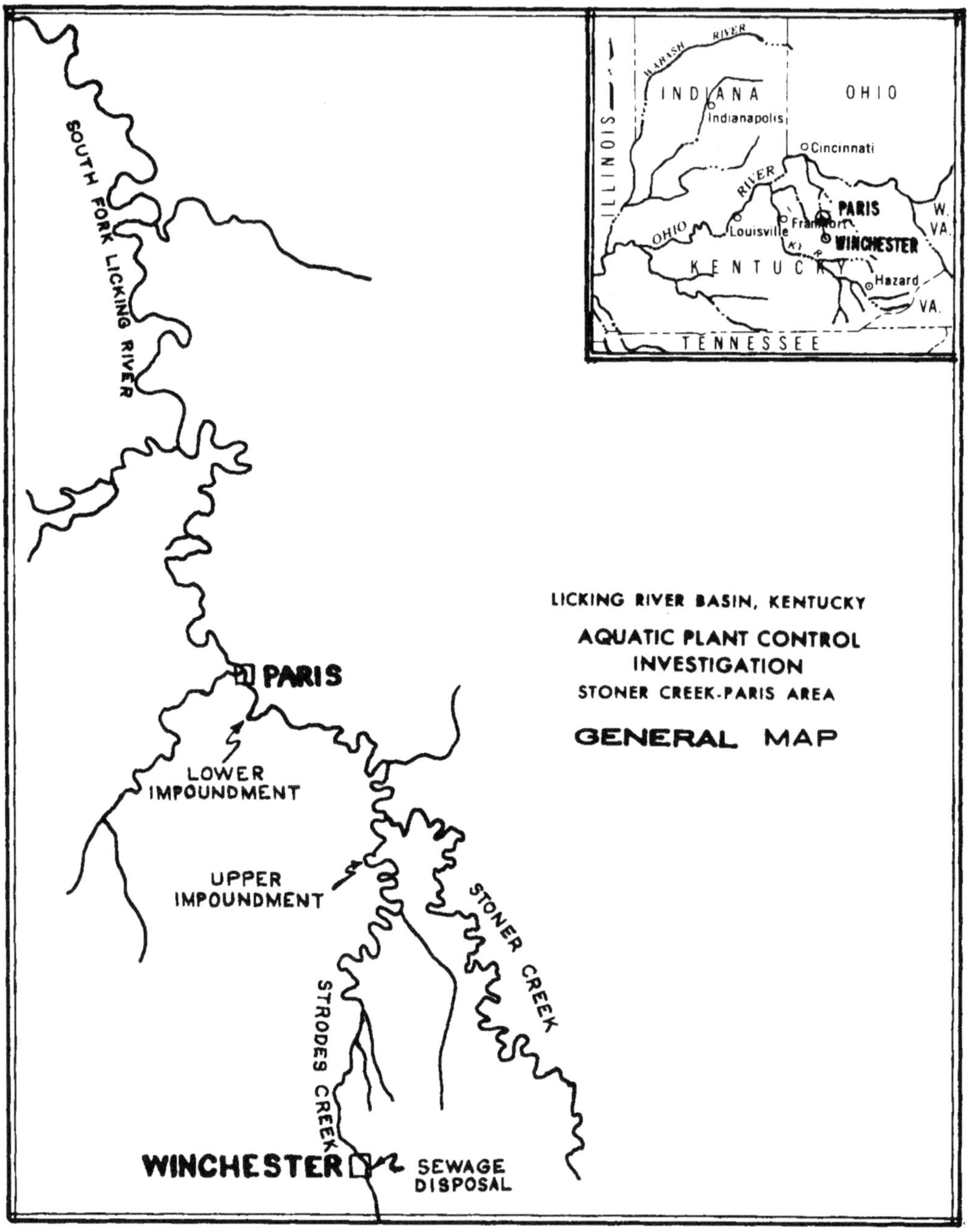

FIGURE 1. Licking River Basin showing sewage disposal in the upper and lower impoundments of the Paris water supply system.

nated as navigable in terms of Corps of Engineers operations.[1] However, this criteria did not preclude exploring the problem providing local authorities with available information on suitable methods or procedures for dealing with it.

TECHNICAL INVESTIGATIONS

The results of an aquatic plant survey are summarized as follows:

1. At that time, the water treatment plant and accompanying lake inspection exhibited neglible taste and odor problems, and only minor infestations of duckweed in the lake were observed (an inspection on September 16, 1970, indicated that serious taste and odor problems did exist).
2. Algal growth in the lake is controlled by standard applications of copper sulfate.
3. The bottom of the lake was inspected by a scuba diver and found to be relatively unsedimented, indicating that sediments contained in the runoffs from agricultural land were not significant.
4. Relatively high levels of chlorination were indicated by the plant manager.
5. Treatment with activated carbon was being used to reduce odor and taste problems, and this control was relatively effective.
6. An inspection of an upper lake, about 5 mi upstream, indicated the water was relatively clear, but some growth of duckweed and algae were noted.
7. No serious odor or taste problems were immediately apparent.
8. Strodes Creek was inspected near the Clark-Bourbon County line, and little or no duckweed was present, but a relatively heavy bloom of algae was noted.
9. Strodes Creek originates within the City of Winchester, Ky. and is the recipient of the effluent of the city sewage treatment plant.

During the inspection, water samples were obtained from the lake adjacent to the plant, the upstream lake, and from Strodes Creek in the immediate vicinity of the sewage plant discharge at Winchester. The three water samples, which were obtained from Stoner Creek and tributaries, were analyzed by a professional testing laboratory.[3] The results of this analysis are shown below (Table 1).

Water Quality

As a result of past water supply problems, the city of Winchester has gone to the Kentucky River for water and created a transbasin diversion. Subsequent to secondary treatment, sewage effluent is released into the Licking River Basin. In the vicinity of Strodes Creek, where Winchester discharges its treated sewage effluent, design flow is negligible. It is estimated that total flows in the order of 12 cfs is needed at the present, increasing to 15 cfs by 1980 and 40 cfs by 2020. Since Winchester is in the headwater region, sites are not available for the storage of water to regulate the stream flow. Sewage plant effluent could be returned to the Kentucky River Basin, but this would deny reusage of the water to downstream users in the Licking River Basin. Therefore, intensive treatment in holding ponds for "polishing" the effluent or transbasin movement of water for dilution may be solutions to the problem.

The chief operator of the water treatment plant at Paris, furnished data on the threshold odor test from the year 1965 to present. This threshold odor data was correlated with the month and in some cases, stream flow. The information depicted on these graphs was subject to considerable interpretation due to the quality of the raw data. However, the graphs did indicate that the higher thresholds occurred at time when algae, rather than duckweed, would be suspect for taste and odor problems.

A follow-up investigation was made on the upper lake on Stoner Creek using diving techniques. That investigation revealed that the bottom of the lake was overlayed with about 2 in. of very black sediment, indicating that this upper is functioning as a "sink" for materials introduced with runoff. In view of the abundant plant life which flourishes within its environment, it may also be functioning as a purification basin. A

Table 1
WATER QUALITY PARAMETERS (PPM)

Parameter	Location 1	Location 2	Location 3
NH_3-N	12.45	0.61	0.50
Nitrite (NO_3)	30.80	26.40	28.60
Nitrite (NO_3)	0.37	0.00	0.01
Total nitrogen (TKN)[a]	1.12	0.16	0.16
Phosphates (PO)	33.5	0.60	0.70

Note: 1 = Strodes Creek, slightly downstream from Winchester, Ky., sewage treatment plant outfall; 2 = Stoner Creek, upstream impoundment owned by City of Paris, 5 mi upstream from water treatment plant; 3 = Stoner Creek, water supply impoundment treatment plant.

[a] Total nitrogen TKN (Total Kjeldahl Nitrogen) includes ammonia and organic nitrogen values.

recheck of the lake in the vicinity of the Paris water treatment plant revealed that distinct taste and odor problems were apparent in raw and finished water. As the lake had recently been treated with copper sulfate and provided good diving visibility, several dives were made to the bottom; these confirmed the original observation that little sediment was present. The water was a bluish color and suspended therein were myriad small whitish particles of algal detritus. The color and suspension were not noticeable in the finished water.

Results of this study indicate that the presence of duckweed is symptomatic rather than casual. The majority of factors creating taste and odor problems appeared to be the result of an overabundance of nutrients in the stream; however, the introduction of domestic waste could cause a taste and odor problem of its own. The amount of nutrients is indicated by a laboratory analysis of water samples obtained at three various locations on Stoner and Strodes Creeks. These nutrients are probably the primary cause of excessive algal growth, which was observed during visual inspections of the lakes and upper reaches of the stream and is recognized by various leading authorities as causing both taste and odor problems. Another area which could cause such problems is associated with the decay processes. Therefore, it may be assumed that treatment of aquatic plant nuisances can be at best considered an interim or stopgap measure; ultimately, as problems and user demands increase, such treatment will lose its effectiveness.

Control Measures

In was concluded that implementation of the measures listed below could have a beneficial effect on taste and odor problems in the Paris, water supply system:

1. A floating boom 1000 yd upstream from the Paris water plant to prevent layering of duckweed in the immediate vicinity
2. Some method of mechanical removal of accumulated growth of duckweed and other aquatic plants during periods of extensive growth
3. Aeration of the lower lake to improve oxygenization by pumping air into an air-diffusing apparatus floated near the bottom level
4. Modifying the treatment operation to allow for greater aeration of the treated

water, thus driving off such gases and other volatile compounds which could contribute to the taste and odor problem in the finished water

Additionally, it was noted that herbicide treatment of duckweed in a potable water system might be obtained by use of Aqua-Kleen® pellets. Improved control of algae could be obtained by use of Cutrine®. Both of these are commercially available products. Prior to incorporating any of these measures, however, the city should engage the services of a firm possessing special expertise in the field of water treatment.

Biological Control of Duckweed

Basic studies have been initiated by various research groups, including the Corps of Engineers in cooperation with the U.S. Department of Agriculture at Fort Lauderdale, Fla. to determine the feasibility of the white amur *(Ctenopharyngodon idella)* for control of Hydrilla and duckweed.[5] The results indicate that the white amur is capable of ingesting large amounts of duckweed and converting it into fish flesh (Figure 2). The biochemical oxygen demand was very high in small aquaria and substantially lower in concrete tanks and ponds. It was apparent that the size of the container had a highly significant effect (Figure 3). Thus, an appropriate stocking rate would be required to establish a satisfactory balance of the biological systems involved. It is recommended that such a system be established to control nutrient pollution.[6]

As indicated in the above reports, southern naiad and *Chara* sp. appear to be among the submersed aquatic plants most preferred by the white amur. However, their preference appears to depend on several factors; one of these factors is fish size and another is the stage of growth of the plants. This herbivorous fish seems to prefer the young, tender plant parts. More research is needed in larger bodies of water to adequately determine the aquatic plants preferred by the white amur. Some troublesome aquatic plants such as duckweed appear to be fairly high in the order of preference.

The white amur appears to have fairly high conversion rates during certain times of the year. The type of vegetation also affects the amount of plant material required to increase the growth of this herbivorous fish. Although approximately half of the vegetation consumed by the white amur passes through the fish, the other digested material appears to be efficiently utilized. In this study, the white amur converted 1.26 to 4.64 g dry of hydrilla into 1.0 g of fish during certain periods of the year; this is fairly high for a herbivorous animal. Data on conversion of duckweed to fish flesh also indicate that the white amur is more efficient at certain periods of the year. Additional research is needed to adequately evaluate the factors involved in the conversion of aquatic plants to fish protein. This information will be useful in evaluating the potential of the white amur as a possible cultural fish to help increase protein supplies.

The white amur appears to be one of the most promising biological agents for the control of many species of aquatic plants, particularly for those which are submersed and floating. This fish is also promising as an efficient and effective agent for converting troublesome aquatic plant growth into a useful product such as fish protein. Proper management will be required in order to utilize this fish for its greatest potential.[5,6]

SUMMARY AND CONCLUSIONS

Based on studies made for this situation it is concluded that the underlying cause of the taste and odor problems which appear in the Paris, Ky. water supply is nutrient-bearing material (or materials) introduced in Stoner Creek upstream from Paris, i.e., the sewage treatment plant. In addition to inherent taste and odor, these materials contribute to frequent proliferations of various aquatic flora, including duckweeds.

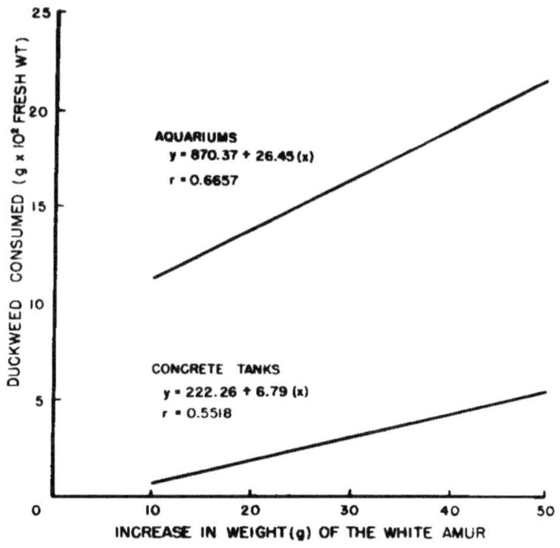

FIGURE 2. Correlation of duckweed consumption to growth of white amur. (From Michewicz, J. E., Sutton, D. L., and Blackburn, R. D., *Hyacinth Control J.*, 10, 22, 1972. With permission.)

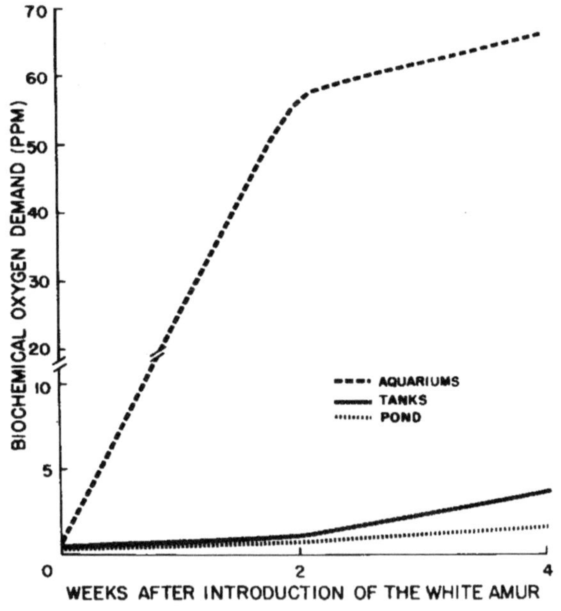

FIGURE 3. Biological oxygen demand of water from aquaria and concrete tanks containing white amur feeding on duckweed. (From Michewicz, J. E., Sutton, D. L., and Blackburn, R. D., *Hyacinth Control J.*, 10, 22, 1972. With permission.)

During their life cycle and decomposition, these flora produce other various tastes and odors. On-site treatment of such proliferations would be complicated by existing state and federal regulations restricting the use of chemicals in potable water supplies, the intrusion of upstream untreated water into the treated water, and the probable continued or increased introduction of nutrient-bearing material into the stream. Duckweed is a result of the nutrient enrichment, not the cause of the problem. Mechanical removal of duckweed is a very temporary solution to the problem. Introduction of the white amur to control vegegative growth is a potential solution and should be investigated further.

REFERENCES

1. **Gangstad, E. O.**, Aquatic plant control program, *Hyacinth Control J.*, 9, 46, 1971.
2. **Mackenthun, K. M. and Ingram, W. M.**, Biological Associated Problems in Freshwater Environment, U.S. Department of the Interior and Federal Water Pollution Control Administration, U.S. Government Printing Office, Washington, D.C., 1967.
3. U.S. Geological Survey, Techniques for Water Resource Investigations of the U.S. Geological Service, Government Printing Office, Washington, D.C., 1970.
4. **Michewicz, J. E., Sutton, D. L., and Blackburn, R. D.**, Water quality of small enclosures stocked with white amur, *Hyacinth Control J.*, 10, 22, 1972.
5. **Hesser, E. and Gangstad, E. O.**, Nuisance aquatic macrophyte growth, *J. Aquatic Plant Manage.*, 16, 11, 1978.
6. **Gangstad, E. O.**, *Weed Control Methods for River Basin Management*, CRC Press, Boca Raton, Fla., 1978.

Chapter 9

AQUATIC PLANT PROBLEMS IN LAKE TANEYCOMO

INTRODUCTION

Lake Taneycomo, located in the Missouri Ozark Hills only a few miles from the Arkansas state line, was formed in 1913 by Powersite Dam. A private impoundment built by The Empire District Electric Company impounds water to a depth of 40 ft and backs the lake 19½ mi up the narrow White River Valley. Excessive siltation from frequent and severe floodwater has reduced the surface area of the lake from its original 2414 acres to 1730 acres. The lake is long and quite narrow — less than 2000 ft wide at its widest. Current is visible at the cities of Branson and Hollister, which are located about 13 mi upstream from Powersite Dam. Due to these features, it is frequently referred to as a "river lake". Lake Taneycomo has for many years been a favorite recreation area, particularly for fishing, boating, and swimming.

Construction of Table Rock Dam, located 2½ mi above the upper end of Lake Taneycomo, began in 1954. The U.S. Army Corps of Engineers built this project to produce hydroelectric power, curtail floods in the White River Valley, and provide other beneficial water uses. Upon completion of the dam, the aquatic environment was abruptly altered by the cold hypolimnion water discharged almost directly in Lake Taneycomo.

The outlet structures of the dam consist of four penstocks which draw water from 140 ft below the surface at the top of the power pool. Daily maximum discharges average about 2000 ft/sec through one or more penstocks, and flows of up to 15,000 ft/sec may be discharged through the four penstocks. During the generation of hydroelectric power, the discharged average water temperature is 48°F.

Local interests first brought the aquatic plant problem in Lake Taneycomo to the attention of the Corps of Engineers in 1966, primarily to determine if the Corps could provide assistance in controlling the infestation. After an investigation, it was determined that this project was not eligible for financial assistance under existing legislative authority for aquatic plant control.[2]

AQUATIC PLANT CONTROL STUDIES

Anderson[1] reported that aquatic weeds in the lake hindered the operation of excursion, pleasure, and fishing boats during the summers of 1964 and 1965. The weeds were also aesthetically undesirable for the tourist industry. During the study conducted in 1964 and 1965, aquatic weeds observed to be a problem or nuisance included leafy pondweed *(Potamogeton foliosus)* and slender naiad *(Najas flexilis)*. The probem was determined to be caused by the nutrients from the fertility of the watershed and the inflow of treated and untreated domestic wastes. Temporary control of these weeds was obtained in a ¾-acre dike-in swimming area with Hydrothol® 47 and 191. However, the concentration needed for an effective 30-min contact time caused fish mortality.

Whitley identified ten species of higher aquatic plants during studies conducted during 1968 and 1969. The following six species were abundant — slender naiad *(N. flexilis)*, small pondweed *(P. pusillus)*, curlyleaf pondweed (P. crispus), American elodea (*Elodea canadensis*), coontail moss *(Ceratophyllum demersum)*, and horned-pondweed *(Zannichellia palustris)*. Limited growths of leafy pondweed *(P. foliosus)* and sago pondweed *(P. pectinatus)* were also present. Duckweed *(Lemna* sp.) and waterweed *(Wolffia* sp.) developed in dense weed beds in late summer and were associated with effluent from a waste stabilization lagoon. Filamentous algae also developed in mid-summer in many weed beds. Other algae (*Nitella* sp. and *Chara* sp.) grew extensively in the upper half of the lake but were not serious problems.

Four methods of controlling excessive aquatic vegetation were evaluated during this study period. They were chemical herbicide treatments, light restriction, dredging, and bottom sealing. Whitley concluded that herbicide di-(N, N-demethylalkylamine) salt of endothall gave 95 to 100% control. Light restriction by plastic sheets was determined to be an impractical control method because of the frequent strong current. Dredging was predicted to be effective in the lower part of the lake and ineffective in the upper area while bottom sealing proved totally ineffective.

POTENTIAL METHODS OF CONTROL

Winter Drawdown

A winter drawdown is probably the best potential method of controlling aquatic plant growth in Lake Taneycomo. Preliminary investigations indicate that the maximum depth of aquatic weed penetration is 8 ft. Therefore, a drawdown of 8 ft for a period of 10 days during the winter period should be effective in achieving a high degree of control and alleviating weed problems. Not all aquatic weeds would be eradicated by the drawndown, and significant adverse effect on the fish population is anticipated. Some damage would occur to private boat docks due to the drawdown; however, this could be used as a period to perform normal maintenance of the docks. Using this technique, control could be achieved by lowering the lake level and exposing the weed beds and shoreline to freezing temperatures. The time selected would be based on a climatological record and coincide with the period having the lowest average temperatures and, specifically, on the basis of short- and long-range forecasts for the period selected for the drawdown. To achieve maximum results, the drawdown should coincide with a period of freezing temperature, preferably as low as 10°F.[4-13]

A general acceptance of a drawdown program from local interests and the public would be necessary to effectively accomplish a weed kill. This could be effected through the public dissemination of information explaining what is planned, why the program is being carried out, and who will benefit. This will require participation by all interested parties including private power companies and state and federal agencies. To gain the acceptance of private dock owners and the general public, these groups must be convinced that the drawdown will improve the lake for fishing, boating, and recreational use. Available information indicates that a drawdown to the extent proposed would create some problems with the anchorage of boat docks. Due to the relatively stable lake level of Lake Taneycomo, many boat dock anchorages were not built to permit flotation of docks under the proposed drawdown conditions. Therefore, temporary anchorages would have to be provided, and many facilities or docks would have to be allowed to stand on the shoreline.

Power generation at Table Rock would have to be curtailed or stopped during the drawdown period. After the lake has been drawn down 8 ft, it may be possible to resume limited power generation at Table Rock and Lake Taneycomo. Table Rock discharges would have to be kept within the capacity of the lower power plant thereby maintaining the lake level in Taneycomo. Details of the drawdown and loss of power revenues would have to be coordinated closely with Southwestern Power Administration, Corps of Engineers, Empire District Electric Company, and lake administrators. As the drawdown would be accomplished during a period of low recreational activity in which the lake is used primarily for trout fishing, recreational interests should not be a serious hindrance in carrying out the program. The limiting factor for successful operation of this method may be the location of Lake Taneycomo in an area of relatively mild winters. Also the need for increased power generation during unusually cold weather may restrict flexibility in operation which would be required at Table Rock Dam.

Dredging

This is practical in the vicinity of Rockaway Beach. An estimated 40 acres could be controlled by this name. Advantages to be gained by dredging are

1. Elimination of accumulated nutrients
2. Deepening of desirable waterways
3. Probable creation of valuable landfill
4. Fish-kill unlikely
5. Effects that would be relatively long-lasting

Problems are

1. Considerable costs will be involved in preparing diked disposal ponds with control structures for dewatering and impounding solids. The impounded area must be large enough to settle the fine solids such that the runoff would meet state and federal water quality standards and will be ecologically acceptable. Also, the height of the dikes must be sufficiently above the high water elevation to prevent overtopping during flood.
2. Odors from decaying vegetation and other materials in the impoundments would create an odor problem that might require treatment with lime or other chemicals. Dredging during winter probably would prvent odor problems.
3. A specially designed dredge with cutter blades would be used.
4. Rockaway Beach, a motel and cottage resort area, will be practically unusable during dredging operations; however, use is limited in winter.

Mechanical Cutting

This involves manually pulling, cutting, and raking the weeds through the use of chain-dragging, hand-operated underwater weed saws as power-driven underwater weed cutting and removal units are usually inefficient and expensive. Additional problems of disposing of the cut weeds would have to be overcome. However, mechanical removal has advantages from an ecological standpoint because:

1. No foreign substances are introduced into the water.
2. Some of the excess nutrients are removed from the water.
3. Organic filling rate of the lake will be reduced.
4. Immediate relief from nuisance weed infestation would be accomplished.

Herbivorous Fish

The white amur *(Ctenopharyngodon idella)* is one of the most promising herbivorous fish for the control of rooted aquatic vegetation. However, the Missouri Department of Conservation is presently reluctant to permit the introduction of exotic fish species in state waters and has only approved the introduction of the white amur for scientific purposes.[19]

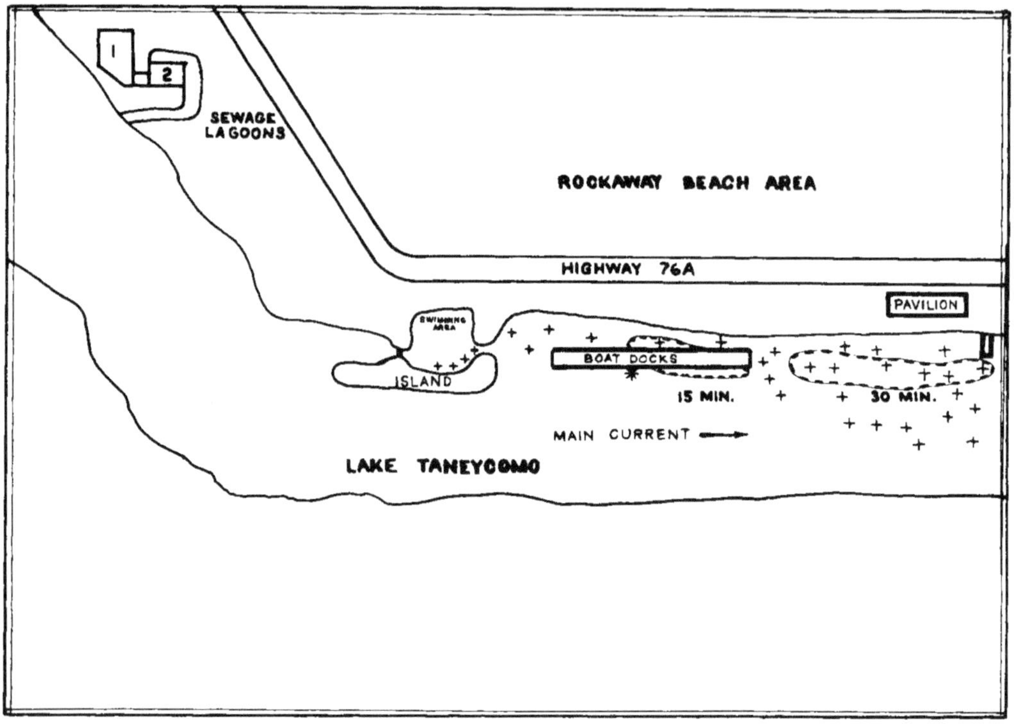

FIGURE 1. Showing the location of sewage lagoons, Rockaway Beach, and the treatment areas in Lake Taneycomo. Lake Taneycomo and the Rockaway Beach area showing water movement of dye placement from point * (by boat docks) and movement 15 and 30 min after application.

RESULTS AND DISCUSSION

The objectives of the study in 1968 and 1969 (Figure 1) were to determine and evaluate methods of controlling excessive vegetation in cold flowing water by chemical herbicide treatments, light restriction, dredging, and bottom sealing.

The most effective herbicide tested was the di-(N,N-dimethylakylamine) salt of endothall which gave 95 to 100% control of all plant species. The most resistant plant was *E. canadensis*. Preemergence applications of dichlobenil controlled *P. pusillus* but was not effective against *N. flexilis, C. demersum, E. canadensis,* and *P. crispus*. Potassium endothall applied during a lake drawdown gave good control of *N. flexilis, P. pusillus, P. crispus,* and *Z. palustris,* partial control of *C. demersum,* and no control of *E. canadensis*.

Dredging was judged to be an effective method of weed control in the lower part of the lake but this method was not successful in the upper area, where there was a greater interchange of clear water and few deposits of fertile sediment.

Light limitation by plastic sheets was not a practical method of weed control in Lake Taneycomo due to the difficulties created by strong currents. Application of a bottom sealant did not inhibit growth of vegetation the following spring.

Aquatic Plant Species

Prior to this study, it was thought that the weed problem was due to only two species of pondweeds, *N. flexilis* and *P. foliosus*. In 1968, ten aquatic plants were identified in addition to several kinds of algae. Listed below are location of the major growths of aquatic vegetation along with a few comments on their manner of reproduction and contribution to the weed problem (Table 1).

Table 1
AQUATIC-VEGETATION IN LAKE TANEYCOMO IN 1968

Month	*Najas flexillis*	*Potamogenton pusillus*	*P. crispus*	*Zannichellia palustris*	*Elodea canadensis*	*Ceratophyllum demersum*	*P. foliosus*	*P. pectinatus*	*Lemna minor*	*Wolffia columbiana*	*Nitella*	*Chara*	*Filamentous algae*
February	d	d	—	—	1	d	—	—	—	—	1	1	1
March	d	1	—	—	1	1	—	—	—	—	2	1	1
April	d	1	1	—	1	1	—	—	—	—	2	2	2
May	1	2	4	—	1	2	—	—	—	—	2	2	2
June	1	3	4	—	1	3	—	—	1	—	2	3	2
July	2	4	4	2	2	4	1	1	2	1	2	3	3
August	3	4	4	3	3	4	1	1	2	2	2	3	3
September	4	4	1	1	4	3	—	—	2	2	2	2	2
October	2	2	—	—	4	1	—	—	—	—	2	1	1
November	d	d	—	—	4	d	—	—	—	—	1	1	1
December	d	d	—	—	2	d	—	—	—	—	1	1	1

Note: d = plant fragments appeared dead; 1 = trace; 2 = light growth; 3 = moderate growth; 4 = heavy growth; and — = not found.

1. Algae, *Nitella, Chara,* and filamentous algae can be found all year in the colder water in the upper part of the reservoir above Bull Creek. They do not create a significant problem in this region at present. Filamentous algae add to the undesirable appearance of the weed beds in the lower half of the lake after vascular aquatic plants reach the surface.
2. *P. crispus,* curly-leaf pondweed, grows early in the season in cold water and at depths to 12 ft. This exotic plant is spreading rapidly in the lake below Branson, may become the major problem plant in the near future, and develops from seed and winter buds.
3. *P. pusillus,* fine-leaf pondweed, grows early in cold water and flowers in warm water. There are extensive growths from ½ mi above Bull Creek to the lower reservoir area. Heavy growth occurred in the Rockaway Beach area. It develops from seed and vegetatively from previous season's growth.
4. *E. canadensis,* waterweed, grows in dense beds from Branson to Cedar Point. It is present all year and grows rapidly in cold water. *Elodea* is the major problem plant in the Youth Camp area and the fishing camp dock area between Branson and Rockaway Beach. It is the most resistant to herbicides of all the plants present in the lake, however, it fluctuates some years due to severe dieback from disease.
5. *C. demersum,* coontail, grows in the lower half of the reservoir from ½ mi above Bull Creek to Powersite. Coontail grows in both cold and warm water and was the major problem plant below Rockaway Beach in 1968. It develops primarily from winter buds in this lake.
6. *N. flexilis,* bushy pondweed, was prevalent below Bull Creek. *Najas* grew mostly in the warm, quiet water areas and was a major problem at Edgewater Beach and Powersite. It develops primarily from sprouting at the joints of previous season's growth.

Table 2
SOLUBLE PHOSPHATE CONCENTRATIONS IN LAKE TANEYCOMO

Location	May	June	July	August	September	October	November
Table Rock tailwater	2.5	0.6	2.6	0.5	1.2	0.5	0.4
Branson	4.4	2.6	1.2	0.5	2.2	1.6	2.2
Below Branson	4.2	3.0	3.0	3.0	2.2	2.0	0.5
Below Rockaway Beach	2.2	3.0	0.6	0.6	—	0.5	—
Forsyth	1.4	3.0	1.2	0.5	—	0.5	—
Powersite	6.3	3.6	0.5	2.1	0.8	0.5	6.8
Bull Creek	0.8	3.0	0.5	3.6	1.2	0.6	3.4
Branson Lagoon effluent							3060.0

Note: Data is expressed as parts per billion phosphate.

7. *L. minor,* duckweed, was found from the Branson Lagoon effluent downstream to Powersite. Heavy growths develop on the water surface in the denser beds of higher aquatic plants, particularly in the Rockaway Beach area. Recent improvements in the Branson-Hollister Sewage Treatment Systems should reduce the growth of duckweed.
8. *W. columbiana,* watermeal, was found primarily in the Rockaway Beach area trapped in heavy growths of higher plants.
9. *Z. pectinatus,* sago pondweed, and *P. foliosus,* leafy pondweed, were found in the Branson area in trace amounts and are not a problem at present.

Phosphorus Nutrition

Soluble phosphorus levels were measured since phosphorus usually is considered the first limiting major plant nutrient. The results are shown in Table 2. A low level of soluble phosphorus was usually present at all stations. The phosphate levels increased at Branson and continued to increase further below the town. A reduction occurred below Rockaway Beach and at Forsyth which could be due to uptake by aquatic vegetation in these heavy growth areas. A very high level of phosphate was measured in the Branson Lagoon effluent at its point of entry into the lake. These measurements were made prior to recent improvement in sewage treatment.

Temperature

In 1968, the temperature measurements ranged from 49 to the upper part of the lake between Table Rock Dam and Branson. A warm surface layer of water usually started at Bull Creek just above Rockaway Beach, and the highest temperature measured in this area was 76.5°F. In August, a maximum surface temperature of 81° was measured at the lower end of the lake at Powersite; however, at 4 ft the temperatures was 64°F. Once dense weed growth reaches the surface, the current is slowed, and higher water temperatures occur in these areas. Manipulation of water levels or turbine intake levels to maintain a lower surface sufficient to affect aquatic plant growth appears impractical.

Table 3
PERCENT CONTROL OF AQUATIC VEGETATION IN LAKE TANEYCOME BY SEVERAL HERBICIDES IN 1968

Herbicide	*Najas flexillis*	*Potamogeton pusillus*	*P. crispus*	*Zannichellia palustris*	*Elodea canadensis*	*Ceratophyllum demersum*	*P. foliosus*	*P. pectinatus*	*Lemna minor*	*Fulamentous algae*
Hydrothol® 47 liquid	95	100	100	100	95	95	0	0	100	95
Hydrothol® 47 granules	10	75	100	0	10	10	0	0	0	0
Casoron® granules	0	95	0	0	0	0	0	0	0	0
Potassium endothall liquid[a]	90	90	100	100	25	75	100	100	0	0
Endothall-Silvex® pellets	0	10	0	10	0	0	0	0	0	0

[a] Potassium endothall was applied to vegetation on the exposed bottom and in shallow water during a drawdown of the lake.

Effects of Herbicide Treatments

Sixty herbicide test plots were applied and evaluated in 1968. A brief summary of these results are shown in Table 3. Casoron® was very effective against the dominant pondweed, *P. pusillus;* however, it was ineffective against *Najas, Ceratophyllum, Elodea,* and *P. crispus.* These results were disappointing because these species can usually be controlled in ponds and lakes with this herbicide. Apparently, the greater water interchange in Lake Taneycomo reduces Casoron® residues in the bottom mud below effective levels.

Hydrothol® 47 was the most effective herbicide tested, and 95 to 100% control of all plant species was obtained in some of the test plots. The most resistant plant was *Elodea.* The use of Hydrothol® 47 is limited in large applications by its high toxicity to fish for a short period following application. Excessive fish mortality was not a problem in these test applications. However this would be significant if large areas were treated.

Potassium endothall was applied during a 5-ft drawdown of the lake. There was little or no current at this time, and it was very effective against all species of plants except *Elodea,* which was affected to only a slight degree. A cationic bottom sealant was applied to five plants. The plants growing in the test plots at the time of application did not appear to be affected, and later observations in 1969 revealed that this was not an effective method of control.

Water Quality Studies

During the summer of 1972, an experimental course entitled The Biology of Water Pollution was taught in the Biology Department College of the Ozarks, Clarksville, Ark. One of the course objectives was to provide biology students with a relevant and practical experience in pollution studies by collecting and analyzing water samples to determine the nature and levels of pollutants entering Lake Taneycomo.[14-18]

The sites were at Rockaway Beach, Branson Sewer Plant, Roark Creek-Branson, and Turkey Creek-Hollister. Sampling was carried during the last 2 weeks of July and the first 2 weeks of August. Determinations of several factors were made on each sample, but those of greatest significance included the coliform counts and the phosphate determinations. The data collected from these determinations are summarized in Tables 4 to 7 using standard methods.

Coliform refers to a group of bacteria which usually originate in the intestinal tracts of warm-blooded animals and man. The presence of relatively large numbers of the bacteria in a body of water is considered a reliable indication of recent sewage contamination. Under optimum conditions in cool clean water, coliform bacteria can survive no longer than 2 to 3 days, and most die within the first few hours of exposure to open water. The presence of coliform bacteria is therefore both an indication of degree of sewage contamination and the relative recency of such pollution.

The contamination of a body of water with sewage effluent indicates the possible presence of other enteric organisms (in addition to the coliform) which may be pathogenic such as those responsible for typhoid fever, cholera, and dysentery, etc. A potential health hazard may therefore exist under these conditions. A direct correlation has been established between coliform populations in the range of 2300 to 2700 coliforms per 100 mℓ of water and the incidence of infectious diseases. Perhaps even more alarming than the potential health hazard is the accelerated eutrophication of bodies of water, which is caused by the input of sewage and sewage effluent. Eutrophication is characterized by prolific algal and aquatic weed growth. This growth is stimulated by the fertilization of the lake with nutrients carried in sewage effluent.

Scientific reports indicate that under most natural circumstances, phosphorus is the limiting element in the stimulation of aquatic plant growth. Therefore, the addition of phosphorus will usually result in a proportional increase in plant growth. Three to 10 mg of phosphorous per liter (ppm) is capable of yielding 300 m to 1 g of organic matter per liter. Effluents from properly functioning sewage treatment plants generally average 3 to 10 mg phosphate per liter of effluent. Translated into English equivalents, approximately 112 gal of sewage effluent carrying 10 ppm phosphate will stimulate the growth of about 1 lb organic matter.

The following chart shows the amount of phosphorous commonly encountered in certain types of water.

Effluents from secondary treatment plants	3.000—10.00 ppm
Agricultural drainage	0.05—1.0 ppm
Unpolluted lakes	0.01—0.04 ppm
Eutrophic (polluted) lakes	0.03—1.5 ppm

Interpretation of phosphate determination is not always easy because phosphates may be taken up by actively growing plant life almost as rapidly as they enter a lake. Low phosphate readings, therefore, will not always indicate low phosphate pollution.

The data as observed in Tables 4 to 7 indicate Lake Taneycomo's problems are to a large extent directly attributable to the lack of adequate sewage treatment by the communities and resorts which lie on the banks of Taneycomo and within its watershed. During the past several years, the aquatic plant problems have been somewhat reduced due to higher flows from the Table Rock Reservoirs and lower visitation of the public use areas.

Table 4
COLIFORM COUNTS (PER 100 M*l*) AND PHOSPHOROUS (PPM)e FOR ROCKAWAY BEACH SITES

Date	Site 1	Site 2	Site 3	Site 4
7/13	1,233	137	72,566	114,020
7/16	6,300	2,600	533	4,900
7/20	2,133	2,033	366	1,233
7/21	2,533	2,166	5,536	25,866
7/23	100	6,066	14,666	18,000
8/3	4,566	12,100	8,133	24,400
8/5	33	4,700	22,966	8,566
Av/100/m*l*	2,400	4,300	14,500	27,000
PO_4 (ppm)				
7/23	8.39	1.68	0.54	0.47

Courtesy of the Biology Department, College of the Ozarks, Clarksville, Ark. With permission.

Table 5
COLIFORM COUNTS (PER 100 M*l*) AND PHOSPHORUS (PPM) FOR BRANSON SEWAGE PLANT SITES

Date	Site 1	Site 2	Site 3	Site 4
7/18	405,450	408,500	121,200	53,000
7/24	252,665	477,587	384,195	40,777
8/8	344,250	12,043	173,000	7,440
Av/100/m*l*	344,122	300,373	226,131	33,872
PO_4(ppm)	30	25	9	0.36

Courtesy of the Biology Department, College of the Ozarks, Clarksville, Ark. With permission.

Table 6
COLIFORM COUNTS (PER 100 M*l*) FOR ROARK CREEK-BRANSON CAMPGROUND SWIMMING AREA

Date	Site 1	Site 2	Site 3	Site 4
7/12	10,087	42,473	10,667	407
7/17	38,400	115,000	11,750	5,950
7/19	55,333	79,000	667	2,100
8/4	31,700	34,850	76,000	21,150
8/8	48,500	21,500	11,000	10,000
8/11	9,000	9,000	4,000	1,000
Av/100/m*l*	32,170	50,304	19,014	6,768

Courtesy of the Biology Department, College of the Ozarks, Clarksville, Ark. With permission.

Table 7
COLIFORM COUNTS (PER 100 M*l*) FOR TURKEY CREEK-HOLLISTER SITES

Site No.	Count[a]	Site No.	Count	Site No.	Count
1	3,300	8	26,820	15	2,700
2	3,817	9	37,823	16	4,220
3	22,978	10	11,166	17	8,220
4	10,810	11	3,802	18	15,966
5	178,810	12	4,600	19	4,266
6	87,600	13	3,033	20	5,666
7	12,750	14	1,842	21	933
Av/100/m*l*	45,720		12,726		5,993

[a] Average of July and August sampling dates.

Courtesy of the Biology Department, College of the Ozarks, Clarksville, Ark. With permission.

SUMMARY AND CONCLUSIONS

Rooted aquatic vascular plants are found to inhabit shallow parts of lakes and rivers, particularly those which are highly eutrophic from enrichment with sewage effluents. The biomass of plant growth cannot be materially reduced except by adequate treatment of the sewage to remove nitrogen and phosphorous that enters the ecosystem.

REFERENCES

1. **Anderson, R. O.**, Investigation of Aquatic Vegetation in Lake Taneycomo at Rockaway Beach, Missouri, and Tests with Herbicides Hydrothol 47 and 191, Missouri Cooperative Fishery Unit, University of Missouri, Columbia, 1966.
2. **Whitley, J. R.**, Control of Undesirable Aquatic Vegetation in Lake Taneycomo, Missouri, Missouri Cooperative Fishery Unit, University of Missouri, Columbia, 1970.
3. U.S. Geological Survey, Techniques for Water Resource Investigations of the U.S. Geological Service, Government Printing Office, Washington, D.C., 1970.
4. **Eschmeyer, R. W., Manges, D. E., and Hasbauer, O. F.**, Trends in fishing on T.V.A. storage waters, *J. Tenn. Acad. Sci.*, 22(1), 45, 1947.
5. **Eschmeyer, R. W.**, Fish Conservation Fundamentals, Bulletin No. 38, Sport Fishing Institute, 1954.
6. **Grizzel, R. A., Jr.**, Fish and wildlife management on watershed projects, in Trans. 25th N. Am. Wildl. Conf., Washington, D.C., 1960, 186.
7. **Heman, M. L., Campbell, R. S., and Redmond, L. C.**, Manipulation of fish populations through reservoir drawdown, *Trans. Am. Fish. Soc.*, 98(2), 293, 1969.
8. **Husley, A. H.**, Effects of a fall and winter drawdown on a flood control lake, *Proc. 10th Annu. Conf. S.E. Assoc. Game Fish Comm.*, Vol. 10, Little Rock, Ark., 1957, 285.
9. **Husley, A. H.**, A proposal for the management reservoirs for fisheries, in *Proc. 12th Annu. Conf. S.E. Assoc. Game Fish Comm.*, Vol. 12, Louisville, Ky., 1958, 132.
10. **Jenkins, R. M.**, Reservoir Fish Management-Progress and Challenge, Sport Fishing Institute, Washington, D.C., 1961.
11. **Lantz, K. E., Davis, J. T., and Hughes, J. S.**, Water level fluctuation — its effect on vegetation control and fish population management, in *Proc. 18th Annu. Conf. S.E. Assoc. Game Fish Comm.*, Clearwater, Fla., 1964.

12. **Pierce, P. C., Frey, J. E., and Yawn, H. M.,** An evaluation of fishery management techniques utilizing winter drawdowns, in *Proc. 17th Annu. Conf. S.E. Assoc. Game Fish Comm.*, Hot Springs, Ark., 1963, 347.
13. **Wood, R.,** The significance of managed water levels in developing the fisheries of large impoundments, *Tenn. Acad. Sci.*, 26(3), 214, 1951.
14. **Field, R. and Lager, J. A.,** Urban runoff pollution control — state of the art, *J. Environ. Eng. Div., ASCE*, 101, 107, 1975.
15. **Hayden, J. W., Chisman, J. A., and Macauly, H. H.,** Simulation of water quality management policies, *J. Environ. Eng. Div., ASCE*, 101, 623, 1975.
16. **Haith, D. A.,** Land use and water quality in New York rivers, *J. Environ. Eng. Div., ASCE*, 102, 1, 1976.
17. **Haith, D. A. and Dougherty, J. V.,** Nonpoint source of pollution from agricultural runoff, *J. Environ. Eng. Div., ASCE*, 102, 1055, 1976.
18. **Freedman, P. L. and Canale, R. P.,** Aquatic macrophytes in White Lake, Michigan, *J. Environ. Eng. Div., ASCE*, 103, 431, 1977.
19. **Gangstad, E. O., Raynes, J. J., and Burress, R. H.,** Herbivorous fish for aquatic plant control, Tech. Rep. 4, U.S. Army Waterways Experiment Station, Vicksburg, Miss., 1973.

Chapter 10

WATER QUALITY EVALUATIONS OF THE LOWER MYSTIC LAKE, MASSACHUSETTS*

INTRODUCTION

The Lower Mystic Lake in Arlington-Medford, Mass. is a meromictic lake. Approximately 24% of the lake's 3,115,790 m^3 volume is brackish water of marine origin. The mixolimnion is thermally stratified seasonally causing stratification of some chemical species. The chemocline is a narrow, transitional zone occurring between 9.1 and 10.7 m. The highly reduced anaerobic monimolimnion is divided into two portions located in the cryptodepressions of the lake. Some dissolved gases are stratified throughout the water column indicating that they penetrate through the chemocline to the mixolimnion. The northern cryptodepression appears to be undergoing less anaerobic decomposition than the southern cryptodepression. A mean hydrogen sulfide concentration of 152 mg ℓ^{-1} in both cryptodepressions has been the source of occasional odor problems near the urban area along the lakeshore.

The meromictic characteristics of the Lower Mystic Lake in Arlington-Medford, Mass. were identified during a planning study funded by the U.S. Army Corps of Engineers' Aquatic Plant Control Program.[10] The study was conducted by the Corps' New England Division (NED) in conjunction with the Boston Metropolitan District Commission (MDC) to determine the feasibility of ridding the lake of its occasional odor problems.

At the outset of the study it was necessary to determine whether the odors occur because the lake is eutrophic or whether the lake is meromictic, characterized by an anaerobic monimolimnion. The lake was determined to be a chemically stratified (meromictic) lake. The MDC is presently considering whether mechanical destratification to remove the hydrogen sulfide from the noxious bottom layer of water is warranted.

MORPHOLOGY

The Lower Mystic Lake is located 11 km upstream on the coastal Mystic River, which empties into Boston Harbor from the northwest. It was separated from its holomictic companion, the Upper Mystic Lake, by a masonry dam built at the natural isthmus between the two lakes in 1864.[6]

The morphometric characteristics of the 36.4-ha Lower Mystic Lake are presented in Table 1. The lake consists of two kettle holes; the northern hole is 17 m deep and the southern hole is 24 m deep, while the maximum depth between the holes is 8.8 m. The lake is situated in a 189-km^2 urban drainage basin of mature topographic development with elevations generally less than 60 m above mean sea level (msl). The mean annual temperature of the area is approximately 10.5°C with a mean annual precipitation of approximately 109 cm/year. Potential ice-forming conditions on the lake are achieved on approximately 90 days/year; however, the mean duration of ice cover has not been determined.

The area-capacity vs. depth relationship of the lake is presented in Table 2. The lake contains approximately 3,115,790 m^3 of water, 740,050 m^3 of which is brackish. This is approximately 24% of the lake's total volume 9% in the northern hole and 15% in

* By Richard A. Cassidy, Hydraulic Engineer, U.S. Army Corps of Engineers, Waltham, Mass.

Table 1
LOWER MYSTIC LAKE MORPHOMETRY

Location	42° 25′ N, 71° 08′ W
Elevation	1 m (msl)
Total drainage area	70 km^2
Surface area (V)	36.4 ha
Volume (V)	3,115,790 m^3
Length (1)	975 m
Breadth (b_x)	490 m
Mean breadth (b = A/1)	375 m
Maximum depth (z_m)	24 m
Cryptodepression depth (z_c)	23 m
Mean depth ($\bar{z}$ = V/A)	9 m
Lake form ratio ($z0_m$)	0.4
Shoreline (L)	2,875 m
Shoreline development (Dl − L/2 $\sqrt{\pi A}$)	1.3
Development of volume (Dv = $3\bar{z}/z_m$)	1.1
Mean slope (tanε)	14°
Chemocline zone	9.1—10.7 m
Chemocline volume	199,650 m^3
Mixolimnion volume	2,375,740 m^3
Monimolimnion volume (excluding chemocline)	540,400 m^3

From Hutchinson, C.E., *A Treatise on Limnology*, Vol. 1, John Wiley & Sons, New York, 1957. With permission.

Table 2
LOWER MYSTIC LAKE AREA — CAPACITY TABLE

Depth (m)	Area (m^2)	Capacity (m^3)
0	364,210	3,115,790
3.0	291,500	2,132,221
6.1	228,000	1,326,996
9.1	163,300	740,046
12.2	94,310	340,750
15.2	41,560	136,945
18.3	17,000	46,177
21.3	7,130	9,982
24.1	0	0

the southern hole. Since the surface elevation of the lake is at 1 m above msl, both holes are considered cryptodepressions. The chemocline occurs between the depths of 9.1 and 10.7 m.

According to Chute[7] the lake is of glacial origin, formed approximately 14,000 years ago. It became subject to tidal influence when the post-glacial ocean rose to form the present Boston Harbor. Until 1909, its waters fluctuated semidiurnally with the brackish waters of the Mystic River Estuary. A sanitary survey conducted in 1861 disclosed that saline water occurred at a depth of 5.5 and high sulfur concentrations at approximately 6 m. No saline water was detected in the Upper Mystic Lake at that time. In 1909 the Cradock Dam was constructed 3 km downstream of the lake to prevent the polluted waters of Boston Harbor from reaching the lakes. However, slugs of brackish water continued to be admitted upstream toward the lake whenever the navigation locks were opened. In 1951, the water level of the lake was lowered on weekdays to allow the placement of utility conduits under a portion of the lake. On weekends the lake was refilled with saline water from the Mystic River. In 1961 an MDC sanitary survey revealed that considerable leakage through the dam's navigation locks carried polluted saline water into the lake. Anaerobic conditions were observed below the depth of 4.6 m. In 1966 the Amelia Earhart Dam was built by the MDC near the mouth of the Mystic River transforming it into a fresh-water basin, thus stopping all brackish water exchange in the lake. The lake's saline water, high in sulfates, became trapped.

METHODS AND MATERIALS

Physicochemical measurements obtained during a preliminary field investigations by the NED on December 5, 1973, and June 6, 1974[19] indicated that the Lower Mystic Lake displayed unusual limnological attributes atypical of holomictic lakes. Because there are about 32 meromictic lakes in North America[17] peremptory data from a Cambridge, Mass. consulting firm were obtained to verify the suspected meromixis.

Data were collected by the firm on September 25 to 26, 1974. *In situ* dissolved oxygen and temperature measurements were performed utilizing a Delta® Scientific Meter (model 85) and results were corrected for salinity interference. Water samples for laboratory analyses were collected with a hand piston — pump and calibrated sampling line. Each water sample consisted of 2-ℓ volumes collected from required depths and brought to the firm's laboratory approximately 10 km from the lake. Sample preservations and analyses were performed according to the methods outlined by the American Public Health Association,[2] American Petroleum Institute[1] and Brown et al.[5]

Analyses for specific gravity and chloride, sodium, calcium, magnesium, and sulfate ions in the mixolimnion were performed according to the American Public Health Association[2] methods while tests for these parameters in the monimolimnion were performed according to the American Petroleum Institute[1] methods. Analyses for all alkalinity values were performed electrometrically according to Brown et al.[5] Analyses for pH, specific electrical conductance, total sulfide, ammonia nitrogen, organic nitrogen, chemical oxygen demand (COD), biochemical oxygen demand (BOD), orthophosphate, total phosphorus, nitrate, and potassium at all depths were performed according to the American Public Health Association[2] procedures. Samples for methane analyses were collected in an evacuated 3.8-ℓ bottle fitted with inlet and outlet tubes and analyzed by gas chromatographic methods.

RESULTS AND DISCUSSION

The data indicate that the lake is divided into two major zones. The mixolimnion occurs from the surface to 9.1 m and the monimolimnion from 10.7 m to the lake bottom. It appears that the thermally stratified mixolimnion induces chemical stratifications of some ionic species while some dissolved gases are stratified throughout both zones. A narrow, distinct transitional zone, the chemocline, occurs between 9.1 and 10.7 m and is treated as being limited to a single plane at 9.1 m. Physicochemical data are presented in Tables 3 and 4.

Temperature

According to Ruttner[15] the year-round temperatures near the bottom of some meromictic lakes approach the mean yearly air temperature, staying relatively constant throughout the seasons. The mean annual air temperature at Boston, Mass. is 10.5°C

Table 3
LOWER MYSTIC LAKE PHYSICOCHEMICAL DATA

Northern Hole (September 25 to 26, 1974)

	Depth Below Surface (m)								
Parameters	0.3	3.0	6.1	7.6	9.1	10.7	12.2	14.3	14.6
Temperature (°C)	16.0	16.0	16.0	14.5	13.1	12.0	11.0	10.0	9.5
Dissolved oxygen (mg/ℓ)	9.0	8.8	7.6	0.1	0.0	0.0	0.0	0.0	0.0
pH	7.4	7.4	7.3	7.0	7.3	7.1	7.1	6.8	6.7
Specific gravity	1.006	1.007	1.006	1.003	1.008	1.016	1.025	1.023	1.020
Specific electric conductance (μmhos/cm)	520	726	616	870	1,210	11,220	12,760	13,420	15,620
Chloride (mg/ℓ)	120	120	120	220	400	9,000	9,980	10.050	10.200
Potassium (mg/ℓ)	8.0	8.1	8.0	12.3	18.2	230	242	246	246
Sodium (mg/ℓ)	30	50	50	100	220	5,000	5,900	6,000	6,000
Calcium (mg/ℓ)	18.9	19.2	19.3	20.1	20.1	200	250	248	250
Magnesium (mg/ℓ)	9.6	10.0	9.9	14.0	25.9	645	680	720	720
Total sulfide (mg/ℓ)	0.2	0.2	0.3	0.4	0.5	100	150	185	170
Sulfate (mg/ℓ)	33	30	38	42	55	57	350	150	159
Alkalinity to pH 4.2, $CaCO_3$ (mg/ℓ)	97	95	100	142	190	2,040	2,500	2,900	3,000
Ammonia nitrogen (mg/ℓ)	1.0	1.2	0.5	0.9	0.5	1.0	1.8	2.0	3.0
Nitrate nitrogen (mg/ℓ)	<0.2	<0.2	0.2	0.2	<0.2	<0.2	<0.2	<0.2	<0.2
Organic nitrogen (mg/ℓ)	1.4	1.0	2.2	0.8	5.5	13.4	8.6	9.0	12.7
Total phosphorus (mg/ℓ)	0.08	0.08	0.06	0.08	0.10	13.8	17.0	20.0	20.5
Orthophosphate P(mg/ℓ)	<0.05	0.05	<0.05	0.06	0.06	10.5	14.3	20.0	19.9
Methane (mg/ℓ)	0.27	0.05	0.11	1.0	1.5	3.8	5.7	5.6	4.2
BOD (mg/ℓ)	2.0	1.6	1.9	1.2	1.0	—	200	240	260
COD (mg/ℓ)	35	32	40	45	42	275	286	300	320

Table 4
LOWER MYSTIC LAKE PHYSICOCHEMICAL DATA

Southern Hole (September 25 to 26, 1974)

Depth Below Surface (m)

Parameters	0.3	3.0	6.1	7.6	9.1	10.7	12.2	15.2	18.3	21.3	22.9
Temperature (°C)	16.9	16.5	16.5	15.2	14.0	12.0	10.1	9.5	8.9	8.0	8.0
Dissolved oxygen (mg/ℓ)	8.9	8.5	4.2	0.0	0.0	0.0	0.0	0.0	0.0	0.0	0.0
pH	7.2	7.2	7.1	6.9	7.0	6.6	6.6	6.7	6.7	6.7	6.8
Specific gravity	1.010	1.006	1.017	1.001	1.010	1.012	1.014	1.013	1.027	1.019	1.009
Specific electric conductance (μmhos/cm)	470	445	640	730	1,120	13,090	12,650	13,860	13,530	13,860	14,740
Chloride (mg/ℓ)	116	118	139	216	450	10,150	9,900	10,650	10,500	10,580	10,380
Potassium (mg/ℓ)	4.0	4.0	8.0	10.0	20.1	252	250	256	253	258	258
Sodium (mg/ℓ)	48	50	52	102	220	6,030	6,030	6,060	6,030	6,000	6,080
Calcium (mg/ℓ)	15.1	15.8	20.0	19.8	20.3	250	248	251	255	253	255
Magnesium (mg/ℓ)	6.0	6.0	6.0	16.1	29.1	720	718	720	725	730	725
Total sulfide (mg/ℓ)	0.06	0.15	0.85	0.96	2.6	128	162	160	159	145	155
Sulfate (mg/ℓ)	27	25	48	44	53	481	500	427	452	380	360
Alkalinity to pH 4.2, $CaCO_3$ (mg/ℓ)	95	95	94	149	187	2,230	2,420	2,600	2,630	2,600	2,760
Ammonia nitrogen (mg/ℓ)	1.3	1.5	1.4	0.6	10	27	37	29	30	25	28
Nitrate nitrogen (mg/ℓ)	<0.2	<0.2	<0.2	<0.2	<0.2	<0.2	<0.2	<0.2	<0.2	<0.2	<0.2
Organic nitrogen (mg/ℓ)	1.1	1.1	1.1	4.1	1.0	1.2	1.2	5.5	4.0	1.0	7.6
Total phosphorus (mg/ℓ)	0.05	0.07	0.06	0.08	0.08	15.0	16.0	17.5	17.8	17.8	17.8
Orthophosphate (P mg/ℓ)	<0.05	0.06	0.05	0.07	0.06	14.3	14.2	15.0	15.3	16.9	17.0
Methane, (mg/ℓ)	<0.01	0.05	0.09	0.78	1.4	—	7.7	12.7	13.5	13.2	4.6
BOD (mg/ℓ)	1.8	2.1	1.4	1.2	1.5	210	160	240	200	180	250
COD (mg/ℓ)	30	30	27	35	25	280	210	290	300	300	350

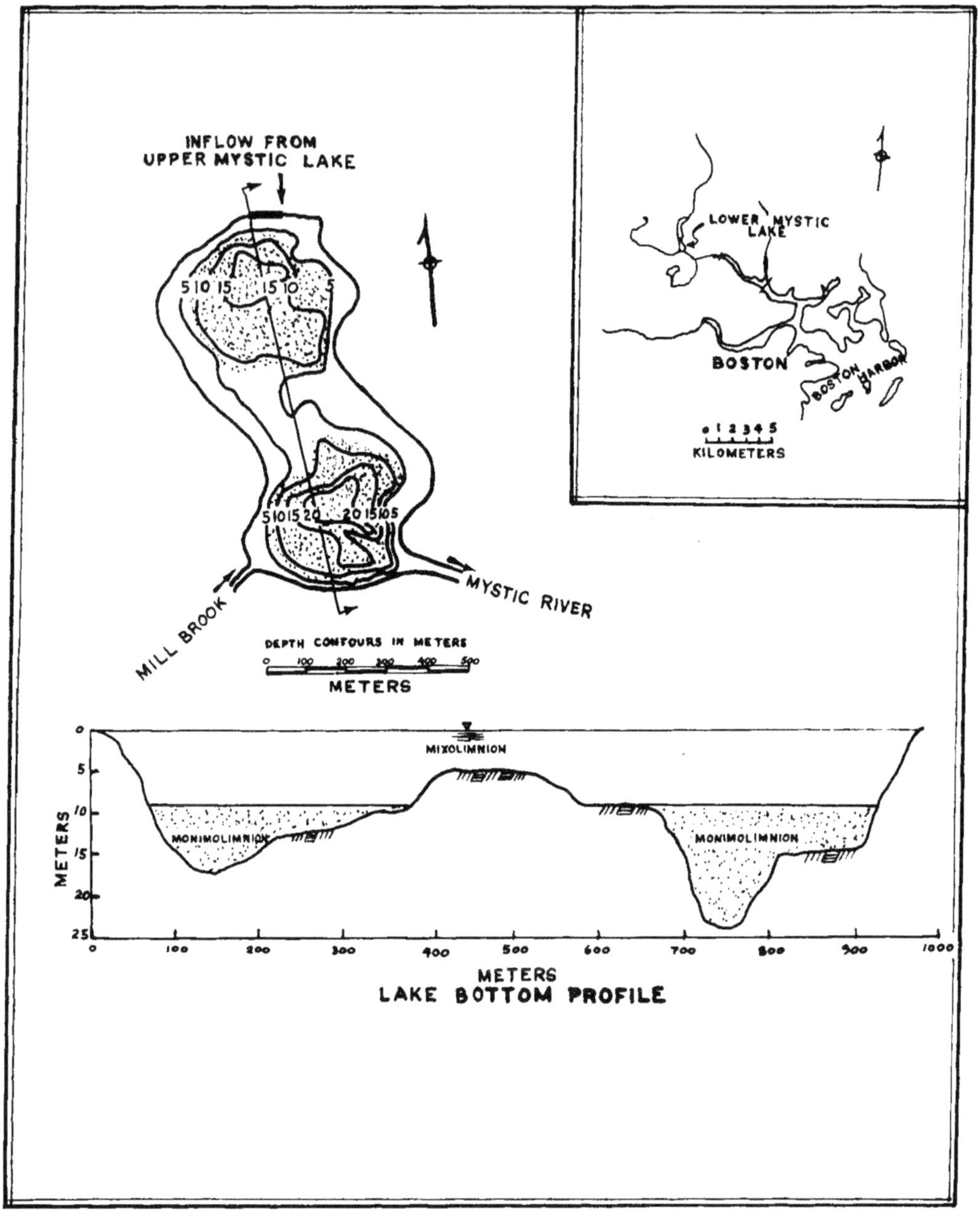

FIGURE 1. Morphometric characteristics of Lower Mystic Lake.

while the mean temperature of the monimolimnion is approximately 10°C. Field data collected on December 5, 1973, and June 6, 1974, indicate little seasonal variation. Figure 1 presents the temperature profile in the southern hole on September 25, 1974. The major source of heat in the monimolimnion appears to be conduction from subsoil geothermal origins and conduction from the mixolimnion, especially in the summer. The September 1974 data indicate that the upper portions of each monimolimnion is stratified. According to Hutchinson[11] additional heat is transported into the lower zone by warm dissolved and suspended material which follow descending density currents along the slope of the lake bottom. The major source of heat loss from the monimolimnion is probably through conduction to the soil and to the mixolimnion in the winter.

Specific Electrical Conductance

Specific electrical conductance was used to locate the chemocline.[9] It occurs between 9.1 and 10.7 m at each location. The lower layer of the stratified mixolimnion had a 2-fold increase, but the chemocline increased nearly 20-fold over upper level mixolimnion values.

Specific Gravity

Specific gravity was measured to determine the density characteristics of the brackish water trapped in the monimolimnion.

Total Sulfide

The hydrogen sulfide concentration is of prime interest in this study because it has been the source of occasional odor problems for the urban area near the lake. The total sulfide measurements include the dissolved hydrogen sulfide gas, the HS ions, and acidsoluble metallic sulfides. Dissolved H_2S gas and HS ions were not identified apart from acid-soluble metallic sulfides because testing must begin within 3 min of sampling. In meromictic lakes characterized by considerable amount of saline water of marine or geologic origin, a large amount of hydrogen sulfide gas can be produced in the monimolimnion from the high sulfate concentrations.[11] Lesser amounts are produced by the organic decomposition of precipitates from the upper layer.

The total sulfide concentration in the monimolimnion of each hole is approximately 152 mg ℓ^{-1} totaling nearly 113 × 103 kg. The distribution of total sulfide in the monimolimnion is partially due to the internal circulation of this lower zone.[9] However, the monimolimnion may be considered a reflux system diffusing some H_2S gas through the chemocline to the overlying mixolimnion.[8] Total sulfide concentrations were stratified in the mixolimnion. Figure 1 illustrates the increasing concentration with mixolimnion depth and the extremely high monimolimnion concentration in the southern hole.

Hydrogen Ions

According to Anderson[3] little vertical variation of the pH concentration is indicative of the highly buffering nature of the monimolimnion. Hydrogen sulfide, a weak acid, is approximately 80% unionized at pH levels near 7 and is therefore a potential source of serious odor problems. Most of the 113 × 103 kg of hydrogen sulfide in the monimolimnia are in an unionized state.

Methane

According to Cassidy[6] the watershed upstream of the lake has a long history of domestic and industrial pollution dating back to the early 1870s. The main tributary to the Mystic Lakes, the Aberjona River, received discharges from gelatin, dyeing, and tanning industries during the late 1800s. Even today material of domestic origin continues to precipitate into the monimolimnion. Anaerobic decomposition is the major source of methane production and a minor source of hydrogen sulfide production. Methane appears to diffuse through the chemocline to the surface. The mixolimnion of each hole has concentrations of less than 1 mg ℓ^{-1}, while the northern monimolimnion mean is 4.8 mg ℓ^{-1}, and the southern monimolimnion is 10.3 mg ℓ^{-1}. The methane concentrations nearest the bottom in both holes are less than those at other levels within the monimolimnion.

Dissolved Oxygen

According to Hutchinson[11] the monimolimnia of meromictic lakes have little or no dissolved oxygen present. Numerous meromictic lakes do not exhibit any dissolved oxygen in the lower layer.[12-18] No dissolved oxygen occurred in the lower mixolimnia of the Lower Mystic Lake and none in the monimolimnion of each hole.

Sulfate Ions

The sulfate concentrations in the trapped, anaerobic brackish water of the lake is the major factor contributing to hydrogen sulfide production. Sulfate ions are the second or third most abundant anions in fresh water; however, brackish water contains considerably more sulfate ions than fresh water. According to Ramm and Bella[14], hydrogen sulfide is anaerobically produced by bacteria when the sulfate ion concentration reaches a kinetic threshold of approximately 300 mg ℓ^{-1}. Theoretically 3 mg of sulfate will produce 1 mg of sulfide. The lesser concentrations of both sulfates and methane in the northern hole compared to the southern hole appear to indicate a decrease in anaerobic biochemical and chemical activity in the northern hole. It was not determined, however, whether the concentration differences between the monimolimnia of each hole is seasonal.

Alkalinity

The alkalinity in the lake is a measure of bicarbonate ions because at pH values less than 8.3 the water is below the bicarbonate to carbonic acid equivalence point. According to Weimer and Lee[18] the high alkalinity in the monimolimnion of merocictic lakes, acting as a buffer, is partially attributed to the release of CO_2 from the anaerobic mud. It is observed that 1.9 mol of CO_2 is formed by the oxidation of organic carbon for each mole of hydrogen sulfide gas produced. When the production of CO_2 is accompanied by the simultaneous release of ammonia, removing free hydrogen ions from the water, the alkalinity is increased. The mean value in the northern monimolimnion was 2610 mg ℓ^{-1} as $CaCO_3$ and 2540 mg ℓ^{-1} in the southern monimolimnion.

Ammonia Nitrogen

Ammonia nitrogen is a measure of the ammonia gas, ammonium, and ammonium hydroxide forms of nitrogen. Ammonia gas produced by the decomposition of organic compounds in lake sediments reacts with water to form ammonium hydroxide, liberating heat. This reaction tends to increase the pH of the water; however, it is partially buffered because other ammonium ions dissociate from their salts of chlorides and sulfates and combine with hydroxyl ions thus freeing hydrogen ions.[13] Like the total sulfide and methane values, the ammonia-nitrogen appears to diffuse through the chemocline to the mixolimnion. The mean ammonia nitrogen concentration of 2.0 mg ℓ^{-1} in the northern monimolimnion was lower than the mean southern monimolimnion concentrations of 29 mg ℓ^{-1}. This difference, like the methane and sulfate concentrations, indicates that ammonification in the northern monimolimnion is less than the southern monimolimnion.

Organic Nitrogen

The complex proteins in the organic matter on the lake bottom anaerobically decompose through various amino acid forms degrading to a completely mineralized state. The anaerobic decomposition of suspended organic matter in the water column is different between the mixolimnion and monimolimnion of the northern and southern hole.

Other Chemical Parameters

BOD, COD, organic nitrogen, orthophosphate, and total phosphorus concentrations indicate the poor quality of the lake, especially the monimolimnion. Future consideration for the elimination of the hydrogen sulfide odor problem must involve the evaluation of the environmental effects of these and other chemical species.

SUMMARY AND CONCLUSIONS

Two cryptodepressions containing brackish water of marine origin have caused meromixis in the Lower Mystic Lake. The chemocline occurs between 9.1 and 10.7 m. Meromixis was accentuated when the Boston Metropolitan District Commission Dam transformed the tidal area into a fresh-water basin, thus stopping brackish water exchange in the lake. The high sulfate anion concentration in the trapped brackish water became the source of high (152 mg ℓ^{-1}) hydrogen sulfide production causing occasional odor problems near the urban area along the lake shore.

REFERENCES

1. API *API Recommended Practices for Analysis of Oil-Field Waters,* 2nd ed., American Petroleum Institute, Dallas, 1968.
2. APHA, *Standard Methods for the Examination of Water and Wastewater,* 13th ed., American Public Health Association, Washington, D.C., 1971.
3. **Anderson, G. C.,** Seasonal characteristics of two saline lakes in Washington, *Limnol. Oceanogr.,* 3, 51, 1958.
4. **Anderson, G. C.,** Some limnological features of a shallow saline meromictic lake, *Limnol. Oceanogr.,* 3, 259, 1958.
5. **Brown, E., Skougstad, M. W., and Fishman, M. J.,** Methods for Collection and Analysis of Water Samples for Dissolved Minerals and Gases: Techniques of Water-Resources Investigations of the U.S. Geological Survey, Vol. 5, Superintendent of Documents, U.S. Government Printing Office, Washington, D.C., 1970.
6. **Cassidy, R. A.,** The Distribution of Chromogenic and Other Heterotrophic Bacteria in the Upper Mystic Lake Including a Brief History of the Lake, M.S. thesis, Tufts University, Medford, Mass., 1971.
7. **Chute, N. E.,** Glacial Geology of the Mystic Lakes — Fresh Pond Area, Massachusetts, U.S. Geol. Surv. Bull. 1061-F, Superintendent of Documents, U.S. Government Printing Office, Washington, D.C., 1959.
8. **Deevey, E. S., Nakai, N., and Stuiver, M.,** Fractionation of sulfur and carbon isotopes in a meromictic lake, *Science,* 139, 407, 1963.
9. **Edmondson, W. T. and Anderson, G. C.,** Some features of saline lakes in central Washington, *Limnol. Oceanogr.,* 10(Suppl.), R87, 1965.
10. **Gangstad, E. O.,** Aquatic plant control program, *Hyacinth Control J.,* 9(1), 46, 1971.
11. **Hutchinson, C. E.,** *A Treatise on Limnology,* Vol. 1, John Wiley & Sons, New York, 1957.
12. **Likens, G. E. and Hasler, A. D.,** Movement of radioisotopes in a chemically stratified lake, *Science,* 131, 1676, 1960.
13. **McKee, J. E. and Wolf, H. F.,** *Water Quality Criteria,* 2nd ed., Publ. No. 3-A, Resources Agency of California, San Francisco, 1963.
14. **Ramm, A. E. and Bella, D. A.,** Sulfide production in anaerobic microcosms, *Limnol. Oceanogr.,* 19 L110, 1974.
15. **Ruttner, F.,** *Fundamentals of Limnology,* 3rd ed., University of Toronto Press, Canada, 1966.
16. **Timms, B. V.,** A meromictic lake in Australia, *Limnol. Oceanogr.,* 17, 917, 1972.
17. **Walker, K. F.,** The stability of meromictic lakes in central Washington, *Limnol. Oceanogr.,* 19, 209, 1974.
18. **Weimer, W. C. and Lee, G. F.,** Some considerations of the chemical limnology of meromictic Lake Mary, *Limnol. Oceanogr.,* 18, 414, 1973.
19. U.S. Army Corps of Engineers, New England Division, unpublished data, December 5, 1973 and June 6, 1974.

Chapter 11

THE PROBLEM OF AMOEBIC MENINGOENCEPHALITIS IN WATER RESOURCE PLANNING*

INTRODUCTION

Any body of water used for recreational purposes, be it a swimming pool, pond, or lake, has the potential for being a source of contagion, not only of bacteria, viruses, and various enteric protozoan parasites, but also by the soil amoeba *Naegleria graberi* and related species. The Corps of Engineers has developed criteria and standards over the years to minimize public health hazards resulting from recreational developments and indeed all phases of water resource developments by compliance with all applicable health and building codes.

THE DISEASE

The first doumented case of primary amoebic meningeoencephalitis (PAM) caused by *N. gruberi* was reported from Australia in 1965.[3] Since Fowler and Carter's paper there has been recognition of an increasing number of cases of the very virulent disease, PAM. Duma,[6] in an excellent review paper, discusses this disease and summarizes the state of knowledge regarding it. Other studies[7-9] discuss the occurrence of the disease with attention to epidemiology, clinical problems associated with amoebeosis caused by *Naegleria* and related genera of hartmanellid amoebae. PAM is a disease confined to the central nervous system. According to Duma[6] there are three major clinical varieties:

1. Acute, in which case the onset is sudden, the course fulminant and rapidly fatal (1 to 4 days), and the responsible organism an amoeboflagellate belonging to the genus *Naegleria*
2. Chronic, in which case the onset is relatively insidious, the course prolonged (1 to 2 months), and the responsible organism probably the *Hartmanella* or *Acanthamoeba*
3. Benign aseptic, in which case the disease picture simulates viral aseptic meningitis, but the responsible organism is again probably *Hartmanella* or *Acanthamoeba* (although *Naegleria* has not been excluded)

Duma (op. cit.) notes that most cases of acute PAM occur during the warmest times of year probably due to increased swimming and water contact activity and elevated temperature which are conducive to amoeba proliferation. Natural and man-made lakes as well as warm or heated chlorinated swimming pools (indoor or outdoor) or warm mineral pools have been sources of infection. One of the largest epidemics of PAM occurred in Chesterfield County, Va. in 1969. Three lakes in the James River Basin are believed to be responsible for 16 deaths.

PAM has come up as a factor in both Corps of Engineers planning facility management.[1,4] Ahearn, in a study sponsored by the Corps on Algal Amebic Bloom in Lake Sidney Lanier, discusses a Corps of Engineers impoundment. He also mentions an identical bloom occurring in nearby Lake Jackson in 1971. During the preliminary

* By Dr. Philip Phillips, North Atlantic Division, U.S. Army Corps of Engineers, New York.

stage of Ahearn's investigation, a fatal case of PAM was reported from a Marine on Lake Jackson. No cases of PAM have, however, been reported for Lake Lanier. A fish kill was also reported temporally associated with the *Anabena* — ameba bloom in Lake Lanier. Whether the fish kill resulted from amoebic related pathogenesis is conjectural. Ahearn[1] reports that tests demonstrated no toxic effects of amoebae on the common guppy *(Lebistes reticulatus).* The amoeba bloom reported by Ahearn occurred in conjunction with a bloom of the blue-green algae *Anabaena circiniais* which is known to be toxic to mammals and birds. Ahearn also noted increased densities of the alga *Gonium* sp., rotifere *(Notholia),* copepodes *(Daplinia),* and numerous species of protozoans of the genus *Paromecian.* Unidentified biflagellated algae were also reported.

Ahearn's[1] account of the amoeba blood is as follows: "As the densities of *A. circinalis* decreased, the surface of the water became coated with a white scum composed mainly of amebae . . ., amebic precysts . . . and amebic cysts . . . The amebae occurred profusely among the *A. circinalis* and were observed to ingest single trichomes on filament. The heterocysts of *A. circinalis* did not appear to be digested and were accumulated within the food vacuole of the ameba. The complex of amebic cysts and heterocysts were erroneously considered to be a brown alga in a preliminary report. The amebae were not observed to ingest cells of *Diastridium* sp. The algal bloom was most intense in the vicinity of high concentrations of phosphate, nitrates, iron, and organic matter in areas which receive waste. The high nutrient loading of the lake as noted by Ahearn will result in eutrophication of greater lake areas and interfere with recreational usage. It also appears circumstantially that should nutrient reduction take place there will be a reduction in algal and amoebic blooms as well as a decrease in the rate of eutrophication.

PATHOGENICITY

There are major questions to be addressed brought up by Ahearn's study. What are the toxicities of the algae? What is the actual and potential pathogenicity of the amebae and the actual relationship of the algal blooms to the proliferation of amoebae? Can amoebic blooms be precluded by effective control of algal proliferation and nutrient loading?

In a study of amoebic symbionts of anemones[13] I found that when the anemone *Bunodosoma cavernata* is maintained in the laboratory, if the temperature is consistantly elevated and the aquarium water is not periodically flushed, large populations of protozoans develop on the anemones and ultimately within the water in the aquaria. The hartmonellid amoeba *Valkampfia* in addition to being found in normal and necrotic anemone tissues, was found to be a dominant organism in fouled aquarium water. Apparently, high temperatures and medium enrichment due to putrefication in a closed aquarium system are conductive to a protozoan bloom dominated by *Vahlkampfia.* Three morphological types of *Vahlkampfia* were found. The following description of *Vahlkampfia* is given:

> Most amoebae were 7 to 10 μm in length and formed only one broad, fan-shaped pseudopod which, moreover, consisted mainly of a large hyaline area. Formation of a new pseudopod was preceded by flooding the old pseudopod with granular endoplasm and the formation of a new hyaline bulge. Rate of pseudopod formation was hastened by the heat of a substage illuminator. The second observed trophozoite stage was considerably larger (25 to 30 μm) and had numerous pseudopodia. Their morphology approached that of the former type after being on a microscope slide (above an illuminator) for several minutes. These larger amoebae underwent division and "reverted" to the smaller, monopseudopodial type. On two occasions a nucleus (2 to 3 μm in diameter) with a large endosome was observed in the large amoebic variant. Large numbers of *Vahlkampfia* cysts were found in decaying and moribund anemones as well as in putrefying

media. After being on a microscope slide for about 5 min above a substage illuminator, these cysts released one amoeba each. Cysts were rounded, considerably flattened, had wrinkled surfaces, and were between 5 to 9 μm in diameter. *Vahlkampfia* invades anemone tissue and will emerge from tissue fragments when heated by substage illumination for a few minutes. The amoebae can be demonstrated in both normal and necrotic tissues, the greatest concentration of amoebae occurring in necrotic tissues with a large bacterial population. The amoebae were observed to ingest bacteria, indicating that greater degree of association with necrotic tissue may be due to greater availability of food organisms in necrotic as opposed to healthy tissue.

There are certain very distinct parallels between Ahearn's observation on the Naeglerild bloom in Lake Lanier and my own work using an aquarium system. Circumstantially, amoebic blooms were noted in conjunction with (1) elevated temperature, (2)nutrient enrichment, and (3) increased densities of other protozon and infusoriform organisms.

Kadlec et al.[12] found that a reservoir of virulent *N. fowleri* was contained in a cracked wall of a swimming pool which was the source of several outbreaks of PAM between 1961 an 1965. According to their investigation, a layer of free air about 10 cm thick developed between the concrete wall of the swimming pool and the tiled interior surface wherein the temperature fluctuated to between 27 and 30°C. The water in this space was relatively isolated from the influence of disinfectants and extensive population of *Naegleria* developed on the inner walls of the pocket in the wall. Kadlec et al.[12] theorizes that periodic fluctuation of the water level up on occasion flushed pathogenic amoebae from the pocket behind the tile layer resulting in infection of swimmers. They also state that reconstruction of the wall should remove the risk of the occurrence of PAM among future pool users.

With the contest of swimming pool safety it would appear mandatory the interior surfaces of the pool be maintained so as to preclude the development of stagnant water pockets which can serve as reservoirs for pathogens.

ETIOLOGY

Jamieson and Anderson[11] describe a method for the isolation of *Naegleria*. Their findings in Table 1, which show the sources of *Naegleria* spp. isolates, give testimony to the ubiquitous nature of these amoebae. Isolates were tested for pathogenicity by intranasal inoculation of mice.

A very basic problem in dealing with PAM is the assessment of its occurrence under various conditions, in various locales. As previously noted, the question of this type of amoebiasis has come up in Corps of Engineers recreational facility planning. Gathright Lake, Va., now almost complete, had become a locus of some controversy over PAM because of its incidence in the James River Basin. In a review of the Draft Environmental Statement on Gathright Lake, the "Citizens Against Pollution, Inc." of Norfolk, Va. (Corps of Engineers op. cit.) expressed considerable concern over possible adverse impacts on human health and safety. The following paragraphs are from comments supplied to the Corps of Engineers and published in the Final Environmental Impact Statement (FEIS) of January 19, 1973:

No consideration is given to direct adverse effects of the dam and lake on human health and safety. There will be an increased incidence of drownings due to steep banks leading into deep water. People unaccustomed to lakes and boats will fall from their boats and drown. People will fall from the dam and be injured or killed. The increased automobile traffic in the area will lead to accidents and injuries. The impoundment itself will be a threat to those living downstream. The dam could be breached by a variety of influences ranging from meteorite impact to human error.

Since Culbertson's pioneering article in 1959[8] and medical literature has accumulated on the subject of PAM,[9] it is now virtually certain that

Table 1
SOURCE OF 130 ISOLATES OF NAEGLERIA SPECIES

Water source	Number of isolates	Number of samples
Swimming pools	50	100
Rain water tanks	10	50
Piped water supplies (unfiltered but chlorinated)	20	40
Surface fresh water (rivers, lakes, and dams)	46	100
Thermal spring	4	10
Total	130	300

1. The disease is contracted by swimming in river water which has been warmed — especially though not always — warmed in contact with the soil.
2. Once contracted, the disease is uniformly fatal (one questionable case report to the contrary).
3. There is no effective treatment for the disease.
4. The causative organism, a free-living soil amoeba, enters through the noses of those swimming in contaminated water and makes its way through the nasal mucosa, along olfactory nerve fibers, through the cribriform plate into the subarachnoid space at the base of the brain, hence, to multiply and cause a fulminating meningoencephalitis which is rapidly fatal.
5. An incubation period of 7 days or less with some mild upper respiratory symptoms exposure.

Onset of the illness itself is dramatically abrupt with severe headache, fever, anorexia, nausea, vomiting, and malaise. The sense of smell and taste (olfactory not lingual) is often impaired early in the course of the acute illness. Usually the patient is hospitalized on the 1st to 3rd day of acute illness, and then dies 24 to 48 hours later.

As of July 1971, 54 cases (53 deaths) of PAM had been reported in the world medical literature. Sixteen of those cases, 30%, occurred in the Commonwealth of Virginia. One occurred in Franklin, Va. The other 15 cases were traced to 3 tepid fresh water swimming lakes in Chesterfield County, Virginia. Nine of those cases were traced to a lake 2 mi from the banks of the James River. Five were traced to a lake situated on the banks of the James. The remaining cases came from a lake about 6 mi from the banks of the James. All three lakes were inside a circle of 5 mi radius.

Any complete 102 statement on the Gathright Project must deal effectively with the difficult questions raised here. Further study of this distressing problem is desperately needed and must be done before any further riverine impoundments are built for recreational purposes. To continue blithely spending public funds to create such impoundments for recreational purposes without first making an exhaustive study of the incidence of PAM contracted from such impoundments already in operation would be unthinkable. In this connection I wish to emphasize that knowledge of this disease is very new and has not penetrated the ranks of private physicians who no doubt have been caring for such cases. Therefore, the true incidence of this disease is completely unknown since physicians are unfamiliar with it and therefore cannot make the diagnosis. To date, retrospective study of autopsy materials has been spotty and woefully incomplete. In short, the true incidence of PAM contracted from swimming in publicly financed impoundments in the U.S. remains an unknown though a highly suspicious quantity.

This letter sparked a considerable effort by the Corps of Engineers to address this area of contention. This resulted into the incorporation of a section on "environmental health" into the FEIS.[4]

Environmental Health

Another aspect of project management deals with insect vector control, principally mosquitoes, which could carry viral encephalitis. For the most part, problems develop with leftover borrow areas, marshes within the conservation pool elevation, and drawdown which could leave temporary ponds. None of the above should pose a problem at the Gathright project. No borrow pits will remain nor any marsh area be adjacent to the maximum or minimum pool. The natural bottom contours of the reservoir will prodrainage as the full conservation pool recedes.

The control of terrestrial arthropods and rodents is largely a matter of thorough maintenance of recreation areas including garbage disposal, pumping and treatment of chemical toilets, and control of nuisance vegetation.

Although lakes, ponds, and other waterways can and do serve as vectors of many significant diseases, there is no reason to assume that the proposed Gathright Lake will be any different or have a greater probability of carrying waterborne disease than any other comparable body of water in the state of Virginia. Assuming proper sewage disposal procedures are practiced in the vicinity of the project, pathogenic organisms associated with septic discharges should be minimized.

It will be the responsibility of the reservoir manager to ensure proper maintenance of recreation areas and facilities. He will also request and coordinate routine field surveys and epidemiological surveillance with the Virginia Department of Health and/or U.S. Public Health Service.

Researchers have shown some concern on the incidence of waterborne PAM in a localized area of Virginia. This disease is now known on a global basis from 57 cases, 17 of which (30%) have occurred in Chesterfield County, Va. with 7 more throughout the U.S. From available medical literature, the disease in those cases diagnosed to date is invariably fatal. A recent article by Carter[3] reviews the current state of knowledge regarding the disease. It has been recognized as a public health concern only in the last decade, and, although a rather extensive medical literature has developed regarding this protozoan infection, virtually nothing is known regarding the ecology of the almost ubiquitous naegleriid and hartmanellid amoebae involved in the epidemiology of the disease. The infection is lethal; death ensues within 3 to 5 days from the onset of symptoms. The symptoms initially are those of a respiratory infection and progress very rapidly to coma, severe neurological disturbance, and death due to cardiopulmonary failure.

Since extremely little is known about the ecology of the infective organisms, it is impossible to predict an outbreak of the disease. Little can be stated other than the disease occurs during warmer months of the year and is linked to contact with tepid water. Carter notes that several cases have been traced to heated swimming pools and at least one in Australia to chlorinated tap water.

With respect to the Gathright Lake it would be simplistic to ignore the possibility of an outbreak of this disease; however, the probability of an outbreak in Gathright is no greater than anywhere else. Of the three lakes in Chesterfield County where the disease has been traced (Moores Lake, Lake Chester, and Lake Manchester) all are relatively small, shallow natural impoundments used for recreational purposes. Simply on the basis of size and anticipated low water temperatures in Gathright there is little in common with the three lakes in Chesterfield County.

In only one instance has the Corps of Engineers had first-hand experience with amoebae, and this was on Lake Lanier in Georgia. When studying a blue-green algal bloom, it was determined that a species of amoeba, not of the same group that causes meningoencephalitis, consumes that algae and thereby serves as a biological control. During that same summer (1969), on nearby Lake Jackson, a youngster contracted fatal amoebic meningoencephalitis; however, the causative organism was different from that which preyed upon the algae in Lake Lanier. To date, there have been no cases of the disease traced directly to a Corps' impoundment.

Certainly the incidence of the amoebic infection in Virginia demands public health vigilance, especially with respect to any fresh-water area used for water contact activities. The present state-of-the-art can allow little more than this. In view of the concern expressed on the possibility of PAM being contracted at the Gathright Lake Project, the matter was brought to the attention of the Office of the Surgeon General, Department of the Army. The letter reply dated December 18, 1972, is appended in the Correspondence Section. However, two paragraphs are quoted as follows:

> Species of the genus *Naegleria* are ubiquitous; only a few species (perhaps only *N. fowleri*) have been demonstrated to cause disease. Documented epidemics have been notable for their localization, implying that pathologic species are either (1) limited in distribution, (2) pathogenic only under specific conditions, or (3) present in sufficient numbers only in specific circumstances. Most cases have occurred in warm lake water to date. Prolonged hot spells or bacterial contamination may be contributing factors, but not enough is known to elucidate the ecology of the organism or the epidemiology of the disease.
>
> In the light of the present understanding of the disease, a risk may indeed exist that primary amebic meningoencephalitis would occur in the water to be impounded by the dam. However, there is no reason to believe that the risk is any greater for the impoundment than for any other natural or man-made body of fresh water in Virginia, the U.S., or, for that matter, the world. Nor can risks be predetermined by the presence of *Naegleria* since they are so common and disease incidence is low, even allowing for misdiagnosis.

The following appear to be the most important measures to be taken to prevent an outbreak of PAM.

Swimming pools:

1. Adequate cleaning of all surfaces, filters, and water lines
2. Elimination of cracks and crevices
3. Adequate flushing and cycling of pool water
4. Proper levels of chlorination
5. Careful regulation of thermal regimen so as to preclude over-heating or maintenance of a tepid condition; development of more stringent standards as to maximum acceptable limits
6. Prevention of contact of soil with pool water by limiting access of swimmers to earthen areas and site design which includes concrete or tile walkways and paved areas adjacent to pool
7. Require disinfective foot rinse and shower prior to immersion in pool

With respect to impoundment construction and management:

1. Elimination of stagnant backwaters
2. Control of thermal regime
3. Adequate oxygenation of water column
4. Elimination of sources of pollution and nutrient enrichment
5. Design features to minimize cul-de-sac and stagnant shallows.

Since the technology of identifying pathogenic *Naegleria* is still quite complex and exotic with respect to "normal" laboratory analysis of water, development of an unequivocal test that can be routinely performed by technicians and possibly by facilities management personnel would go far in controlling this public health menance.

SUMMARY AND CONCLUSIONS

Primary amoebic meningoencephalitis (PAM) is an extremely virulent disease of the central nervous system. The actual organism is a soil amoeba called *Naegleria graberi*. It is associated with fresh bodies of water including lakes, ponds, and swimming pools. The *Naegleria* is associated with amoebic blooms which occur when there is an elevation of temperature, a nutrient enrichment, and an increase of densities of other protozoan and infusoriform organisms. There has been a reported increase in the number of cases of PAM, and because of this, the U.S. Army Corps of Engineers has a strong interest in minimizing this health hazard.

PAM is a fatal disease. The individual comes into contact with the organism while swimming in tepid water. The organism is first associated with the nasal passages and eventually to the subarachnoid space at the base of the brain causing a case of meningoencephalitis. The initial symptoms are those of a respiratory infection. Death is due to cardiopulmonary failure.

The state of Virginia has had 30% of all reported cases. This has been brought to the attention of the Corps with their Gathright Lake project in Virginia. Because of this the Corps has incorporated an "environmental health" section into the FEIS.

The evaluation of Gathright Lake concerning PAM shows little concern for the disease. Although the disease has been found in three lakes in Virginia, those lakes have little in common with Gathright. In the conclusion of the FEIS it was said that the presence of *Naegleria* is everywhere and not all strains are virulent. The virulent strains are "(1) limited in distribution, (2) pathogenic only under specific conditions, or (3) present in sufficient numbers only in specific circumstances." There are no conclusive reasons to beware that the presence of *Naegleria* at Gathright Lake causes a greater risk than at other bodies of water.

Since the FEIS was released, there has been a great deal of new information concerning PAM. There are now several measures used in the prevention of an outbreak. There are regulations for swimming as well as to construction on fresh water.

There is a great deal more to learn about *Naegleria* and the disease of PAM. Although the Corps has never had a case of PAM on one of its projects, they have taken the responsibility by taking measures to prevent an outbreak of it.

REFERENCES

1. **Ahearn, D. G.**, Investigation and Report of Brown Dinoflagellate Occurring in Lake Sidney Lanier, Georgia and Terminal Report, Army Corps of Engineers, Mobile District, Alabama, 1971.
2. **Ahearn, D. G., Hal, E. T., Jr., and Rheinhardt, D. J.**, Hypnodinium-like algal blooms in Georgia lakes, *BioScience*, 21(3), 115, 1971.
3. **Carter, R. F.**, Primary amoebic meningo-encephalitis, *Trans. R. Soc. Trop. Med. Hyg.*, 66, 193, 1972.
4. Corps of Engineers Final Environmental Impact Statement, Gathright Lake, Virginia, Army Engineer Division Norfolk, Fort Norfolk, Va., 1973.
5. **Das, S. R.**, Isolation of *Naegleria* and *Hartmanella* amoebae from benchkenham (London) soils and their pathogenicity in mice, *Trans. Soc. Trop. Med. Hyg.*, 66, 663, 1972.
6. **Duma, R. J.**, Primary amoebic meningoencephalitis, *CRC Crit. Rev. Clin. Lab. Sci.*, 3, 163, 1972.
7. **Duma, R. J., Ferrell, H. W., Nelson, E. C., and Jones, M. M.** Primary amebic meningoencephalitis, *N. Engl. J. Med.*, 281, 1315, 1969.
8. **Duma, R. J., Rosenblum, W. I., McGehee, R. F., Jones, M. M., and Nelson, E. C.**, Primary amoebic meningoencephalitis caused by naegleria: two new cases, response to amphotericin B, and A review, *Ann. Intern. Med.*, 74(6), 861, 1971 .
9. **Duma, R. J., Shumaker, J. B., and Callicott, J. H., Jr.**, Primary amebic meningoencephalitis, *Arch. Environ. Health*, 23, 43, 1971.
10. **Griffin, J. L.**, Temperature tolerance of pathogenic and nonpathogenic free-living amoebas, *Science*, 178, 869, 1972.
11. **Jamieson, A. and Anderson, K.**, A method for the isolation of *Naegleria* species from water samples, *Pathology*, 5, 55, 1973.
12. **Kadlec, V., Cerva, L., and Skvarova, J.**, Virulent *Naegleria fowleri* in an indoor swimming pool, *Science*, 201, 1029, 1978.
13. **Phillips, P. J.**, Protozoan symbionts from the anemone banodosoma cavernata from Galveston Island, Texas, *Gulf Res. Rep.*, 4, 186, 1973.

Part III: Special Studies on Large Scale Recreation Water Areas

Chapter 12

AQUATIC PLANT CONTROL IN KITTY HAWK BAY

INTRODUCTION

For a solution to the Eurasian watermilfoil *(Myriophyllum spicatum* L.) problem in Kitty Hawk Bay, four different types of action were considered, namely (1) herbicide application, (2) salinity control, (3) mechanical control, and (4) do-nothing.*

Herbicide Application

There has been a fair amount of scientific research on herbicide application for milfoil control and its effects. Experimentation in tidewater areas of Chesapeake Bay with the butoxyethanol ester of 2,4-dichlorophenoxy (2,4-D) attached to attaclay granules at 22 kg ae/ha indicated no lethal effect to caged animals and did not kill the valuable native plants.[1,2] In 1968, an experiment using 22 kg ae/ha of the butoxyethanol ester of 2,4-D was conducted on 320 ha in Currituck Sound. No acute adverse effects on fish or shellfish were observed.[5] The Tennessee Valley Authority (TVA) has used the same herbicide in some of its reservoirs at levels of 44 kg ae/ha without any observable detrimental effects on fauna or water quality.[3,4]

Salinity Control

Research on the control of Eurasian watermilfoil by increasing salinity levels has shown positive results. The level of salinity required is somewhere between one-third to one-half sea strength. Three types of activities utilizing salinity control have been proposed:

1. Creation of salt-water overwash areas of the outer banks which would allow overwash of sea water into Currituck Sound during spring and fall high tides
2. Creation of an inlet in the outer Banks, resulting in the change of Currituck Sound from a fresh-water to salt-water environment
3. Construction of a salt-water pumping facility to increase the salinity level in Currituck Sound

The first of these three actions would not provide enough salt water to remedy the milfoil problem throughout the entire area of infestation. It would probably result in control of milfoil in a limited area around the overwash site.

Consultation with the U.S. Army Corps of Engineers indicated that constructing such overwash areas might result in increased erosion on the Currituck Sound side of the Banks, since whatever water went in must come back out, with resulting flooding on the Currituck Sound areas. For these reasons, such an action must be considered as, at best, a partial solution to the milfoil problem, and not feasible for eradication of milfoil.

Estimated cost of overwash — There is no good estimate on the cost of establishing such salt-water overwash areas. In order to raise the salinity of Currituck Sound to the minimal level (30%) for milfoil control would require the influx of 56.5 billion gal of seawater. This is assuming that dilution would be homogeneous throughout the

* From Fish, F., Environmental impact statement for control of Eurasian watermilfoil, Currituck Sound and Kitty Hawk Bay, North Carolina, in Butoxyethanol Ester of 2,4-D for Control of Eurasian Watermilfoil, APCP Tech. Rep. 12, U.S. Army Corps of Engineers, 1976, C3.

Sound and there would be no influx or outflux of fresh or salt water after initial dilution. This, of course, is a faulty assumption. In actuality, the salinity of the Sound would be different in various locations depending on the circulation patterns established by such action. One would have to find an area of high salinity near the site of salt-water influx which would grade outwards to areas of lower salinity.

Control by salt water pumping — The Norfolk District, U.S. Army Corps of Engineers and the city of Virginia Beach were consulted concerning the salt-water pumping program into Back Bay, Va. The program has been in effect for 8 years. The purpose of the project was primarily to reduce the turbidity of the area (43,000 ha with an average depth of 1 m) by flocculation and settling out of particulates due to increased salinity.

Estimated cost of pumping — Initial cost of the pumping station facilities in 1965 was $50,000. Operation costs for the first 7 years of operation was $35,000/annum. Since converting from diesel to electric power in 1971, the cost of operation has been $20,000/annum. In comparing this to the cost of using similar facilities for Currituck Sound, it is worthwhile to note that the surface area of Currituck Sound is 3.6 times that of Back Bay while the volume of water is 5.5 times greater. Therefore, a comparable program would necessitate a much larger investment in terms of capital and operation costs.

Mechanical Control

The most feasible method of mechanical control of Eurasian water milfoil appears to be mowing and the use of water jets to blow the plants out of the substrata. Such actions result in extreme turbidity as a result of the disturbance of bottom materials. This would be detrimental to aquatic life and water quality. The mowing technique employs cutters mounted on a skiff or barge. The equipment moves through the water cutting a path through the aquatic vegetation up to 3 m wide and at varying depths to 2 m. The range of sophistication in such machinery is variable. The simplest cutters are mounted on a small skiff and only cut the milfoil, leaving it free to drift. This is a potential danger due to possible regeneration in new areas from the active floating plants. The most sophisticated machinery employs a "weed harvesting system" approach, utilizing a self-propelled mowing barge, a collection barge for accumulating the mowed weed and carrying it to shore, and a conveyor system for transferring the weed from water to trucks for removal.

The primary limitation of such equipment is that it cannot cover large areas rapidly. Around 0.3 ha/hr is the average harvesting rate. Mowing without harvesting can move up to 0.6 ha/hr. For this reason, such actions could not be applied to the entire Currituck Sound.

Estimated Costs of Mechanical Harvest

The boat-towed cutting barge with a cutting rate of 1 ha/hr costs about $1000. The self-propelled cutting barge with a cutting rate of 0.6 ha/hr costs about $12,500. The weed-harvesting system for cutting and removing weeds is estimated at $59,500 (harvester $31,900, transporter $19,500, and conveyor $9,000). These estimates do not include costs for maintenance and labor. Unless a practical use of the harvested material is found, the disposal cost appears to be prohibitive.

The disadvantages of using this alternative are

1. The high cost of capital investment

2. The slow rate of cutting limits the area of control
3. The possibility of spreading milfoil infestation by fragmentation of plants from cutting operations
4. Possible deoxygenation of waters by decaying plants cut, but not harvested

Do-Nothing

The "do-nothing" alternative is difficult to evaluate because there seems to be no certainty as to the future of the milfoil problem. The advantages of the do-nothing program are (1) no cost (except for monitoring of infestation areas by aerial photograph), (2) it does not require action which might have a large scale environmental impact, and (3) it does not eliminate the option for implementing another alternative at some future date. The disadvantages of using the do-nothing alternatives are (1) it does not generate an immediate solution to the problems now associated with milfoil infestation and (2) it may allow the milfoil to spread to other areas.

It is thought that the short-term uses of the environment involved in this project do not compromise the maintenance and enhancement of long-term productivity in an environmental, social, or economic sense. This agency is still in the process of evaluating the long-term feasibility of the mechanical mowing alternative and it is possible that either limited herbiciding or mechanical mowing may be utilized in the future.

PROGRAM OF STUDY*

The Kitty Hawk Bay area is located in northeastern coastal North Carolina at the confluence of several large, relatively shallow bodies of water which includes Currituck Sound, Albemarle Sound, and Kitty Hawk Bay (Figure 1). The area is separated from the Atlantic Ocean by a barrier strand which extends in a southward direction from southeastern Virginia. Oregon Inlet, which marks the end of the barrier strand approximately 10 km to the south, is the nearest direct connection between this area an the ocean. This is a shallow inlet with a low exchange volume. Thus, the Kitty Hawk Bay-lower Currituck Sound area has extremely low salinity (less than 1 part per thousand) and is generally considered to be fresh water. At times the salinity may increase due to sea spray and infrequent ocean flooding across the barrier strand. In the areas treated with 2,4-D, the water depth ranged from under 1 m to slightly under 2 m. Water turbidity was low and the bottom was easily visible in most of the area. The bottom consisted of a firm sand-mud mixture which became soft and more organic in the coves. All of the areas contained a high density of aquatic macrophytes, primarily dominated by Eurasian watermilfoil. The entire area is affected by strong wind tides which can cause water levels to fluctuate.

Kitty Hawk Bay is a protected body of water with an irregular shoreline and numerous marsh islands. It is bound on the east and north by the barrier strand and on the south by Colington Island. The western end of the bay opens into Albermarle Sound. A study plot was selected within Kitty Hawk Bay on Colington Creek, and approximately 725 acres on Kitty Hawk Bay. The Martin Point study plot was located at the mouth of Jean Guite Creek, which is surrounded on three sides by the barrier strand and on the north by Currituck Sound.

* From Hall, A. B., Dgetsinger, K., Brinson, M. M., and Davis, G. J., Changes in structure and metabolism in a Eurasian watermilfoil community following 2,4-D treatment, in Butoxyethanol Ester of 2,4-D for Control of Eurasian Watermilfoil, APCP Tech. Rep. 12, U.S. Army Corps of Engineers, 1976, G3.

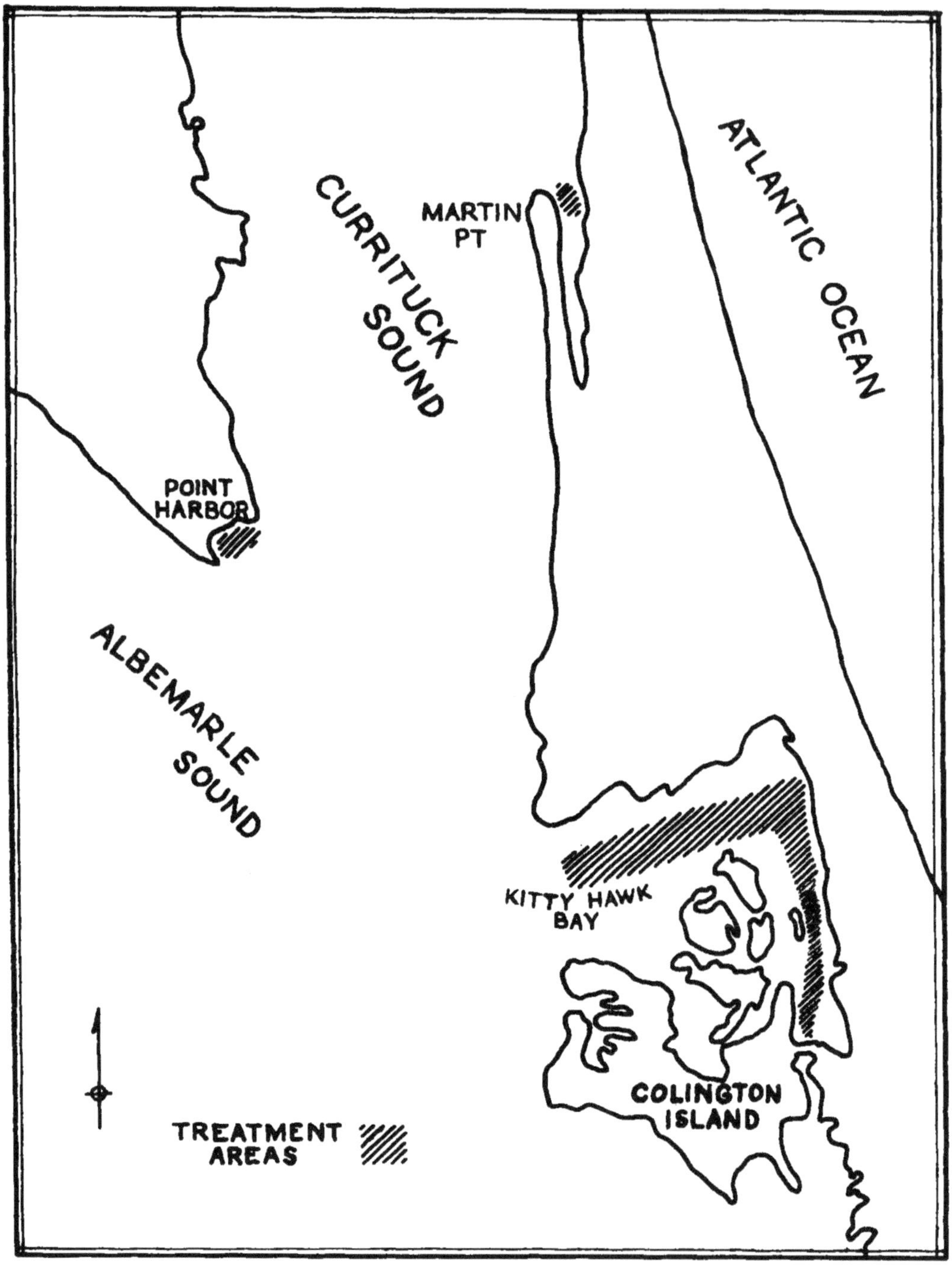

FIGURE 1. Kitty Hawk Bay study area.

Methods and Materials

On July 13 and 14, 1974, a 20% acid-equivalent (ae) granular formulation of the butoxyethanol ester of 2,4-D was applied at the rate of 22 kg/ae/ha by helicopter on three treatment areas. In Kitty Hawk Bay a boundary line was established to divide the study plot into a treatment plot and the adjacent control plot. Midpoints of 10 randomly selected transects were permanently established along this boundary line.

The bouys used in marking the transects were placed into position by boat prior to sample collections. The total study plot in Kitty Hawk Bay in which the transects were randomized was approximately 1000 by 500 m. Late in the study, it was necessary to establish a new control plot, behind Burnt Island.

Quadrants along the transects of all three study areas were sampled during June and early July and the third week of August 1974. Since the harvest method was employed, the post-treatment transects were positioned 2 to 3 m north of the initial (pretreatment) transects to prevent an overlapping of sample areas. Water depth was measured at all plots.

Sample collection was facilitated by the use of skin diving apparatus or scuba gear, depending on water depth. A 0.1 m^2 iron quadrant frame was dropped on the bottom near the buoy and all plants rooted within the frame were harvested above the substrate. The stems were brought to the surface and counted to determine the density of each species. Biomass determinations were based upon the random selection of two 0.1-m^2 quadrants along each transect in Kitty Hawk Bay. The plants from each quadrant were washed thoroughly, species were separated and counted, and the roots and shoots were separated. A 10- to 25-g aliquot was selected randomly from each sample and dried at 70°C for 24 hr to a constant dry weight.

Detritus samples were obtained with an Ekman dredge. The dredge was hand pushed approximately 10 cm into the substrate, and the trapped material was carried to the surface and washed through a fine mesh screen. Wet and dry weights of the organic material were determined in the laboratory with techniques used on the plant samples.

Results and Discussion

The treated Eurasian watermilfoil in all three study areas began to exhibit symptoms 24 hr after the 2,4-D application The flower spikes, which are normally several centimeters above the water surface in July, fell flat upon the water, and the tips of the stems began to twist and curl in abnormal positions. The upper stems became bloated and puffy. After 72 hr, the stems became brittle and continued to deteriorate, and after 96 hr the stems began to brown and bend toward the bottom (native macrophytes appeared healthy during the same period). The stems continued to sink and deteriorate, and by the 15th day the Eurasian watermilfoil had been essentially eliminated in all of the treated areas. A few stem fragments remained attached to the roots, while others separated from the root systems and formed mats along the surface. Most of the mats drifted onto the shorelines of the treated ateas and decomposition continued. Leaf deterioration could not be followed once the leaves fell from the stems. Leaves were not observed in the stem mats, nor were they found on the bottom. Apparently leaf decomposition was rapid and/or the leaves were dispersed by wind tides. Stem decomposition would be expected to take longer than leaf decomposition since the proportions of lignin and cellulose are higher in the stems. General observations and analysis of the Ekman dredge samples gave no evidence of particulate detrital accumulation on or in the sediment.

Once the Eurasian watermilfoil disappeared from the study areas, wind effects became noticeable. Wave action increased along with the water turbidity as the bottom was stirred up. The Secchi disc transparency, which was greater than 1.5 m before treatment, was less than 0.3 m after the watermilfoil decline. Native macrophytes began to deteriorate during the 3rd posttreatment week. By the 7th posttreatment week, a phytoplankton bloom (which also contributed to turbidity) was observed in Kitty Hawk Bay, during which the milfoil in Control Area II and surrounding coves began to die. This delayed dieback did not exhibit the typical 2,4-D symptoms, but a much slower type of deterioration occurred.

Table 1
RELATIVE ABUNDANCE OF AQUATIC MACROPHYTES IN KITTY HAWK BAY

	Relative abundance at indicated date (%)				
Species	1963	1966	1968	1973	1974
M. spicatum	—	—	25	75	58
R. maritima	50	90	37	—	23
P. pectinatus	45	Trace	37	25	2
P. perfoliatus	5	1	—	—	—
N. guadalupensis	—	Trace	—	—	17
V. americana	—	5	—	—	—

From Hall, A. B., Dgetsinger, K., Brinson, M. M., and Davis, G. J., in Butoxyethanol Ester of 2,4-D for Control of Eurasian Watermilfoil, APCP Tech. Rep. 12, U.S. Army Waterways Experiment Station, Vicksburg, Miss., 1976.

Table 2
PERCENT FREQUENCY OF AQUATIC MACROPHYTES AT POINT HARBOR AND MOUTH OF JEAN GUITE CREEK (MARTIN POINT) 1959—1962

	Percent frequency at indicated date					
	Martin Point			Point Harbor		
Species	1959	1960	1962	1959	1960	1962
R. maritima	100	80	0	30	0	80
P. pectinatus	0	40	20	0	0	0
P. perfoliatus	0	60	40	20	20	50
Najas sp.	80	80	0	20	0	0
V. americana	70	20	0	70	50	70
M. spicatum	0	0	0	0	0	0

From Hall, A. B., Dgetsinger, K., Brinson, M. M., and Davis, G. J., in Butoxyethanol Ester of 2,4-D for Control of Eurasian Watermilfoil, APCP Tech. Rep. 12, U.S. Army Waterways Experiment Station, Vicksburg, Miss., 1976.

Four species of aquatic macrophytes were found during the pretreatment study at Kitty Hawk Bay, Point Harbor, and Martin Point: *Myriophyllum spicatum* L., *Ruppia maritima* L., *Potamogeton pectinatus* L., and *Najas guadalupensis* Magnus. Earlier observations of Kitty Hawk Bay macrophytes indicated a shift in species composition from native species in 1963 to a 75% Eurasian watermilfoil and 25% native species community in 1973 (Table 1). A U.S. Fish and Wildlife Service Report shows that no beds of Eurasian watermilfoil were found at Point Harbor and Martin Point between 1959 and 1962 (Table 2).

In 1974 pretreatment densities of aquatic macrophytes in Kitty Hawk Bay and Martin Point were found to be approximately 1000 stems/m^2, while they totaled 300 stems/m^2 at Point Harbor (Table 3). Eurasian watermilfoil contributed most of the density (58% at Kitty Hawk Bay and at Martin Point and 100% at Point Harbor). Although

Table 3
PRETREATMENT DENSITIES AND BIOMASS (DRY WEIGHT) OF AQUATIC MACROPHYTES IN STUDY AREAS OF KITTY HAWK BAY, MARTIN POINT, AND POINT HARBOR

	Pretreatment density and biomass					
	Pretreatment density (stems/m)			Biomass (g/m)		
Species	Kitty Hawk Bay	Martin Point	Point Harbor	Kitty Hawk Bay	Martin Point	Point Harbor
M. spicatum	626.3	584.0	308.0	274.0	255.5	134.7
R. maritima	248.7	255.6	0.0	8.1	8.3	0.0
N. guadalupensis	188.0	165.0	0.0	4.5	3.9	0.0
P. pectinatus	17.9	0.0	0.0	5.6	0.0	0.0
Total[a]	1080.9	1004.6	308.0	292.2 ±91.9	267.7 ±121.3	134.7 ±45.9

[a] Biomass (g/m) totals represent the mean ± 1 SD.

From Hall, A. B., Dgetsinger, K., Brinson, M. M., and Davis, G. J., in Butoxyethanol Ester of 2,4-D for Control of Eurasian Watermilfoil, APCP Tech. Rep. 12, U.S. Army Waterways Experiment Station, Vicksburg, Miss., 1976.

Table 4
6-WEEK POSTTREATMENT DENSITIES AND BIOMASS (DRY WEIGHT) OF AQUATIC MACROPHYTES IN STUDY AREAS OF KITTY HAWK BAY, MARTIN POINT, AND POINT HARBOR

	Pretreatment density and biomass					
	Pretreatment density (stems/m)			Biomass (g/m)		
Species	Kitty Hawk Bay	Martin Point	Point Harbor	Kitty Hawk Bay	Martin Point	Point Harbor
M. spicatum	0.0	0.0	0.0	0.0	0.0	0.0
R. maritima	110.5	24.8	0.0	3.6	0.8	0.0
N. guadalupensis	7.2	10.2	0.0	0.1	0.1	0.0
P. pectinatus	4.8	0.0	0.0	0.7	0.0	0.0
Total[a]	122.5	35.0	0.0	4.4 ±3.2	0.9 ±1.4	0.0

[a] Biomass (g/m) totals represent the mean ± 1 SD.

From Hall, A. B., Dgetsinger, K., Brinson, M. M., and Davis, G. J., in Butoxyethanol Ester of 2,4-D for Control of Eurasian Watermilfoil, APCP Tech. Rep. 12, U.S. Army Waterways Experiment Station, Vicksburg, Miss., 1976.

no native species were found within the immediate treatment area at Point Harbor, beds of *P. perfoliatus* were observed outside of the treatment area on sandbars which separate Currituck Sound from the Point Harbor cove). At Kitty Hawk Bay and Martin Point 23% of the macrophyte communities consisted of *R. maritima*, while 17% was *N. guadalupensis* and 2% *P. pectinatus*.

Six weeks after treatment (Table 4), the macrophyte density had decreased in Kitty Hawk Bay to 122.5 stems/m^2 and in Martin Point to 35 stems/m^2. No macrophytes remained in the Point Harbor study plot. No Eurasian watermilfoil was present in any of the study areas. In Kitty Hawk Bay the remaining macrophytes consisted of 90% *R. maritima,* 6% *N. guadalupensis,* and 4% *P. pectinatus.* At Martin Point 70% was *R. maritima,* 29% *N. guadalupensis,* and less than 1% *P. pectinatus.* Total pretreatment biomass (dry weight) of the aquatic macrophytes ranged from 134.7 g/m^2 at Martin Point. Although Eurasian watermilfoil only accounted for 58% of the macrophyte density (with the exception of the Point Harbor study plot), it comprised over 95% of the biomass in the pretreatment study plots. The remainder consisted of various native macrophytes. Posttreatment biomass showed an elimination of the Eurasian watermilfoil standing crop and a decline in the native macrophytes in Kitty Hawk Bay (from 18.2 to 4.4 g/m^2) and in Martin Point (from 12.2 to 0.9 g/m^2). It also indicates an absence of macrophyte biomass in the Point Harbor study plot.

SUMMARY AND CONCLUSIONS

A joint decision was made by the U.S. Army Corps of Engineer and the Department of Natural Economic Resources of North Carolina to treat Eurasian watermilfoil in certain areas of infestation with a 20% ae formulation of the butoxyethanol ester of 2,4-D using impregnated attaclay granules. Surface applications of 2,4-D at this level were found selective for control of Eurasian watermilfoil with a subsequent release of growth of native waterfowl food plants, such as *R. maritima* (widgeon grass), *P. pectinatus* (sago pondweed), *P. perfoliatus* (redhead grass), and *N. guadalupensis* (a southern naiad).

REFERENCES

1. **Blakey, H. L.,** Outlook for weed control activities, *Hyacinth Control J.,* 5, 18, 1965.
2. **Gangstad, E. O.,** Aquatic plant control program, *Hyacinth Control J.,* 9, 46, 1971.
3. **Smith, G. E. and Isom, B. G.,** Investigation of effects of large-scale applications of 2,4-D on aquatic fauna and water quality, *Pest. Monitoring J.,* 1, 16, 1967.
4. **Wajtalik, T. A., Hall, T. F., and Hill, L. O.,** Monitoring ecological conditions associated with wide-scale applications of DMA 2,4-D to aquatic environments, *Pest. Monitoring J.,* 4 (4), 184, 1971.
5. **Whitney, E. W., Montgomery, A. B., Martin, E. C., and Gangstad, E. O.,** The effects of 2,4-D application on the biota and water quality in Currituck Sound, *Hyacinth Control J.,* 11, 13, 1973.
6. **Hall, A. B., Getsunger, K., Brinson, M. M., and Davis, G. J.,** in Butoxyethanol Ester of 2,4-D for Control of Eurasian Watermilfoil, Aquatic Plant Control Program, Tech. Rep. 12, U.S. Army Waterways Experiment Station, Vicksburg, Miss., 1976.

Chapter 13

AQUATIC WEEDS IN LAKE SEMINOLE*

INTRODUCTION

The importance of aquatic weeds has increased significantly in recent years, especially in the southeastern and southern coastal areas of the U.S. Excessive amounts of certain aquatic weeds interfere with commercial navigation and pleasure craft, reduce streamflow, contribute to flooding, interfere with fishermen and fish production, interfere with shoreline developments, interfere with swimming and other aquatic recreation, and contribute to mosquito reproduction.

In this chapter, plant species are classified into four types: submerged, emergent, floating, and marginal. Some plants are listed under two or more categories. The distribution and approximate acreage of selected plant species are shown on separate maps. Other species in the nine-lake survey portions are listed in table form.

The shoreline, water depths, stream flows, plant nutrients in the water, minor variations in pool elevations, and area climatic conditions are very favorable for growth and spread of a variety of obnoxious aquatic plants in Lake Seminole.

The significance of aquatic plants in the maintenance of commercial navigation waters, drainage canals, artificial impoundments, and naturally ponded areas has increased decidedly in recent years, especially in the southeastern and southern coastal areas of the U.S. Excessive amounts of certain aquatic plants interfere with or prevent movement of commercial boats and pleasure craft, reduce streamflow, and contribute to flooding, increase costs of processing water for municipal purposes, interfere with fish production and harvest, interfere with swimming, contribute to mosquito production, and/or increase costs of mosquito control for public health purposes.

Under favorable climatic conditions, water temperatures, and water levels, some aquatic plants grow rapidly. This occurs especially where the biological balance has been upset drastically and the waters are clear, shallow, quiet to slow moving, and rich in nutritional substances. Furthermore, some plants (because of their growth characteristics or locations) are almost impossible to control by mechanical means or by approved chemicals applied at rates sufficient to produce satisfactory kills without being very harmful to other desirable aquatic organisms. Also, cost of the chemicals or the required application method may prevent or limit the extent of control work.

Planned plant surveys are essential to determine the plant species present, define the problems, plan programs, and prepare cost estimates and program funds. Although satisfactory control or eradication of some aquatics appears almost impossible at the present time, early detection of foci of infestations may be very important. Beginning control before plants become widely or firmly established sometimes enables maintenance forces to keep certain plants subdued or controlled.

A general systematic aquatic plant survey covering a variety of the more objectionable aquatic plants in Lake Seminole was performed because of the rapid rate at which several of the more objectionable aquatic plants established themselves and spread within and around shallow water areas and numerous lake islands. Also, it was performed because of the costs and effectiveness of the mosquito control program, the costs of aquatic plant control (water hyacinth), and increases in light-trap collections

* By Gholson, A. K., Jr., Assistant Reservoir Manager, U.S. Army Corps of Engineers, Lake Seminole, Chattahoochee, Fla. From Gholson, A. K., Jr., Aquatic weeds of Lake Seminole, *Hyacinth Control J.*, 7, 18, 1968. With permission.

of a mosquito, *Mansonia perturbans,* a vector of encephalitis. Increases in prevalence of this species appear to have been very closely correlated with increases in cattail believed the principal host plant of the aquatic stages of the Lake Seminole mosquito.

Prior to August 1960, somewhat limited plant survey work on Lake Seminole was performed mainly in connection with other scheduled work; therefore, subsequent reports usually pertained to one plant or to a few aquatic plants that apparently were increasing rapidly and/or at a rate sufficiently important to recommend initiation of control measures as soon as practicable. However, water hyacinth is the only plant on which any specific or special attack was made before August 1960.

PHYSIOGRAPHY OF THE REGION

Lake Seminole is located entirely in the Dougherty Plain, one of four subdivisions of the Gulf Coastal Plain. The Dougherty Plain is a broad lowland stretching from central Georgia into southeastern Alabama and adjacent Florida. It was formed by solution of a bolt of soft Eocene and Oligocene limestones.

The more striking topographic and drainage features of the Dougherty Plain adjacent to Lake Seminole are mainly the result of solution. The solution of the limestone has lowered the surface of the region to a level at which the use and value of extensive areas are controlled by the water table. Also, it has transformed the upper part of the rock into a cavernous mass full of cavities and communicating passages through which flows a major portion of the drainage. The resulting topography is notably flat, having numerous shallow saucer-shaped depressions or ponds which vary from a fraction of an acre to many acres in size. Some authorities have called part of this area the Lime Sink Region. Many of the sinks have gently sloping sides. Some contain water throughout the year; others dry up, except during rainy periods. Normally, there are comparatively few small surface streams. Rain water sinks rapidly into the sandy soil, finds its way along channels in the porous limestone, and issues as blue springs along banks of the larger creeks and rivers.

The area above the ordinary flood plain of the rivers, known as the second bottom, contains ponds of various sizes. They vary in size from a fraction of an acre to many acres. The most notable because of their size are Ray's Lake and Lewis' Pond.

AQUATIC PLANT SURVEY

The general aquatic plant survey of Lake Seminole was performed in August 1960, during a period when a large percentage of the plants were actively growing, flowering, and/or producing seeds. Three representatives of the Corps of Engineers, Mobile District, participated in all phases of the field work. These were the Assistant Reservoir Manager (botanist), one boat operator,* and a biologist of Engineering Division, Mobile, Ala. Two representatives of the Georgia Game and Fish Commission, Atlanta, contributed to the work.

For convenience and record purposes, the lake was divided into nine survey portions (Table 1). The survey divisions were designated as follows: (1) Chattahoochee River portion, extending from the dam to the upstream limit of the survey on that river; (2) game management area of the Florida Game and Fresh Water Fish Commission; (3) the areas between Fish Pond Drain and the Chattahoochee River; (4) all of Fish Pond Drain; (5) channel north of Big Island from Fish Pond Drain to Spring Creek; (6) Spring Creek Arm upstream to Old Lake Decatur; (7) Old Lake Decatur, formerly an

* Operations Division, Jim Woodruff Dam, Chattahoochee, Fla.

Table 1
AQUATIC PLANTS RECORDED IN LAKE SURVEY PORTIONS (1—9)

Plant names: Common	Plant names: Scientific	Chattahoochee River (1)	Game Management Area Florida (2)	Fish Pond Drain to Chattahoochee River (3)	Fish Pond Drain (4)	Spring Creek Fish Pond Drain Channel (5)	Spring Creek (6)	Old Lake Decatur (7)	Flint River to Spring Creek (8)	Flint River (9)
		Survey portion names and numbers								
Alligator weed	*Alternanther philixeroides*									X[a]
Aquatic grass	*Paspalum* spp.	X						X		X
Bald cypress	*Taxodium distichum*									
Banana lily	*Nymphoides aquaticum*				X	X				
Black willow	*Salix nigra*	X	X		X	X	X	X	X	X
Bladderwort	*Utricularia inflata*		X		X					
Buttonball	*Cephalanthus occidentlis*	X	X	X	X	X	X	X	X	X
Buttonweed	*Diodia virginiana*							X		
Cattail	*Typha latifolia*	X	X	X	X	X	X		X	X
Chara	*Chara fragalis*		X			X				X
Cowlily	*Nuphar advena*		X		X	X	X			
Cutgrass	*Leersia hexandra*	X	X	X	X	X			X	
Evening primrose	*Ludwigia* spp.				X	X		X		
Fanwort	*Caboma caroliniana*					X	X			
Frogfruit	*Lippia nodiflora*		X				X	X		
Giant cutgrass	*Zizaniopsis miliacea*	X								X
Hydrolea	*Hydrolea quadrivalvis*	X	X						X	X
Hydrotrida	*Hydrotrida caroliniana*		X		X	X	X	X	X	X
Lizard's-tail	*Saururus cornuus*	X								X
Lotus	*Nelumbo lutes*	X	X		X					
Maidencane	*Panicum hemitomum*		X		X	X	X		X	X
Micranthemum	*Micranthemum imbrosum*		X							
Myriophyllum	*Myriophyllum heterophyllum*				X	X			X	
Naiad	*Najas conferta, N. guadalupensis*	X	X	X	X	X	X		X	X
Parrot's-feather	*M. brasiliense*	X	X	X	X	X	X		X	X

Table 1 (continued)
AQUATIC PLANTS RECORDED IN LAKE SURVEY PORTIONS (1—9)

Plant names		Survey portion names and numbers								
Common	Scientific	Chattahoochee River (1)	Game Management Area Florida (2)	Fish Pond Drain to Chattahoochee River (3)	Fish Pond Drain (4)	Spring Creek Fish Pond Drain Channel (5)	Spring Creek (6)	Old Lake Decatur (7)	Flint River to Spring Creek (8)	Flint River (9)
Pondweed	*Potomogeton diversifiolius* spp.	X	X		X	X	X		X	
Primrose willow	*Jussiaea decurrens*	X	X	X			X		X	X
Rushes	*Juncus repens* and *J.* spp.	X		X	X	X	X		X	X
Sedges	*Cyperaceae*	X	X	X		X	X	X	X	X
Smartweed	*Polygonium* spp.	X	X			X		X		X
Southern watergrass	*Hydrochloa caroliniensis*	X	X				X		X	X
Water hyacinth	*Eichhornia crassipes*	X		X			X			X
Water lily	*Nymphaea odorata*		X		X	X	X			
Water penny-wort	*Hydroctyle umbellata*				X	X		X		
Water plan-tain	*Alisma subcordatum*	X	X	X	X					X
	Sagittaria spp.	X	X	X	X					X
	Echinodorus spp.	X	X	X	X					X
	Lophotocarpus spp.	X	X	X	X					X
Water shield	*Brasenia schreberi*	X	X	X	X	X	X		X	
Willowweed	*Justicia americana*	X		X	X		X			X

[a] Presence of aquatic plants at a level of 10% or more of the surface area.

impoundment of the Georgia Power Company; (8) Spring Creek to Flint River Channel; and (9) Flint River portion from the dam to the upstream limit of the survey on that river terminating at a point a few miles north of Bainbridge, Ga.

With the exception of a few areas isolated by dikes or other obstructions and the lake's upper reaches on the Chattahoochee and Flint Rivers where water is confined within the relatively steep old river banks, the survey was conducted in a manner showing a continuous pattern around the shoreline. Shorelines of the numerous round, ovate, or elongate islands were completely covered or spot checked as various factors indicated or permitted.

Species Distribution and Classification

Species and/or groups of aquatic plants were recorded in the survey. Buttonball was the most widely distributed plant, occurring in each of the nine survey portions. Black willow, cattail, najas, and sedges occurred in each of eight portions. Hydrotrida, rushes, and water shield occurred in each of seven portions. Cutgrass, maidencane, pondweed, and primrose willow occurred in each of six portions. Water hyacinth occurred in four of the survey portions. The Flint River Arm had the heaviest infestation. Portion 6 includes a part of uncleared areas on lower Spring Creek adjacent to uncleared Portion 4. The heaviest infestation of parrot's-feather occurred in Portions 1 and 9. Alligator weed occurred mainly in Portion 9, which is one of the larger survey divisions, with only a small amount being found in Portion 8.

Aquatic plants in this survey report are classified into four groups: submerged, emergent, floating, and marginal (see Table 2). According to this classification scheme some plants (under certain conditions) may be and are classified under two or more of these types. Types are defined as follows:

Submerged plants — These rest on or root in the bottom. Their stems and leaves may fill the water or protrude somewhat above the surface. This type includes such plants as najas, certain species of pondweeds, needle rush, and chara.

Emergent plants — These are rooted in the bottom and their leaves and/or stems either stand above or float on the water surface. This type includes such plants as cow lily, water lily, lotus, water shield, cattail, and willowweed.

Floating plants — These are not normally rooted to the bottom, but float freely on the water surface. This type includes such plants as duckweed, water hyacinth, and bladderwort.

Marginal plants — Normally found along the shoreline, this type includes such plants as buttonball, willow, primrose willow, and smartweeds.

Lake Characteristics and Climate

Lake Seminole was formed by the Jim Woodruff Dam, designed and constructed by the U.S. Army Corps of Engineers across the Appalachicola River, west of Chattahoochee, Fla. The dam is located immediately below the union of the Chattahoochee and Flint Rivers which form the two principal arms of the lake. Spring Creek and Fish Pond Drain, two secondary arms of the lake, join the Flint River approximately 6 mi above the confluence of the two rivers. The west end of the dam is located in Florida and the east end in Georgia. Approximately two thirds of Lake Seminole is in Georgia and the remaining one third in Florida. The dam was constructed primarily for navigation with the available head used to produce hydroelectric power. At normal pool elevation, 77 ft mean sea level (msl), the lake has an area of approximately 37,500 acres, extending approximately 50 mi up the Chattahoochee River and 47 mi up the Flint River. The lake has a shoreline of 250 mi. The shoreline mileage does not include the numerous islands, both large and small, located in the lake. Excepting the old

Table 2
LIST OF RECORDED PLANTS CLASSIFIED: SUBMERGED, EMERGENT, FLOATING, AND/OR MARGINAL

Name		Types			
Common	Scientific	Submerged	Emergent	Floating	Marginal
Alligator weed	*Alternanthera philoxeroides*		X[a]	X	X
Aquatic grass	*Paspalum* spp.		X		X
Bald cypress	*Taxodium distichum*		X		X
Banana lily	*Nymphoides aquaticum*		X	X	
Black willow	*Salix nigra*				X
Bladderwort	*Utricularia inflata*	X		X	
Buttonball	*Cephalanthus occidentalis*		X		X
Cattail	*Typha latifolia*		X		X
Chara	*Chara fragalis*	X			
Cowlily or spatterdoc	*Nuphar advena*		X	X	
Cutgrass	*Leersia hexandra*		X		
Evening primrose	*Ludwigia* spp.				X
Fanwort	*Cabomba caroliniana*	X			
Frogfruit	*Lippia nodiflora*				X
Giant cutgrass	*Zizaniopsis miliacea*				X
Hydrolea	*Hydrolea quadrivalvis*		X		X
Hydrotrida	*Hydrotrida caroliniana*		X		X
Lizard's-tail	*Saururus cernuus*				X
Lotus	*Nelumbo lutea*		X	X	
Maidencane	*Panicum hemitomum*		X		
Micranthemum	*Micranthemum imbrosum*				X
Myriophyllum	*Myriophyllum heterophyllum*	X	X		
Najas	*Najas conferta* and *N. guadalupensis*	X			
Parrot's-feather	*M. brasiliense*	X	X	X	
Pondweed	*Potomogeton diversifolius*	X	X		
	P. spp.	X	X		
Primrose willow	*Jussiaea decurrens*		X		X
Rushes	*Juncus repens* and *J.* spp.		X		X
Sedges	*Cyperaciae*		X		X
Smartweed	*Polygonum* spp.		X		X
Southern watergrass	*Hydrochloa caroliniensis*		X		X
Water hyacinth	*Eichhornia crassipes*			X	
Water lily	*Nymphaea ordoratoa*		X	X	
Water pennywort	*Hydrocotyle umbellata*				X
Water plantion	*Alisma subcordatum*		X		X
	Sagittaria spp.		X		X
	Echinodorus spp.		X		X
	Lophotocarpus spp.		X		X
Water shield	*Brasenia schreberi*		X	X	
Willowweed	*Justicia americana*		X		X

[a] Presence of aquatic plant at a level of 10% or more of the surface area.

river, slough and pond areas, and creek channels, a large percentage of the lake is 7 ft or less in depth, having extensive areas too shallow for use of flat-bottom boats propelled by 10-hp outboard motors or larger.

Lake Seminole was impounded in steps or stages as features of the dam and the lock were completed. A partial impoundage was begun in March 1955. The pool was allowed to rise only to elevation 65 msl which occurred during the first week of April. It was then held at approximately that elevation until the final stage of impoundage was begun in January 1957. The lake was allowed to fill to normal top pool elevation (77 msl) in the early part of February 1957. Since that time, the pool elevation has been kept fairly constant, varying from about elevation 76.5 to 77.8 msl.

After impoundment to elevation 77 ft numerous ponds and lime sinks in the second bottom (above the ordinary flood plain of the rivers) became a part of Lake Seminole. Aquatic plants common to the area were found in the ponds and lime sinks. Such plants presented little concern at the time since physical and biological factors have successfully checked them through the years. Impoundment upset the natural biological balance and many of the submerged, emergent, floating and marginal aquatic plants and grasses of the ponds and lime sinks became sources of infestation for many of the shallow water lake areas. The more common of these plants are as follows: fragrant water lily, *Nymphaea odorata;* American lotus, *Nelumbo lutea;* water shield, *Brasenia schreberi;* curly-leaf pondweed, *Potamogeton crispus;* southern naiad, *Najas guadalupensis;* duckweed, *Lemna minor;* maidencane, *Panicum hemitomom;* spatterdock, *Nuphar advena;* bladderwort, *Utricularia inflata;* common cattail, *Typha latifolia;* banana water lily, *Nymphaea meximana;* water willow, *Justicia americana;* chara, *Chara fragalis;* winged water primrose, *Jussiaea decurrens;* and smartweed, *Polygonum* spp.[1]

Extensive areas (approximately 10,000 acres) of Lake Seminole were not cleared of trees or brush prior to impoundment. Practically all flooded trees were dead by 1960, and as a result, natural pruning has reduced the remaining trees to snags of varying height. Waterfowl and fish are attracted to these uncleared areas in great numbers. Floating-type aquatic plants lodge and flourish in parts of these uncleared areas making control from boats hazardous and difficult due to submerged stumps and tall decaying snags. Therefore, control must be accomplished by airplane.

Lake Seminole is located in an area that has a mild climate, a mean temperature of approximately 68°F, and a long growing season from mid March through October. Although winter temperatures sometimes drop below freezing for relatively short periods, they are not sufficiently low to kill or seriously damage aquatic plants in the lake, those in the bordering lowland and marshy areas, and those in lake-affected lime sink areas numerous near the lake between the Chattahoochee and Flint Rivers. Lake Seminole's favorable climate coupled with great expanses of shallow, clear water of the Flint River, Spring Creek, and Fish Pond Drain Arms render conditions most ideal for all types of aquatic plant growth. Apparently, turbid water conditions on the Chattahoochee River Arm have somewhat restricted aquatic weeds in this area. However, impoundments on the Chattahoochee River above Lake Seminole will no doubt, through sedimentation, help clear the waters of the Chattahoochee in the future, rendering them more suitable for aquatic plants.

HISTORY OF AQUATIC WEEDS AND CONTROL

Prior to the intermediate stage of impoundment (April 1955) an aquatic plant problem did not exist on Lake Seminole. Numerous aquatic plants had been noted in the many small ponds (lime sinks) located within the area that was ultimately to become a part of the lake at full impoundment. The plants were largely confined to margins of these ponds and were not considered menaces either by the public or the reservoir management personnel. Since full impoundment (February 1957), however, a few of these plants (maidencane; american lotus; fragrant water lily; bladderwort; water plantain, *Alisma* spp.; water shield; buttonball, *Cephalanthus occidentalis;* banana water lily; fanwort, *Cabomba caroliniana;* winged water primrose; and water willow) have become minor problems in restricted areas. Common cattail, seemingly an unimportant marginal aquatic plant at the time of impoundment, has spread from a small infestation near the old Georgia Power Company dam installation on Spring Creek to much of the shore area of the reservoir. This plant, in addition to presenting problems from a recreational standpoint, has created mosquito control problems. In 1960, an aquatic plant survey revealed that approximately 1500 acres of common cattail were present in Lake Seminole. Another aquatic plant survey in September 1963 revealed approximately the same amount within the reservoir area. Aquatic plants (water hyacinth, *Eichhornia crassipes;* alligator weed, *Alternanthera philoxeroides;* parrot's-feather, *Myriophyllum brasiliense;* and giant cutgrass, *Zizaniopsis miliacea*) presenting serious problems in 1963 were not known to exist in the reservoir area prior to impoundment. It is believed that these plants reached Lake Seminole from impoundments located above it on the Flint River.[2-4]

The initial water hyacinth infestation was observed in the vicinity of Big Horseshoe Bend on the Flint River Arm of Lake Seminole in the early spring of 1955. The herbicide, 2,4-D, at 2 lb/acre was applied from boats in the late summer of 1955, and an excellent kill resulted. Very little control work was attempted after the summer of 1955 until the summer of 1958. At that time, it was estimated that several thousand acres of water hyacinth were on the Flint River Arm of Lake Seminole. Two aerial contracts, specifying use of helicopters, were let in the summer, July and September of 1958. Under the first contract, 1054 acres of water hyacinth were sprayed at a cost of $10,700. This contract partially covered the Flint River Arm from mile 11 to mile 27. At the completion of this contract, inspections indicated that an excellent kill resulted in the areas covered. Calibrations of the equipment prior to beginning spray operations assured us of an application of 2 lb/acre. Under the second contract, 2015 acres of water hyacinth were sprayed at a cost of $12,150. This contract covered the Flint River Arm from mile 37 (Red Bluff) to mile 3 and covered the Chattahoochee River Arm from mile 8 down to the Jim Woodruff Dam. Also covered under this contract were the numerous islands located between the Flint River and the Chattahoochee River Arms of Lakes Seminole. The areas sprayed under the first contract were checked, and all reinfestations, resprouting, and areas untreated during the first contract were sprayed under the second contract. Inspections, both by airplane and boats, made several weeks after completion of the second contract indicated that an excellent kill had been obtained. Herbicides were also applied by boat crews during the summer of 1958 so that all lake areas might be covered.

In the early spring of 1959, water hyacinth were again discovered entering Lake Seminole via the Flint River in the vicinity of Bainbridge, Ga. An investigation revealed that the plants were reaching Lake Seminole from impoundments upstream. The 2,4-D was applied by boat crews during the 1959 and 1960 growing seasons. Boat work during these seasons was limited to shorelines along the main river channel and margins of the uncleared areas. The limitations were necessary because of navigating in the stumps and dead snags of uncleared areas. Keeping the plants sprayed along margins of the infested areas during these years generally prevented further spread of water hyacinth. Acreage of water hyacinth increased considerably during the 1959 and 1960 seasons even though spray operations prevented spread to new areas. This increase occurred in areas not readily reached by the spray boats. Again in early 1960, water hyacinth were observed entering Lake Seminole via the Flint River at Bainbridge, Ga.

During August and September of 1960, a systematic aquatic plant survey was made of the entire lake, and a comprehensive report compiled. At the time of the 1960 survey, it was estimated that 2500 acres of water hyacinth were on Lake Seminole. Herbiciding operations continued during the time of this survey.

Alligator weed was found for the first time on Lake Seminole in the vicinity of Butler Mill Creek on the Flint River Arm during the 1960 aquatic plant survey. This plant was found in small patches from Hutchinson's Ferry to Bainbridge, on both sides of the Flint River Arm. A total of 5 acres was estimated on Lake Seminole.

High inflow into Lake Seminole from the Flint River during the early spring of 1961 greatly reduced water hyacinth acreage in Lake Seminole. Large quantities were stranded above the normal shoreline, in bushes above the water, and large quantities were carried through the gated spillway of the Jim Woodruff Dam into the Appalachicola River. During the growing seasons of 1961 and 1962, alligator weed increased considerably. Water hyacinth and parrot's-feather apparently decreased. During these seasons herbiciding operations were continued by boat crews further reducing the acreage of water hyacinth and parrot's-feather. Alligator weed quickly recovered from the effects of the herbicide applied at low dosages and continued to spread at its usual, characteristic, alarming rate. The growth and development of these plants was similar to observations in Florida[3,4] and Louisiana.[6]

PLANT CONTROL OPERATIONS

An aquatic plant survey conducted in September 1963 revealed that alligator weed had definitely replaced water hyacinth and parrot's-feather in importance and acreage. At that time, it was estimated that 350 acres of alligator weed, 200 acres of water hyacinth and only a trace of parrot's-feather existed on the Flint River Arm of Lake Seminole. The survey also revealed that practically all of the small boat channels (totaling approximately 15 mi) were infested with submersed weeds. This situation can be readily understood when it is known that these channels are shallow (5 to 6 ft) and located in clear-water lake areas. Curly-leaf pondweed was the principal submersed weed in 1963 and continues to be so today. Several experimental attempts have been made to remove the submersed aquatics by applying herbicides in many different rates and combinations to sections of the small boat channels. To date, results have been very poor. Temporary control of this submersed plant has been achieved since 1963 by use of mechanical cutters which traverse the channels twice monthly during the growing season (April to October) (Figure 1). Water recreation problems caused by submersed aquatics will become numerous and acute in the shallow, clear waters of our lakes and streams.

The rapid spread of giant cutgrass on both the Chattahoochee and Flint River Arms of Lake Seminole was also noticeable during the September 1963 aquatic plant survey. In 1963, this plant was becoming well established along the shorelines and in the shallow water areas.

Herbiciding operations were continued during the 1963 growing season. The main objective was to prevent spread of the two major problem plants, water hyacinth and alligator weed. With the exception of the two aerial contracts mentioned earlier, all herbicides have been applied by boat crews. It might be of interest to mention again that boat work on Lake Seminole is limited by shallow water, stumps, and dead trees. In most instances, work is restricted to the deep water edges of the infestations.

Experimental work using various herbicides and combinations of herbicides for alligator weed control was initiated in 1963 by Dr. John Lawrence of Auburn University under the Corps' Expanded Aquatic Plant Control Program. This experimental work has been continued since 1963 and at a gradually reducing rate during 1966 and 1967.

FIGURE 1. Monitoring an aquatic plant control site for obnoxious aquatic species. (Courtesy of the U.S Army Corps of Engineers, Mobile District, Alabama.)

Under prevailing conditions on Lake Seminole, experiments revealed that repeated herbicidal treatments can check the spread of alligator weed.

During the summer of 1964, experimental plots were also established on Lake Seminole by Dr. Lyle Weldon of the Agricultural Research Service, Fort Lauderdale, Fla. for the purpose of comparing experimental results with similar plots established on other impoundments located in the South Atlantic and Gulf Coastal States as part of the environmental testing program under the Expanded Project for Aquatic Plant Control. Several colonies of flea beetles (*Agasicles* n. sp.) were established at Lake Seminole last year and have increased in size and extent to show activity in suppressing alligator weed in the immediate area.[5,6]

Limited aquatic herbicidal control work was done from 1964 through 1967. The purpose of this work was to maintain area control of the known serious aquatic pests. This work has been successful in so far as the emersed and floating species are concerned, but it has been less effective with submersed species. Mechanical cutters have been used continuously during growing seasons to maintain small boatways in navigable conditions. Although mechanical cutters aid the spread of submersed aquatics by vegetative means, temporary relief gained by their use makes it a justifiable operation until a satisfactory control chemical is found.

In 1966, a sizable infestation of Eurasian watermilfoil, *M. spicatum,* was discovered in Spring Creek, a secondary arm located in the midsection of Lake Seminole (Figure 2). Within a 6-month period, an estimated 1500 acres had become infested with this plant. It is speculated that this plant arrived in our area either by fishermen from infested lakes of the Tennessee Valley Authority (TVA) or by collectors of aquatic plants. Whatever the source, this plant renders many hundreds of acres useless during the growing season, and the Lake Seminole infestation may create a possible infestation source for areas downstream from the Jim Woodruff Dam. Presently, Eurasian watermilfoil is found only in the Spring Creek Arm of Lake Seminole. This may be of particular significance because Spring Creek contains a much higher calcium con-

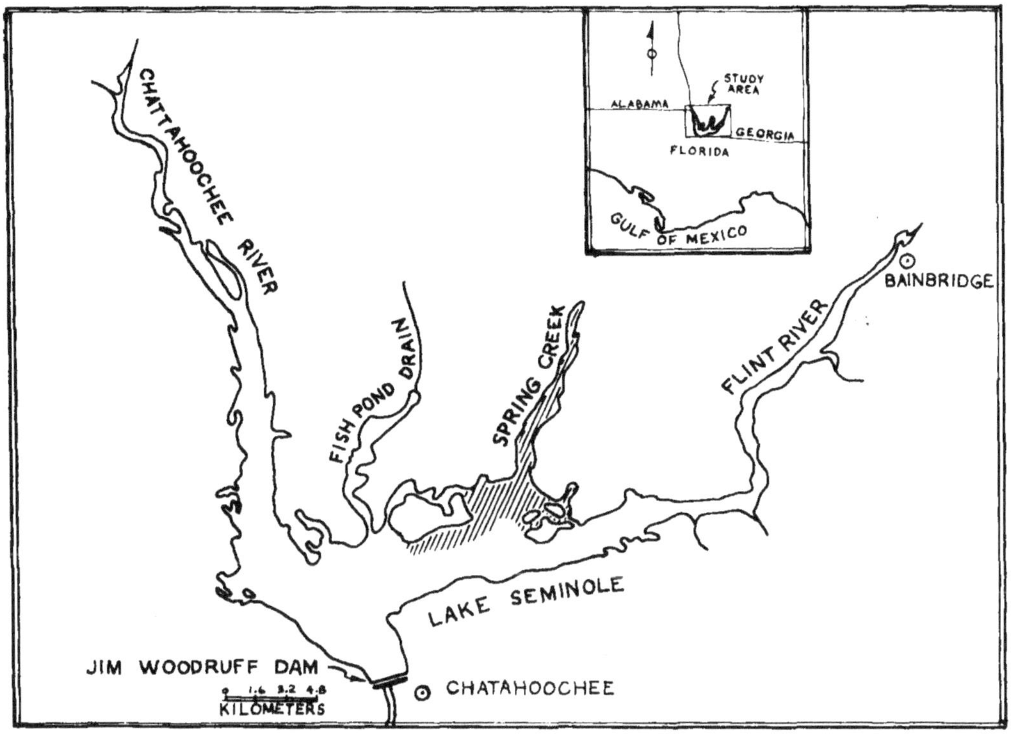

FIGURE 2. Eurasian watermilfoil (cross-hatched) as distributed in the Spring Creek area of Lake Seminole, 1966 to 1970.

centration than other streams coming into Lake Seminole. It is hoped that the lesser calcium concentration found in areas of Lake Seminole other than Spring Creek will prevent its spread to those areas.

Early in 1967, Florida elodea *(Hydrilla verticillata)* and marshweed *(Limnophila indica)* were found in small patches adjacent to a public access area on Lake Seminole. Here again it is suspected that these plants were inadvertently introduced into Lake Seminole by persons collecting aquatic plant material for aquariums. One can only guess what future aquatic problems will be due to introduction of species such as these.

In October 1967, several small patches of marsh pennywort (*Hydrocotyle ranunculoides)* were discovered on the Flint River Arm of Lake Seminole. When discovered, the patches contained less than 3 mi^2 but were very lush in growth. This plant was growing in association with alligator weed and water hyacinth. Pennywort infestations observed during the winter of 1967 and early 1968 were affected little by frost and were actually taking over areas infested with alligator weed. Routine checks during April 1968 revealed that the patches discovered in October 1968 have approached the one-half acre size and have completely taken over areas from alligator weed. In April 1968, two watermilfoils new to Lake Seminole were found on the Fish Pond Drain Arm (secondary). *M. heterophyllum* and *M. laxum* were found in close association with other aquatics in a small embayment of clear water. *M. spicatum* L. (Eurasian watermilfoil) has also been established and continues to persist. In 1 hr Spring Creek employment of Lake Seminole, Jim Woodruff Reservoir, specific conditions appear to favor its growth in this area of the lake.[7]

REFERENCES

1. **Lawrence, J. M. and Weldon, L. W.,** Identification of aquatic weeds, *Hyacinth Control J.*, 4, 6, 1965.
2. **Tabita, A. and Woods, J. W.,** History of hyacinth control in Florida, *Hyacinth Control J.*, 1, 19, 1962.
3. **Timmer, C. E. and Weldon, L. W.,** Evapotranspiration and pollution of water by water hyacinth, *Hyacinth Control J.*, 6, 34, 1967.
4. Expanded Project for Aquatic Plant Control, U.S. Department of the Army, House Document 251, 89th Congress, Washington, D.C., 1967.
5. **Wunderlich, W. E.,** Water hyacinth control in Louisiana, *Hyacinth Control J.*, 3, 4, 1964.
6. **Zeiger, C. F.,** Biological control of alligatorweed with *Agasicles* n. sp. in Florida, *Hyacinth Control J.*, 6, 31, 1967.
7. **Stanley, R. A., Shakelford, E., Wake, D., and Warren, C.,** Effects of season and waterdepth on Eurasian watermilfoil, *J. Aquatic Plant Manage.*, 14, 32, 1976.

Chapter 14

AQUATIC PLANT CONTROL PROBLEMS AT SAM RAYBURN RESERVOIR

INTRODUCTION

Some of the most serious weeds that presently confront human society are those that have evolved successful adaptations enabling them to occupy aquatic ecosystems. In order to provide water supplies, recreation, and electrical energy for urban population centers, we have increased the number of new water impoundments, and created new aquatic weed problems. The explosive development of water hyacinth, *Eichhornia crassipes* (Mart.) Solms., and alligator weed, *Alternathera philoxeroides* (Mart.) griseb., in Sam Rayburn Reservoir and B.A. Steinhagen Lake, is a typical example of this aquatic plant problem.

THE NATURE, SCOPE, AND FREQUENCY OF THE PROBLEM

The proposed management of aquatic plants consists of a spring and fall application of herbicides. Typically, during the spring months 1500 acres of water hyacinth are treated at the Sam Rayburn Project (Figure 1). Concurrently, about 800 acres of alligator weed are treated in the B.A. Steinhagen Lake (Figure 2). In the fall months, treatment areas are usually half of the spring treatment acreages. These biannual treatments have resulted in considerable waters being opened to recreational boating and fishing and should be continued, as in the past, with the use of 2,4-dichlorophenoxy (2,4-D). Aquatic weed problems are annual occurrences due to the warm humid climate of this area which results in very little, if any, disruption of the plants' growth.

B.A. Steinhagen Lake is very shallow, averaging about 7 ft, and is subject to frequent fluctuations. Alligator weed in the lake forms dense hummocks or islands when they break loose from nearby low lying lands. The alligator weed flotage provides harborage for many mosquitoes. Based on mosquito trap collections made each week during the summer months at the lake, the predominant mosquitoes are *Anopheles quadrimaculatus, A. crucians,* and *Uranotaenia sapphirina.* A typical sample usually consists of about 250 *Anopheles* sp. mosquitoes and about 100 *U. sapphirina.* The alligator weed hammocks also group together through wind drift to create larger islands which are capable of supporting a man's weight. These islands can also form enough substrate to support the development of small trees and shrubs. This has resulted in a serious encroachment on recreational use of the lake, interferes with the normal flow of incoming waters, and will, if unchecked, result in more difficult eradication and more usurping of the northern sector of the lake. Control and elimination of alligator weed at B.A. Steinhagen Lake are beneficial in preventing its spreading to other nearby lakes through boaters carrying its debris on their boats. Water hyacinth also impedes flow and provides harborage for mosquitoes at B.A. Steinhagen Lake. They do not form islands but do break loose as mats during high velocity inflows which create annual maintenance problems since the mats have to be manually removed between spraying periods to avoid increased infestation of the lower half of the lake. The mats must be periodically removed from lodging against the U.S. Highway 190 bridge piers to maintain lateral stresses within design limits for the bridge.

Water hyacinth is primarily a problem for recreationalists at the Sam Rayburn Reservoir since they interfere with boaters. Sam Rayburn attracts both the casual and serious sport fishermen as well as summer cottage owners who spend considerable sums

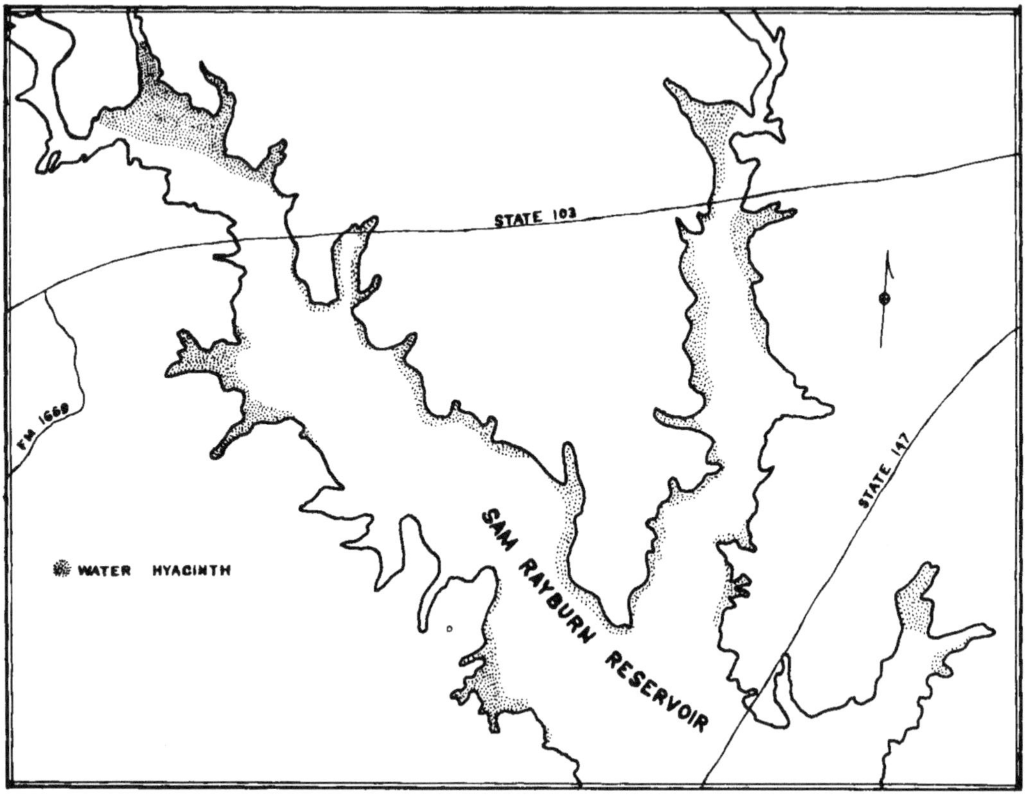

FIGURE 1. During the spring months, about 1500 acres of water hyacinths as shown in the stippled areas, are sprayed with 2,4-D for aquatic plant control.

of money in the local market area. Bass fishing is very popular as witnessed by a recent report of the Texas Park and Wildlife Department telling of 278 Bass Club tournaments held on the lake in 1973, which was the largest number of tournaments held on any lake in the state of Texas for that year. Granted, water hyacinth are good for fish shelter and cover. Because of weather conditions favorable to propagation of water hyacinth they multiply to such an extent that even with chemical spraying there still remain enough in the numerous coves and indentations of the lake to provide fish shelter.

TIME AND PLACE OF PESTICIDE APPLICATION

The optimal time of control for both aquatic plants is the spring season when maximum growth is occurring. However, due to the long growing season of east central Texas, they can be treated at both reservoirs almost all year. Experience has shown a smaller percentage of control when the plants were sprayed during the fall and winter months. Our current program plans for an initial attack in the fall months, with the major part of the program in the spring months and follow-up work thereafter. The most serious area, amounting to about 80% of the water hyacinth infestation at the Sam Rayburn Reservoir occurs in the upper or northern half of the lake with isolated mates throughout the remainder of the reservoir. It is in this area that the Angelina River transitions the lake with numerous shallow areas favorable for sustaining the water hyacinth infestation into the main body of the lake. Approximately 80% of the proposed eradication and/or control treatment is to be carried out in this area. Water

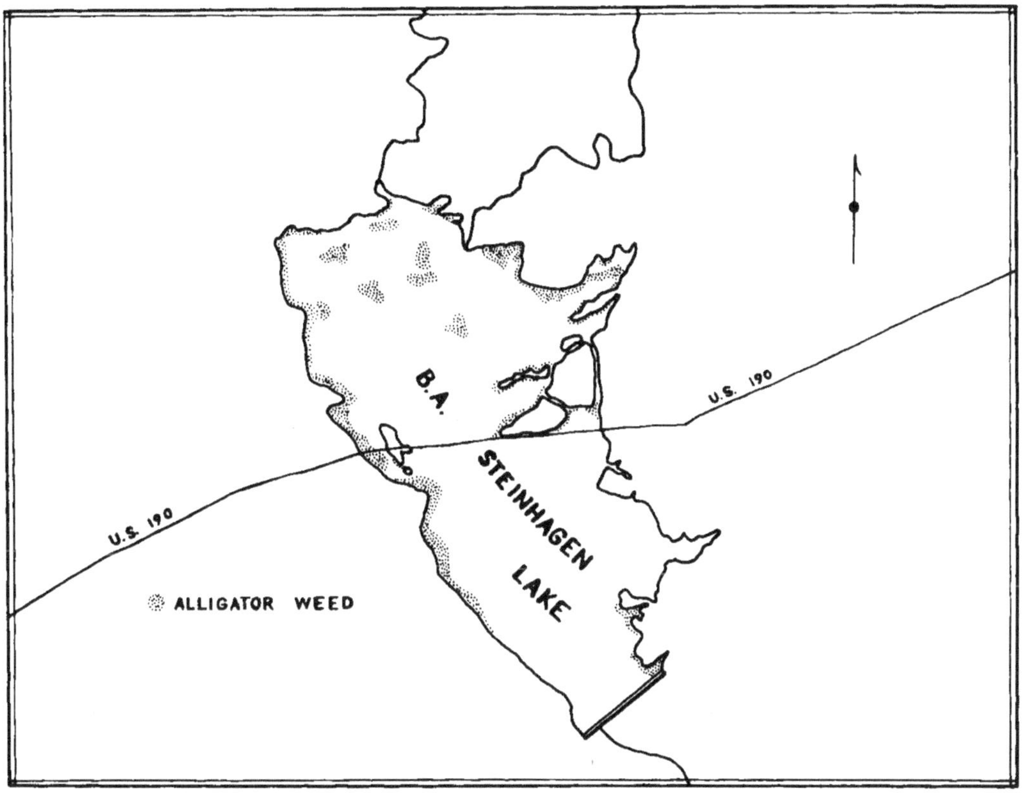

FIGURE 2. About 800 acres of alligator weed (stippled area) are treated in B.A. Steinhagen Lake, for aquatic plant control.

hyacinth and alligator weed are mostly confined to the portion of B.A. Steinhagen Lake above U.S. Highway 190 which crosses the lake at approximately midlake. About 95% of the infestation of both aquatics is found in this area so the program anticipates that 95% of the treatment efforts will be directed toward this area.

DESCRIPTION OF PLANTS

Alligator weed *(Alternanthera philoxeroides)* is an emergent aquatic plant that grows as a mat of vegetation interwoven with stems that have become prostrate. The nodes generally have two buds, each of which may sprout to produce a shoot. Reproduction is by vegetative means from the axillary buds. The leaves are opposite, 5 to 13 cm long, and lance shaped with a distinct midrib. Very few hairs may be found on the leaf surface and margin. The stems are hollow. Flowers occur in long peduncles in a solitary head up to 13 mm in diameter. From 6 to 20 white florets may be found on each flower with each floret having 5 stamens. Viable seed has not been found in the U.S. Roots may form at every node near to or submerged in water, or on moist soil. Upon reaching soil, the small aquatic roots enlarge, thicken, and in general take on the characteristics of other perennial terrestrial roots.

Water hyacinth *(Eichhornia crassipes)* are plants floating and often times rooted in mud, with slender, perennial rootstocks and rosettes of stalked-inflated leaves and fibrous, branched dark roots. Petioles are often times inflated or bladder-like. Flowers on a dense spike-like pedicel are usually quite showy and may be blue, violet, or white.

Individual flowers bearing a yellow marking on the upper portion of the flower are often surrounded by a deep blue margin. Plants reproduce largely by vegetative means and are connected by stolons. Plants range from a few centimeters to nearly a meter in height. Plants infrequently set seed within a large population. The seeds may sink to the bottom and remain dormant until periods of water stress, i.e., droughts. Upon reflooding the seeds may germinate and renew the populations in spite of the conspicuous absence of vegetative matter.

ALTERNATIVE PESTICIDES

The Fort Worth District has not found a more efficient means of controlling water hyacinth than 2,4-D. A letter was also sent to the Regional Administrator, Environmental Protection Agency, Region VI, requesting any available data concerning a more efficient and allowable chemical herbicide for use in reservoirs where the water was destined to be used for irrigation or domestic uses. Their response was negative. No reasonable alternatives, either chemical or mechanical, for water hyacinth control are known other than 2,4-D. Mechanical control, besides being quite costly, would also not be able to adequately harvest the weeds in the shallow backwaters and sloughs of these reservoirs because of the inundated timber. Areas missed would provide numerous foci for reinfestation, and in the case of some aquatics that can reproduce by fragmentation, may well increase the infestation. Biological control for water hyacinth is still in a developmental stage and not available. The alligator weed flea beetle has been introduced into B.A. Steinhagen Lake in recent years for control of alligator weed, but they have never established control of the weed problem. They can be found feeding in isolated areas of the reservoir, but never in sufficient densities to offer the needed control. What control they do exert is good when combined with chemical control for an integrated approach.

The Fort Worth District uses a water solution of either the 40% amine salt of 2,4-D or the butoxyethanol ester of 2,4-D for control of water hyacinth at both reservoirs. The solution would contain not less than 4 lb of active acid equivalent (ae) per gallon mixed to produce not less than 0.25% nor more than 1% concentration by weight of the dilute solution.

DESCRIPTION OF THE ERADICATION PROGRAM

The amount of 2,4-D chemical at the above described rates to treat an estimated total of 3000 acres of water hyacinth in the two reservoirs is approximately 12,000 lb.

The northern half of the Sam Rayburn Reservoir is the primary site for treatment of water hyacinth since this part constitutes about 80% of the problem. The northern half of the B.A. Steinhagen Lake would be the site for about 95% of the alligator weed treatments.

The spraying program is conducted by the Texas Parks and Wildlife Department personnel under the Aquatic Weed Control Program for the state of Texas on a 70 to 30 federal-state cost-sharing basis.

Under the state program the method of application is primarily by conventional spraying from boats. State crew personnel work between the two reservoirs and treat about 20 acres a day.

The Texas Parks and Wildlife Department is in the area of the two reservoirs for several months. No specific dates are given but it is estimated that the state crew requires about 4 months to complete the program. The Texas Parks and Wildlife Department personnel involved in the spraying program are certified by the state of Texas.

No recent study has been made relating to the economic impact of the alligator weed and water hyacinth infestations. However, numerous complaints have been brought to state representatives by local communities and individuals, especially from the northern sector of the Sam Rayburn Reservoir. In addition, there is strong indication that high densities of aquatic weeds contribute significantly to high densities of the Anopheline mosquitoes, especially in the northern sector of the Town Bluff Dam-B.A. Steinhagen Lake. Mosquitoes harbor in the shallow weedy sectors of the lake and use the alligator weed and water hyacinth mats as developmental sites. Control of these weeds would significantly help in decreasing public health hazards and the annoyance factor of the mosquitoes which are possible contributors to loss of work time. *Anopheles quadrimaculatus*, an important vector of malaria in the past decades of medical entomology, is the predominant mosquito in this area. The tremendous reproductive capacity of water hyacinth has resulted in the following summarized economic factors based on observation and conversations rather than specific studies: (1) plant mats have rendered many lakeside areas unusable, leading to severe economic losses; (2) hyacinth mats at confluence of the Angelina River and various tributary bayous lead to frequent flooding upstream on these bayous; and (3) the recreation benefits would be considerably higher through increased people usage of the headwater location of the Sam Rayburn Reservoir if the aquatic weed control program is continued.

ENVIRONMENTAL EFFECTS

As far as environmental considerations are concerned, research studies indicate that the use of the requested chemicals does not result in a safety hazard for water users either for irrigation or for domestic water needs. Storage space in the Sam Rayburn Reservoir has been allocated to the city of Lufkin, Tex. for domestic water supply uses. The enclosed map (Figure 1) shows the geographical relationship between these two projects. No significant adverse effect is anticipated for water users from these projects because of the dilution effects and the fact that no appreciable water quantities are taken directly out of the lakes. There are five owners of nearby small establishments about the Sam Rayburn Reservoir who take water for sprinkling lawns. It is likewise true that waters from Sam Rayburn Reservoir and B.A. Steinhagen Lake are used for rice field irrigation at certain times of the year (primarily spring and summer). Water from Sam Rayburn Dam enters into the Town Bluff Dam-B.A. Steinhagen Lake after a freeflow distance of about 25 river mi. The Town Bluff Dam then releases water downstream. However, it is not until after about 65 mi of river travel that the first use of irrigation water below the Town Bluff Dam is evident. The dilution and dissipation of any chemical sprayed upstream would be negligible at the point where irrigation was needed.

Large-scale spraying in Sam Rayburn Reservoir is confined to the upper northern half of the reservoir with only small isolated areas being treated in the southern portion. As the reservoir is very large (about 55 mi long), the diluent factor should be even greater as a distance of 90 river mi would be realized before first irrigation use. An additional factor increasing the safety potential of 2,4-D applications is the fact that waters from both reservoirs are not sluggish and slow which might allow some concentration of chemicals. Rather, in most all instances, discharges into the river below both dams are in the vicinity of 2000 ft^3 of water per second, and range much higher during the spring and early summer flood periods. These higher rates of release coincide with the best time to spray, thus the dilution of the chemicals should be well maintained. The chemical has been used at these reservoirs since 1963 with no fish kills being reported and no decrease in either fishing participation or success as far as

Table 1
DATE, NUMBER, ORIGIN, AND SITE OF RELEASE OF ALLIGATOR WEED FLEA BEETLE, *AGASICLES HYGROPHIA* (SELMAN AND VOGT), AT B.A. STEINHAGEN LAKE AREA, J.D. MURPHREE WILDLIFE AREA, AND THE HOUSTON AREA, TEX.

Date of release	Number of beetles	Origin of beetles	Site of release	1977 activity
May 1967	1000	Black Creek, Fla.	B.A. Steinhagen Lake	Yes
May 1967	1500	Black Creek, Fla.	J.D. Murphree area	No
June 1968	800	Black Creek, Fla.	B.A. Steinhagen Lake	Yes
June 1968	800	Black Creek, Fla.	J.D. Murphree area	No
September 1969	1000	Savannah River, Ga.	Houston area 1	Yes
September 1969	2000	Flint River, Ga.	Houston area 2	Yes
September 1969	4000	Flint River, Ga.	Houston area 3	Yes
August 1970	300	J.D. Murphree, Tex.	Houston area 4	Yes
October 1970	1000	Flint River, Ga.	Houston area 5	Yes

is known. Bass Club tournaments on Sam Rayburn Dam and Reservoir in 1973 reported a total of 42,000 lb of large-mouth bass being caught. This poundage represented only about 40% of the sport fish poundage taken for the year by tournament contenders.

BIOLOGICAL CONTROL OF ALLIGATOR WEED

Alligator weed is a widespread problem throughout the Gulf Coastal Plain of Texas. The flea beetle has been released and subsequent detailed studies and observations have been made within this region. These areas provide an interesting cross-section since they consist of a natural fresh-water marsh, the J.D. Murphree Wildlife Area, and an artificial water impoundment, B.A. Steinhagen Lake. A summary of flea beetle release and 1977 activity are given in Table 1.

J.D. Murphree Wildlife Area

Field observations of alligator weed at the J.D. Murphree Wildlife Management Area near Port Arthur, Tex. were made in both early November and again in late December 1970. On November 10, 1970, the mats of alligator weed observed were chlorotic. Flea beetles were found everywhere and were observed to be more abundant on some alligator weed mats than on others. Egg masses, as well as all three larval instars, were abundant in some areas and sparse in others. Some plants were supporting two or more egg masses on their apical leaves. First instar larvae that had just hatched were also noted to be abundant on some plants, and the examination of the hollow interiors of some stems showed teneral adult flea beetles present and preparing to exit from the stems.

These observations indicate that the flea beetles have dispersed well and are widely distributed in the J.D. Murphree Area. They have been found up Big Hill Bayou 5 km west of the nearest release site, and significant flea beetle populations have become established in various other areas.

B.A. Steinhagen Lake Area

Observations made in several areas of the Lake near Jasper, Tex. during November and December 1970 indicated the beetle had become well established. A few adult flea beetles and their egg masses were found on some of the remaining alligator weed still present. An area north of the fishing pier located at Walnut Ridge Park was investigated for the first time. High water allowed access to places that normally would be difficult to reach by boat. None of the alligator weed mats in this area were chlorotic but all mats investigated showed strong evidence of heavy feeding damage on the alligator weed, and many mats contained an abundance of beetles. Flea beetles had spread extensively and were found on alligator weed throughout this large area examined north of the fishing pier.

All of the areas north of Walnut Ridge Park contained good to excellent flea beetle populations. Egg masses were numerous and all stages of larvae were abundant. Most of the alligator weed had suffered heavy feeding damage to their leaves, especially the apical clusters. Stem damage was not impressive, although many flea beetle exit holes could be seen. Water level drawdowns have been practiced at B.A. Steinhagen Lake during 1970 to 1972; the flea beetle populations have apparently remained in the creeks and sloughs during that time, following the exposed alligator weed as the water level permitted. An early autumn, followed by continuous warm weather, promoted an abundance of succulent regrowth and the flea beetles strongly attacked the alligator weed. The feeding damage on the alligator weed here was the most impressive of all areas observed at Dam B Reservoir. It appears that the flea beetles have survived during periods of extensive drawdown by staying in the waterways. This survival was facilitated because the shallow backwater areas such as the creeks and sloughs, and the old channel of the Angelina River bed, supported the host plant. The only alligator weed that appears not to be very suitable for the flea beetles are those mats that grow on the flats. Alligator weed that was previously observed in some areas to have spindly stems and to be unattractive to the flea beetles is now observed to have larger stems, and appears more attractive for feeding and capable of supporting developing pupae in the stems.

Factors Limiting Control

The most important natural factors that appear to affect flea beetle populations are climatic in nature. The vulnerability of the host plant under frost conditions eliminates the aerial portions on which the flea beetle must rely, and occasional frost prevents any regrowth until the following spring; hence the size of the flea beetle population is severely limited under these conditions. Other abiotic factors are the rapid changes in water level resulting from heavy and sometimes sudden rainfall and the occurrence of wind velocities of hurricane intensity. Both of these factors have occasionally limited flea beetle effectiveness in those areas where the beetles have been released on the Gulf Coastal Plain.

Combined Control

Results of control operations for the J.D. Murphree Wildlife Area and B.A. Steinhagen Lake are given in Table 2. Spray operations in the J.D. Murphree Wildlife Area were terminated prior to release of the alligator weed flea beetles in 1968. Control operations on water hyacinth in B.A. Steinhagen Lake were continued through 1970. The combined operation was found more effective than either treatment alone, as shown by comparison of J.D. Murphree and B.A. Steinhagen Lake Areas in Table 2.

RESIDUE STUDIES

The Texas Parks and Wildlife Department spraying operations at both reservoirs were under the supervision of the Sam Rayburn Project Engineer, Mr. Aubrey Dry, Jr., and Project Manager, Mr. John Letney. Before initial deployment of the state

Table 2
SUMMARY OF CONTROL SURFACE AND BEETLE POPULATIONS ON ALLIGATOR WEED AT J.D. MURPHREE WILDLIFE AREA AND B.A. STEINHAGEN LAKE TREATMENT SITES

		Alligator weed		Beetle population No./m²	
Date of observation	Control surface percent[a]	Plant height (cm)	Leaf length (cm)	Adults	Larvae
J.D. Murphree Wildlife Area: No Herbicide Applied					
November 1968	59	61	10	92	21
May 1969	27	64	11	102	92
August 1969	71	61	10	11	0
June 1970	6[b]	69	11	11	0
B.A. Steinhagen Lake Area: Sprayed for Water Hyacinth Control					
November 1968	27	38	7	72	62
May 1969	2[b]	51	7	21	102
August 1969	2[b]	43	6	0	0
June 1970	1[b]	51	7	0	0
Summary of Treatment Means					
November 1968	43	51	9	82	41
May 1969	15	56	7	62	92
August 1969	36	53	8	11	0
June 1970	3[b]	60	6	11	0
J.D. Murphree	41	64	10	51	31
B.A. Steinhagen Lake	8[b]	45	8	21	41

[a] Percent total surface leaf area defoliated by insects.
[b] Significant difference (P = 0.05) for interaction error.

work crew force (two spraying crews were assigned to treat both reservoirs), the project supervisory personnel met with Mr. Lou Guerra, Chief of the Noxious Aquatic Vegetation Control Program of the Texas Parks and Wildlife Department, and his work staff. Aerial and topographic maps were checked and priorities set up to define what areas of the reservoir to treat. The main consideration was to place the spraying crews first in the more compacted areas of the reservoirs where complaints were most likely to occur. Complaints generally came from small marinas and shoreline development owners due to occasional blocking off of water access sites by water hyacinths.

A schedule of work days was set up on a tentative basis. However, this could not be realistically followed throughout the course of the year due to wind shifting of hyacinths from site to site, downtime of spray crews due to inclement weather, and mobilization of one or both spraying crews from one assigned area to another when drift of hyacinths caused immediate impacts on accessibility to some shoreline businesses. It was generally agreed that the Texas Parks and Wildlife Department spray crews will check in with the project supervisory personnel on an every-Monday-morning basis to explain or reverify their intended work segment areas for the entirety of the week. This allowed the Corps supervisory personnel to know at all times where the spraying crews were operating. If a shift in the assignment was necessary due to the conditions above mentioned a special call to the project's headquarters was made by the spray crew leader.

It is highly unlikely that residues of 2,4-D will be registered downstream due to the great dilution factor that would be taking place. Flows out of Town Bluff Dam are relatively constant between 450 and 2000 cubic ft^3/sec over most of the year due to upstream releases on the Angelina River out of Sam Rayburn and the normal flow on the uncontrolled Neches River. The Town Bluff Dam—B.A. Steinhagen Lake, being relatively shallow and having no flood control function, must operate primarily on the principle of what flow comes in must be let out as soon as possible.

A breakdown of acres treated and chemical gallonage used per reservoirs is as follows:

Spraying — 1975

Acres Treated	
Sam Rayburn Reservoir	1821
B.A. Steinhagen Lake	663
Total acres treated (1975)	2484
Gallons 2,4-D LV4 used	
Sam Rayburn Reservoir	1223
B.A. Steinhagen Lake	433
Total gallons used	1656

An approximate 95% control of water hyacinths and alligator weed was accomplished by spraying operations using conventional boat spraying gear and LV4 formulation of 2,4-D. The major difference between the previous aerial (helicopter) applications of an invert emulsion of 2,4-D by commercial contract and the present Texas Department Parks and Wildlife Department method of noninvert made by conventional boat spraying gear was the number of stress situations placed on particular water hyacinth density sites. Essentially we have traded off a major one-time stress attack on aquatics over an extensive area by aerial applications (generally over a 3-week time), for an almost continuous day to day application (weather permitting) by conventional boat spraying operations, extending over a 4- to 6-month period.

The first year results by the state using conventional treatment by boat did, however, show us comparable effectiveness (95% control) to that of aerial applications. However, what was noted in subsequent years of state contract work was a steady decline of water hyacinths and alligator weed after consecutive treatment years. With aerial applications we did not note this steady decline after subsequent treatment years. We surmise that repeated stress on plants by follow-up treatments as needed have had a better kill of problem aquatic plants. Apparently, the aerial flyovers during the 3-week period left a significant number of hyacinths and alligator weed patches to reestablish themselves over the year. We therefore believe the long-duration boat spraying operations under the Texas Parks and Wildlife Department spray crews presented the following advantages over aerial applications: (1) a much greater specificity in attacking existing plants (dosing according to density situations), (2) a greater mobility in attacking priority problem areas throughout a greater portion of the year, and (3) reduced costs via the 70 to 30 cost-sharing agreement.

WATER QUALITY STUDIES

Town Bluff Dam—B.A. Steinhagen Lake

A review was made of the U.S. Geological Service (USGS) Water Quality Data for the Angelina — Neches River System. We have extracted data from their 1973 to 1974, 1974 to 1975, and 1975 to 1976 reports and list seven USGS sampling sites and accompanied data in this report. USGS sampling site 1 (the Evadale Stream Monitoring gauge located some 59 mi below the Town Bluff Dam) is important to this report. A comprehensive evaluation of biological, chemical, and physical data was made here each of the above cited years. Of particular importance to this report is the 2,4-D monitoring data reported in micrograms per liter at this station. Based on 2,4-D data per years 1973, 1974, 1975, and 1976, it would appear that very minimal residue of 2,4-D existed at those times, in each year, when active aquatic spraying operations were being conducted upstream at both reservoirs. At no time do residues show 2,4-D exceeding the 0.1 ppm limitation for potable and/or irrigation waters.

This further substantiates somewhat the findings of the Texas Parks and Wildlife Department monitoring. Irrigation water for agricultural uses is picked up from the Lower Neches River below the Town Bluff Dam. However, the first pickup of this water is still about 30 mi below the Evadale Gauge Station (USGS 1 sampling site). We would presume there would be additional dilution involved on 2,4-D before such water was picked up downstream for agrarian uses. The data presented in USGS sampling show the residue of 2,4-D to be very minimal. Project managers at both reservoirs have observed no fish kills as a direct result of chemical spraying of 2,4-D, nor have any official major complaints been directed at project supervisory personnel on the spraying program. There does not appear to be any significant depletion of dissolved oxygen or significantly high biological oxygen demand (BOD) counts made in the USGS sites for the areas in B.A. Steinhagen Lake through die off of aquatic plants.

Sam Rayburn Dam and Reservoir

USGS sampling sites 6 and 7 for the years from October 1973 through September 1975 are good representative sampling points located in the immediate northern sectors of Sam Rayburn Reservoir where high densities of aquatic plants and consequently most treatment application sites are located. Again, no strong deviations on the BOD, or dissolved oxygen was observed in these sampling sites. USGS sampling sites 4 and 5 give the same type information as noted above. These sites are in the immediate area below the Sam Rayburn Dam on the Angelina River.

SUMMARY AND CONCLUSIONS

The Aquatic Plant Control Program at Sam Rayburn Reservoir was found to be effective and environmentally sufficient. Toxic residues were not observed after treatment. Control of water hyacinth and alligator weed is found to contribute to improved recreation development.

REFERENCES

1. Environmental Protection Agency, Decision on request for specific exemption to control alligatorweed and waterhyacinth, *Fed. Reg.*, 40, 29123, 1973.
2. Environmental Protection Agency, Tolerances and exemptions from tolerances for pesticide chemicals, 2,4-D, subparagraph C(f), *Fed. Reg.*, 40, 58287 and 58294, 1975.
3. **Gangstad, E. O.**, Aquatic plant control program, *Hyacinth Control J.*, 9, 46, 1971.
4. **Maddox, D. M. and Hambric, R. N.**, Use of alligatorweed flea beetle in Texas: an exercise in environmental biology, *South. Weed Sci. Soc. Proc.*, 23, 283, 1970.
5. **Maddox, D. M., Andres, L. A., Hennessey, R. D., Blackburn, R. D., and Spencer, N. R.**, Insects to control alligatorweed: an invader of aquatic ecosystems in the United States, *BioScience*, 21, 985, 1971.

Chapter 15

CONTROL OF WATER HYACINTH IN THE ST. JOHNS RIVER, PUTNAM COUNTY, FLORIDA

INTRODUCTION

The Corps of Engineers Aquatic Plant Control Program, initiated in 1899, was accelerated and expanded in 1958 and in 1965, to achieve progressive control and eradication of primary weed infestations of greatest economic importance in the eight South Atlantic and Gulf Coastal States from North Carolina to Texas. Water hyacinth, alligator weed, and Eurasian watermilfoil infestations were the primary problem aquatic weeds. Currently, mechanical, chemical, and biological methods of control are being used. The benefit-to-cost ratio considerably exceeds the 5:1 estimate on the basis of which the program was authorized and sustains the 14:1 ratio determined for the preceding pilot project.

HISTORY OF AQUATIC PLANT INFESTATION

The first authentic account of the presence of water hyacinth in this country was at the Cotton Exposition in New Orleans, La. in 1884. Alligator weed was noted in the U.S. at about that time, but it did not become a real problem for many years. These two plants have spread to many waterways with the greatest infestation in the states of Florida and Louisiana. In 1898, Congress, recognizing the potential hazard to navigation within the waterways, requested that the Chief of Engineers investigate and report on the obstruction of the navigable waters of Florida, Louisiana, and other South Atlantic and Gulf States by water hyacinth. That report (House Document No. 91, 55th Congress, 3rd Session) and the Annual Report of the Chief of Engineers, 1899, pages 1613 to 1623, resulted in authorization of work by the Corps of Engineers "for the construction and operation of suitable vessels, and for the use of log booms as adjuncts to the operation of the vessels, for the removal of water hyacinth in the navigable waters of the States of Florida and Louisiana" by the Rivers and Harbors Act of March 3, 1899.

By 1900, practically all of the main bayous in Louisiana were badly infested with water hyacinth and many of the streams were reported to be completely blocked. In some instances the lower reaches of many of the streams were unsafe for navigation. In Florida, water hyacinth had taken over the St. Johns River at Palatka to such extent that steamboats and other craft were unable to reach the docks or pass through the navigation openings of the bridges. Since then it has spread to many of the remaining waterways in Georgia, Alabama, Mississippi, and Texas and particularly in peninsular Florida and watercourses in Louisiana.

Following the advent of a new herbicide, 2,4-dichlorophenoxy (2,4-D), Congress requested the Corps of Engineers in 1945 to determine whether any expansion of the scope of operations or methods of control was advisable at that time, to determine the nature and extent of the various public benefits that would accrue from extermination and removal, and to determine the amount of local benefits. That report resulted in the establishment of a comprehensive program to provide for control and progressive eradication of obnoxious aquatic plants in the states of North Carolina, South Carolina, Georgia, Florida, Alabama, Mississippi, Louisiana, and Texas; it was designated as the "Expanded Project for Aquatic Plant Control."

Control operations of the project were extended to a nationwide program in 1965 to consider the combined interest of navigation, flood control, drainage, agriculture, fish and wildlife conservation, public health, water quality control, recreation, and related purposes. Most of the control operations are still limited to the Gulf States, Florida and Louisiana specifically, since these states have developed active programs and the need is greatest in these areas.[1]

CORPS OF ENGINEERS PROJECT AUTHORITY

The expanded project for control of different species of obnoxious aquatic plants was initiated July 1, 1967. Section 302 of the Rivers and Harbors Act of 1965 (P.L. 89-298; 33 U.S.C. Section 610) authorized a continuing program based upon the information presented to the 89th Congress on the pilot project as contained in House Document 251,[1] and testimony presented by the Chief of Engineers containing data from the U.S. Department of Agriculture, U.S. Public Health Service, and the U.S. Fish and Wildlife Service to the Subcommittee on Flood Control-Rivers and Harbors of the Committee on Public Works, U.S. Senate in June of 1965. Provisions for research and planning for the program were included which are borne fully by the federal government, and provisions for control operations were made by which local interests must agree to hold and save the U.S. free of claims that may occur from control operations and by which they must agree to participate to the extent of 30% of the cost of such operations.

EFFECTS OF AQUATIC PLANT INFESTATION

The most recent surveys indicate the presence of several types of aquatic infestations in Florida, including water hyacinth, alligator weed, Eurasion watermilfoil, hydrilla, and a number of other surface, bank, and submersed weeds. The long growing season and other environmental elements are particularly favorable for the growth and propagation of aquatic plants in this state.

Water hyacinth and alligator weed often grow sufficiently dense to block or impede boat traffic, damage propellers and marine cooling systems, and increase navigational hazards in the vicinity of bridges, docks, piers, etc. The dense mats formed by alligator weed are particularly difficult for boats to penetrate. The dense above-water foliage is supported on thick underwater mats composed of the stems and laterals of the plants. Without control of these aquatic plants, commercial navigation would cease to exist after a few years in many streams.

Aquatic plants seriously congest or completely blanket slowly flowing natural streams and drainage canals, thus reducing their discharge capacities. Small channels and shallow waterways often become congested, and the resultant retarded runoff increases stages, extent, duration, and frequency of flooding. Effective aquatic plant control will relieve these flood problems.

In areas where irrigation water is pumped from surface supplies, aquatic plants clog pump intakes and reduce flow to the pumps. Normally, land being irrigated is fertile and low. Water in these ditches and canals flows at low velocities and frequently contains fertilizer leached from adjacent croplands. Under these conditions, aquatic plants grow rapidly and seriously retard water flow.

In many areas where masses of floating plants cover smaller waterways and ponds, the infestations have rendered the areas unusable for fish habitat by exhausting the dissolved oxygen in the water, by occluding needed sunlight essential for basic food production, and by rendering shallow water spawning areas unusable. In regards to wildlife, drifting mats of weeds destroy beds of submerged plants defoliate floating-leafed aquatics, and overwhelm marginal food plants, which may be the desirable foods for waterfowl.

Aquatic plant infestations frequently interfere with and practically eliminate recreational use of water areas. In addition to preventing access to and the use of waters for hunting and fishing, they preclude boating, swimming, skiing, and other water-oriented activities. Mosquito production almost invariably occurs in the aquatic plants of waters adjacent to recreation areas and furnishes a source of nuisance to campers and other recreationists.

The production of disease-carrying and pest mosquitos is aided by mats of water hyacinth and alligator weed. Infestation of aquatic plants in the vicinity of population centers occasionally requires mosquito control as a health measure. The leaves of water hyacinth plants trap and hold water which, in the dense mats, provide excellent mosquito-breeding areas.

Where surface waters are used as a source of water supply, heavy growths of aquatic plants are a problem. Studies in Florida have shown the rate of loss of water in water hyacinth infested areas is 3.7 times greater than in areas without hyacinths. Transpiration, when added to evaporation, consumes the water intended for other beneficial uses.[1-5]

CONTROL AND ERADICATION

As early as 1898 tests were conducted to find methods by which the water hyacinth could be effectively eradicated. These tests, and subsequent tests in 1906, utilized a wide variety of chemical sprays. None of these chemical eradicants were found to be both economically and ecologically feasible.

During the early phases of the Corps of Engineers' aquatic weed eradication program, many different mechanical devices were designed to cope with the dense infestations of water hyacinths. Early control operations were based almost exclusively on mechanical procedures such as gang-saw boats to open up paths for navigation, physical removal by derrick and grapple, the use of hand labor to cut blocks of matted water hyacinths with 6-ft timber saws, and the use of special barges equipped with conveyor belts for picking up the plants and depositing them on the shore, as well as crusher boats which crushed the plants into a pulp and then deposited the pulp directly into the water.

While in some cases mechanical methods temporarily open waterways or lakes, they also tend to spread aquatic plants by fragmentation. This, coupled with the presently prohibitive cost of widespread mechanical control measures, has led to current control measures involving the use of chemical and biological agents.

Corps of Engineers' spraying is done by boat and aircraft with an amine salt of 2,4-D. This herbicide is an organic chemical belonging to a group of substances known as plant hormones or growth regulators. It has been widely recommended for aquatic plant control and continues to be the most widely used herbicide. It is nontoxic to humans and animals when ingested in concentrations likely to follow its application as a herbicide. Generally, depending upon size and depth of the body of water and whether it is freely flowing or sluggish, there are no harmful effects from control operations on water supplies. In instances where the supply source is very heavily infested and where there is little discharge, special precautions are taken by treating only portions of the growth at one time. No adverse effects on water supplies have been reported since widespread use of 2,4-D was started by the Corps and state and local agencies in other states for hyacinth control. The spray material contains 4 lb/gal of active acid equivalent (ae) and is applied at the rate of 2 to 4 lb of active ae to an acre of vegetation. The herbicide 2,4-D is toxic to many desirable broadleaf plants. However, its application is controlled at all times to prevent the drifting of spray onto cultivated lands, ornamentals, or desirable aquatics. Aerial spraying is performed only when winds are minimal (3 to 5 m/hr). Spraying by boat or land based equipment is not performed when wind velocities exceed 10 m/hr.[4-6]

Specific Exemption

The chemical treatment of water hyacinth in the St. Johns River, Fla. was declared illegal by the Environmental Protection Agency (EPA) on December 4, 1973 (Figure 1).

On December 23, 1973, an application for a specific exemption from the requirements of the Federal Insecticide, Fungicide, and Rodenticide Act (FIFRA), as amended (86 Stat. 973), was made to the EPA from the Department of the Army, Office of the Chief of Engineers, Director of Civil Works, Washington, D.C. 20314, pursuant to para 166.2(a) of the regulations (40 CFR Part 166) governing exemption of federal and state agencies for use of pesticides under emergency conditions. The regulations were published in the *Federal Register* on December 3, 1973 (38 FR 33303).

The U.S. Corps of Engineers, Department of the Army (hereafter referred to as the "Applicant"), has requested the use of the chemical 2,4-D (dimethylamine salt of 2,4 dichlorophenoxy acetic acid) for control and progressive eradication of water hyacinths from the navigable waters, tributary streams, connecting channels and other allied waters of the St. Johns River, Florida. All interested persons were referred to the application on file with the Registration Division (HM-567), Office of Pesticide Programs, EPA, Washington, D.C., for a statement of the representations contained therein which are summarized below.

I. As to the nature, scope, and frequency of the emergency, On December 4, 1973, the Deputy Assistant Administrator for Pesticide Programs, EPA, sent a letter directing the Corps of Engineers to stop using products containing the chemical 2,4-D in the St. Johns River, since that chemical is not registered for use in flowing/moving waters. The Applicant states that the rapid and extensive growth of water hyacinths in this river is a hazard to navigation and recreation activities. According to the Applicant, there is no known method of control as effective as this chemical.

II. As to description of the pest, water hyacinth, the Applicant states that the plants are floating and oftentime rooted in mud, with slender, perennial rootstocks and rosettes of stalked-inflated leaves and fibrous, branching dark roots. Petioles oftentime are inflated or bladder-like. The plants reproduce largely by vegetative means and are connected by stolons. The water hyacinth shows considerable variation in size, the plants ranging from a few centimeters to nearly a meter in height. Plants infrequently set seed within a large population. The seeds may sink to the bottom and then remain dormant until periods of water stress, i.e., droughts. Upon reflooding, the seeds may germinate and renew the populations in spite of the conspicuous absence of vegetative material.

III. As to pesticide registration, the Applicant states that the use of the chemical 2,4-D would permit them to conduct a program of Aquatic Plant Control by the use of 2,4-D in moving water. However, the chemical 2,4-D is registered for control of aquatic weeds in still water, e.g., lakes, ponds, etc., and a very limited use in the Northwest in irrigation ditches with an established residue tolerance in water of 0.1 ppm.

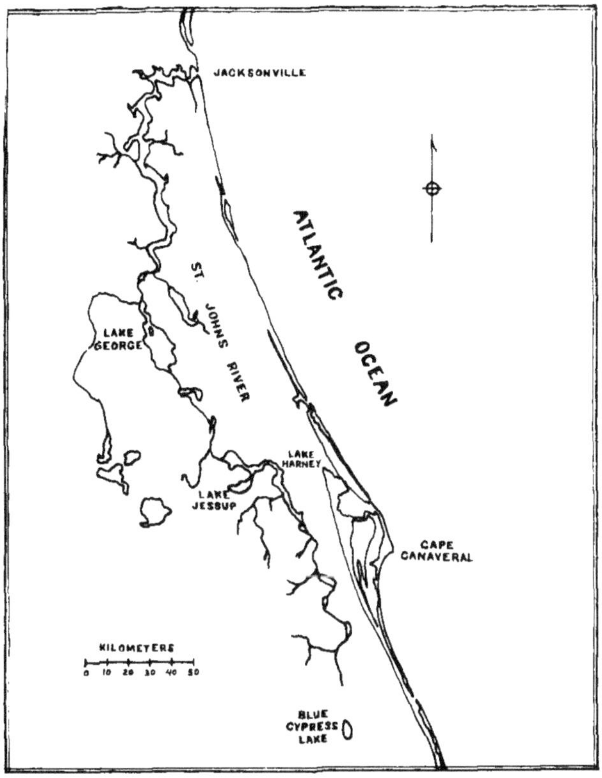

FIGURE 1. Chemical treatment of water hyacinth in the St. Johns River, Florida, was declared illegal by the EPA, December 4, 1973.

IV. As to pesticides to be used, the Applicant states that the chemical 2,4-D has been found to be a very effective and feasible measure of control. He also states that other herbicides such as sodium arsenite or amitrole are effective, but are undesirable from an environmental point of view. It should be noted that sodium arsenite and amitrole are not registered for such use; consequently, before the Corps of Engineers could use these pesticides, it would be required to either apply for registration or request an exemption under Section 18 of the FIFRA, as amended (40 CFR Part 166).

V. Applicants description of the eradication or control program is as follows:

A. The chemical 2,4-D would be applied at a rate of 2 to 4 lb/acre for control of water hyacinths, applied directly to the floating mat. One application is normally sufficient to kill the plant. Follow-up applications as a spot treatment are normally necessary in a control program. For treatment of approximately 3000 acres of water hyacinths, approximately 12,000 lb acid equivalent would be required.

B. The chemical would be applied on floating mats of water hyacinths in the St. Johns River. The program in this request for specific exemption covers only the area of the St. Johns River from Lake Harney to approximately Jacksonville.

C. Chemical application would be made from air boats as a direct spray at 75 to 100 lb pressure in 100 to 200 gal of water per acre.

D. Treatment would involve the spring-summer season, April through September.

E. The chemical would be applied by trained personnel of the Corps of Engineers or by contracts under the direct supervision of the Corps of Engineers. Engineer Regulation ER 1130-2-412, Planning and Engineer Regulation ER 1130-2-413, Operations, and a manual for the personnel training program are available from the Corps of Engineers upon request.

VI. As to the statement of economic benefits and losses anticipated with and without the exemption, the Applicant states that annual benefits in the interest of commercial navigation and recreation are estimated at $1,092,750 with a benefit-to-cost ratio of 3.0:1.

VII. As to the analysis of possible adverse effects on man and the environment, the Final Environmental Impact Statement (FEIS) for the Hyacinth Control Program in Florida (which includes the St. Johns River) was filed with the Council for Environmental Quality on September 11, 1973.

Residual 2,4-D Levels

In response to conditions of the specific exemption, the U.S. Army Corps of Engineers conducted extensive studies to determine the residual levels of 2,4-D in flowing water, i.e., the St. Johns River, Fla. The levels of 2,4-D observed were well below established tolerance limits and there was no accumulation of 2,4-D in blue crabs (*Callinectus sapidus* Rathbun) in the treatment area.[7]

The Corps of Engineers applied for a 2,4-D tolerance for aquatic application to the Department of Health, Education, and Welfare, Food and Drug Administration on August 7, 1970. The functions of the Food and Drug Administration were transferred to the EPA under Public Law 92-516.

A series of meetings were undertaken with the EPA and members of the Corps of Engineers in an effort to finalize the applications. Reports were submitted and reviewed, and new studies were initiated in coordination with EPA to obtain additional data.[8-17]

On October 28, 1975, the EPA announced (40 FR 50124) that the Department of the Army, Office of the Chief of Engineers (DAEN-CWO-R), Washington, D.C., had submitted a petition (FAP6H5104) which proposed that 21 CFR 123.100 be amended by establishing a food additive regulation permitting the use of the herbicide and plant regulator 2,4-D in potable water with a tolerance of 0.1 ppm, residues that would result from the application of the methylamine salt of 2,4-D in water hyacinth control programs conducted by the Corps of Engineers or other federal, state, or local public agencies in ponds, lakes, reservoirs, marshes, bayous, drainage ditches, canals, rivers, and streams that are quiescent or slow moving.

Tolerances were established for residues of the herbicide 2,4-D as follows:

I. 2 ppm in the milled fractions (except flour) derived from barley, oats, rye, and wheat to be ingested as food or to be converted to food. Such residues may be present therein only as a result of application to the growing crop of the herbicides identified in 40 CFR 180.142

II. 0.1 ppm (negligible residue) in potable water; such residues may be present therein only:

A. As a result of the application of the dimethylamine salt of 2,4-D to irrigation ditch banks in the western U.S. in programs of the Bureau of Reclamation; cooperating water user organizations; the Bureau of Sport Fisheries, U.S. Department of the Interior; Agricultural Research Service, U.S. Department of Agriculture; and the Corps of Engineers, U.S. Department of Defense.

B. As a result of the application of the dimethylamine salt of 2,4-D for water hyacinth control in ponds, lakes, reservoirs, marshes, bayous, drainage ditches, canals, rivers, and streams that are quiescent or slow moving in programs of the Corps of Engineers or other federal, state, or local public agencies.

SUMMARY AND CONCLUSIONS

Plants destroyed by herbicides remain in the water and, in decomposing, deplete the oxygen supply and release stored nutrients for support of new plant growth. Oxygen depletion temporarily subjects fresh-water species to physiological stress; however, by treating the lake or body of water in sections at different times, the motile species are able to move to areas of more favorable conditions. The same low-oxygen condition would be present if a large mat of vegetation were to remain untreated, as some plants would be continually dying and depleting the oxygen while the heavy mat would prevent reaeration at the surface. The dying vegetation from chemical treatment will cause a temporary noxious condition. However, once the bulk of the vegetation has been removed, the following maintenance treatments will have little effect on water oxygen levels.

It is found that continued use of 2,4-D will serve the public interest and far outweigh the probable public detriment in discontinuing its use. There have been no observable long-term adverse effects on fish or waterfowl as a consequence of using 2,4-D as an aquatic weed control agent. This chemical is used not only by the Corps of Engineers but by the Game and Fresh Water Fish Commission of the State of Florida, by the Central and Southern Florida Flood Control District, and by numerous counties.

Extensive research and at least 20 years' field operations experience in Florida have shown that chemical control through the use of 2 to 4 kg acid equivalent per hectare of 2,4-D, is the most effective and economical method of control of water hyacinth currently available to the Corps of Engineers. Conversely, if water hyacinths are allowed to grow without control, the damaging effects on the entire public welfare would be very great. Depending on climatological conditions the fresh-water areas of the state would, within a relatively short period, become packed with hyacinths. The oxygen content of the water in such areas would be reduced to a level which would not support fish. Operators of fish camps would not be able to launch boats nor would recreational boaters be able to take advantage of the water areas. Commercial navigation as well as pleasure navigation would be materially impeded. The conveyance capacity of flood control canals would be sharply reduced and evapotranspiration of water would be materially increased. Public health would be affected because of increased mosquito breeding and snake infestations.

The St. Johns River in Florida is an extensive slow-moving river that originates in a marsh area near Vero Beach, Fla. and flows northward approximately 502 km through a series of lakes to Jacksonville, Fla.

The Jacksonville District of the U.S. Army Corps of Engineers along with state and local agencies conducts water hyacinth control operations throughout the St. Johns River Basin, using a chemical spray treatment with the dimethylamine salt of 2,4-D.

The results of a series of studies indicated that the herbicide 2,4-D did not have adverse environmental effects and a registration for this use was approved by the EPA.

REFERENCES

1. U.S. Department of the Army, Expanded Project for Aquatic Plant Control, House Document No. 251, 89th Congress, 1965.
2. **Walker, C. R.**, Toxicological effects of herbicides on fish environment, *Water Sewage Works*, 3, 113, 1969.
3. **Walker, C. R.**, Problems in clearance and registration of herbicide for aquatic areas, *Hyacinth Control J.*, 9, 5, 1971.
4. **Gangstad, E. O.**, Herbicidal control of aquatic plants, *J. Sanit. Eng. Div., ASCE*, 98, 397, 1972.
5. **Holm, L. F., Weldon, L. W., and Blackburn, R. D.**, Aquatic weeds, *Science*, 166, 699, 1969.
6. **Schultz, D. P. and Gangstad, E. O.**, Dissipation of residues of 2,4-D in water, hydrosoil and fish, *Aquatic Plant Manage. J.*, 14, 43, 1976.
7. **Joyce, J. C. and Sikka, H. C.**, Residual 2,4-D levels in the St. Johns River, Florida, *Aquatic Plant Manage. J.*, 15, 76, 1977.
8. **Aly, O. M. and Faust, S. D.**, Studies on the fate of 2,4-D and ester derivatives in natural surface water, *J. Agric. Chem.*, 12, 541, 1964.
9. **Audus, L. J., Ed.**, The Physiology and Biochemistry of Herbicides, Academic Press, New York, 1964.
10. **Averitt, W. K. and Gangstad, E. O.**, Dissipation of 2,4-D residues in static water, *J. Environ. Qual.*, 5, 145, 1976.
11. **Bell, G. R.**, On the photochemical degradation of 2,4-Dichlorophenoxyacetic Acid and the structurally related compounds in the presence and absence of riboflavin, *Bot. Gaz.*, 109, 314, 1956.
12. **DeMarco, J., Symons, J. M., and Robeck, G. G.**, Behavior of synthetic organics in stratified impoundments, *J. Am. Water Works Assoc.*, 59, 965, 1967.
13. **Frank, P. A. and Comes, R. D.**, Herbicides in pond water and hydrosoil, *Weeds*, 15, 210, 1967.
14. **Hemmett, R. B. and Faust, S. F.**, Biodegradation kinetics of 2,4-dichlorophonoxyacetic acid by microorganisms, *Herbicide Rev.*, 29, 191, 1969.
15. **Okey, R. W. and Bogan, R. A.**, Apparent involvement of electronic mechanisms limiting the microbial metabolism of pesticides, *J. Water Pollut. Control. Fed.*, 37, 692, 1965.
16. **Schwartz, H. J.**, Microbial degradation of pesticides, *J. Water Pollut. Control Fed.*, 39, 1701, 1967.
17. **Sikka, H. C., Appleton, H. T., and Gangstad, E. O.**, Uptake and metabolism of dimethylamine salt of 2,4-dichlorophenoxy acetic acid by fish, *J. Agric. Food Chem.*, 25, 1030, 1977.

Chapter 16

AQUATIC WEED CONTROL IN A POTABLE WATER RESERVOIR*

INTRODUCTION

The Walker Dam Impoundment Aquatic Plant Control Project was authorized by Section 302 of the River and Harbor Act of 1965 (Public Law 89-298, 89th Congress). The proposed project is designed to control the profuse infestation of Brasilian waterweed Elodea (*Egeria* sp.) through chemical treatment with a 50/50 mixture of diquat and potassium endothall. The most recent economic analysis (May 1969) indicates a benefit-to-cost ratio of 2.9:1.0.

The Walker Dam Impoundment is a tidal barrier dam located 30 mi southeast of Richmond, Va. on the Chickahominy River (Figure 1).[2] The dam was constructed in 1943 as a war emergency measure. Today, water provided by the reservoir is piped to the nearby cities of Williamsburg and Newport News, Va. The reservoir covers an area of 1100 acres. Maximum dam height is approximately 3 ft above the normal maximum tide level. Extreme high tides from hurricanes or other unusual weather conditions occasionally overlap the tidal barrier.

The Chickahominy River, a tributary of the James River, follows a winding southeasterly path before converging with the James approximately 15 mi west of Williamsburg, Va. Prior to dam construction, the Chickahominy River was navigable in the area inundated by the Walker Dam. Navigability is maintained by a small boat lock which was installed in the face of the dam.

Chickahominy Reservoir is situated in a rural area consisting of lowlying marshland, most of which is below elevation 10 ft, mean sea level (msl). Walkers, a small rural settlement having a population of about 15, is located on the Chickahominy. The nature of the land surrounding the reservoir does not lend itself to widespread agricultural usage. No major industries are associated with the reservoir other than a tree farm maintained by the state.

Aside from its principal service as a portion of the water supply for Newport News and Williamsburg, Chickahominy Reservoir is noted for its fishery. The shallow, enriched, and weed-covered waters provide excellent habitat for large-mouth bass, pickerel, and bluegill. Crappie, flier, bowfin, and other species also inhabit the reservoir in good numbers. Surrounding marshes provide private and commercially guided duck hunting opportunity in fall and winter months. Five separate private fishing marina and duck hunting camps, as well as numerous cottages, cover and control most of the northern shoreline of Chickahominy Reservoir. The State Forestry Department of Virginia maintains a tree farm near the upper end of the reservoir. Nearly all of the southern shoreline is in private ownership as timber tracts, private duck marshes, or summer cabins. One youth camp is also located along the southern shoreline near Walker Dam.

The primary purpose of Chickahominy Reservoir, that of supplying water for domestic purposes, is in jeopardy as are associated recreational activities. Based on information concerning the rate at which the reservoir surface is being reduced by filling (approximately 200 surface acres in 25 years), and on acreage of the original stream channel (around 184 acres), it is predicted that in another 115 years only the stream channel will remain. Long before that time, use of the reservoir as a functional unit will be lost. Fishing and boating waters are not only declining rapidly because of sur-

* By Carl B. Schreck, formerly Assistant Professor of Fisheries, Virginia Polytechnic Institute and State University, Blacksburg.

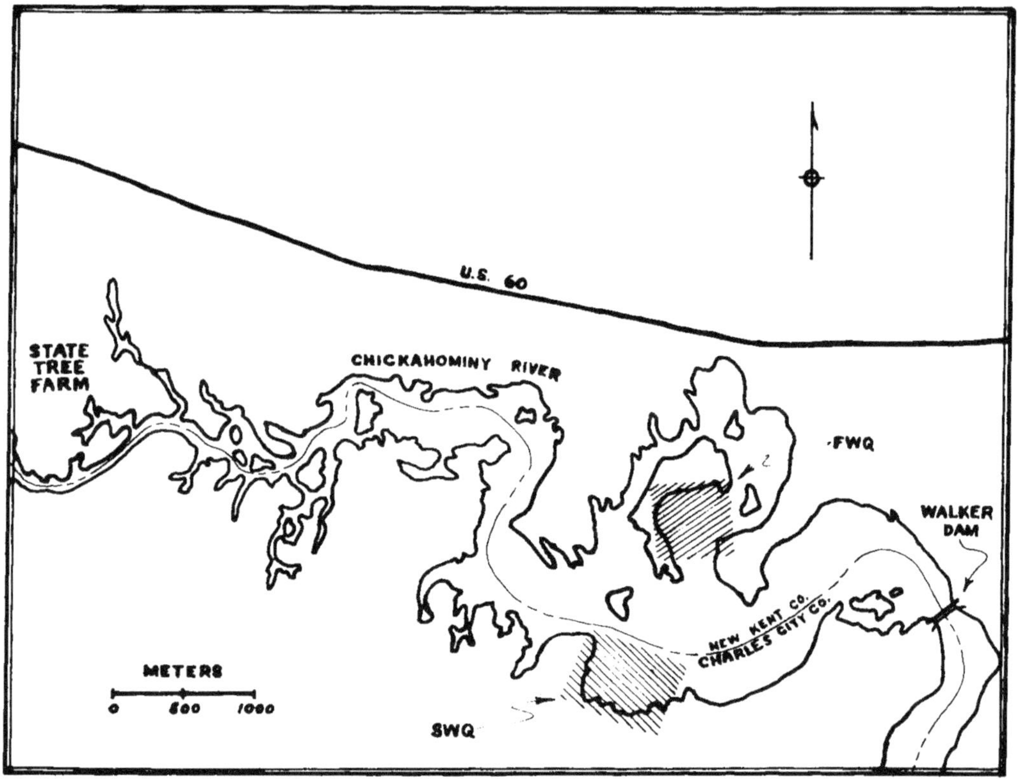

FIGURE 1. Chickahominy River and Reservoir areas of treatments SWQ and FWQ.

face area reduction, but rooted aquatic weeds now restrict the use of much of the reservoir during spring, summer, and fall months. The declining storage capacity and recreational potential contrast sharply with water and recreational needs of Newport News (one of the fastest growing cities in the nation) and the surrounding region.

The reasons for rapid decline in storage capacity and in the restriction of recreational activities are severalfold. A major problem is the profuse infestation of rooted aquatic weeds which provide a constantly increasing accumulation of plant material and which tend to entrap silt in increasing quantities. Shallow areas result. As eutrophication continues, shallow areas are invaded by typical marshland vegetation, and the shallows become intermittently wetted marshes. Large quantities of nitrogen, phosphorus, carbon, and other plant nutrients being supplied from upstream sources compound problems by creating ideal conditions for the growth of plant life. Nutrient inputs into the Chickahominy drainage from sewage disposal systems have reached such high levels that the State Water Control Board has virtually halted further effluent discharges.[5]

Elodea *(Egeria densa)* is the predominant rooted aquatic, although pondweeds (*Potamogeton* sp.), coontail *(Ceratophyllum demersum)*, and several other rooted aquatic plants are common. Green algae (*Pithophora* sp.) and duckweed (*Lemna* sp.) have increased in proportion to rooted aquatics. Shoreline vegetation varies, being composed of rushes, grasses, sedges, water shield *(Brasenia schreberi)*, golden club (*Orontium aquaticum*), spatterdock (*Nuphar* sp.), and many other marsh inhabiting plants.

During the summer of 1967, the Virginia Commission of Game and Inland Fisheries conducted a limited elodea control program. Strips of the reservoir surface, totaling 200 surface acres altogether, were treated with a herbicidal mixture consisting of diquat and potassium endothall. Equal portions of the two herbicides were sprayed at an overall rate of 1.5 gal/surface acre. Treated areas were surveyed during the summers of 1967 and 1968. Chemical treatment was successful in clearing fishing lanes, channels, and docksides. The area treated and cleared was found to total approximately 800 surface acres.

PROPOSED ACTION

Selection of the best herbicides to eliminate the infestation of elodea on the Walker Dam centered around those which biodegrade rapidly, present minimal hazards to humans or other life, are readily accessible in desired quantities, are relatively inexpensive, and have been cleared for use in potable water supplies.

Diquat is known to be an effective herbicide for the control of elodea. As a herbicide diquat stimulates photo-oxidative processes which ultimately destroy plant cells through dessication. The herbicide, a systemic weed killer, has been shown to biodegrade in 10 days or less.

Potassium endothall has also been used in the past to control elodea. The contact nature of this chemical requires a rather long exposure of the plant to the herbicide. Potassium endothall biodegrades in approximately 7 days after initial application.

The proposed chemical treatment will render water in the reservoir unsuitable for drinking purposes for a period of approximately 1 week. During the 1967 chemical treatment, the cities of Williamsburg and Newport News agreed to curtail their use of reservoir waters for a period of 7 days following treatment.

Literature reviews by the commission biologist indicate that application levels well above the proposed rate of 1.5 gal/surface area would be required before concentrations became lethal for fish life. Diquat has been shown to have an acute oral lethal dose (LD) rate of 400 to 440 mg/kg for rats while endothall has an acute oral LD of 35 mg/kg of body weight for rats. Presumably, both diquat and endothall have relatively low toxicities for both fish and mammals. Few fish kills were associated with the initial 200-acre treatment of the reservoir. Fish mortalities were attributed to an oxygen depletion caused by the rapid decomposition of the dead elodea and not necessarily to introduction of the chemical mixture.

Chemical treatment of the entire 1100 acres may present some adverse effects on associated fish and wildife forms. The initial aquatic plant control project was designed to spray approximately 200 acres of the impoundment. Erratic wind and wave conditions caused the herbicides to diffuse at a greater rate than previously anticipated. As a result, approximately 800 acres of the impoundment were effectively treated with diquat and potassium endothall. This treatment represented approximately 73% of the reservoir. If the entire reservoir area is treated, excluding the duck hunting marshes, some minor fish kills can be expected as a result of plant decomposition and an accompanying oxygen drawdown.

A second effect of the chemical treatment is potential damage to the tree farm located near the upper reaches of the lake. Irrigation of the forest nursery with lake waters would have to be suspended until toxicity of the herbicides decreased to an approved level. This problem could be eliminated if the spraying operation purposely avoided the extreme upper reaches of the lake.

A third effect which may be associated with the proposed herbicide treatment is the potential downstream contamination associated with an overlap of the dam by wave action. Possible damage to the downstream fauna and flora would then result; however, this condition would only occur as a result of the high tides generated by a hurricane or some other atypical weather condition.

A fourth effect, contamination of drinking water, is possible if water is pumped from the reservoir before biodegradation of the two herbicides is complete. Literature reviews have indicated both potassium endothall and diquat biodegrade rapidly. No problems were encountered when the chemicals were used previously. However, it is proposed that waters in the vicinity of the existing pumping station, and at selected intervals in the reservoir, be tested for the presence of both chemicals prior to resumption of water supply pumping. Under these conditions, actual contamination of water supplies would be remote.

The Virginia Commission of Game and Inland Fisheries, with the concurrence of the Virginia Department of Health and cities of Williamsburg and Newport News, report that the chemical treatment of the reservoir was relatively successful in light of the other means by which the plant growth could have been controlled.

The first alternative would be to forego the improvement of this reservoir and preserve the present setting of Walker Dam. State and local interests are unanimously in favor of the proposed project since the improvement of the reservoir would be viewed as an economic asset to all who live in the area, to those who depend on the reservoir for at least a partial livelihood, and to those who use the reservoir for recreational purposes. Since there has been no objection to the proposed action, there is no reason to consider the "no-improvement" alternative.[3,9,10]

A second possible alternative would be to alter the reservoir habitat in such a way as to deter the yearly growth of waterweeds. Water level reduction, followed by an interval of exposure to freezing weather, may provide long lasting control of aquatic vegetation. This means of modifying the habitat could result in complete elimination of weed growths. Coordinated flushing actions also would help to reduce the accumulation of silt and humic material in the lake or reservoir. Control of this type can be regarded as permanent, although annual "drawdown" would be required. Unfortunately, no means exist for rapid flushing of Chickahominy Lake, and long-term water level reductions would present problems. Annual exposure of shallows to freezing and dessication, when coupled with flushing actions, if feasible, undoubtedly would be the least expensive, long-term, permanent solution for the control of any rooted aquatic growths in Chickahominy Reservoir. However, manipulation of the reservoir's water level is not recommended by the Norfolk District for the following reasons:

1. The dam is old and would require extensive remodeling, if not replacement, in order to install a new system of gates.
2. At mean sea level, the difference in elevation of Chickahominy River immediately below the impoundment and the impoundment is about 3 ft. During high tides and storms, there is actually an intrusive flow of downstream salt water into the fresh-water impoundment. The slight difference in water levels could make successful lowering of the impoundment, to any extent, an impossibility.
3. Since elodea grows to a depth of about 10 ft and the depth of the impoundment in most areas is less than this figure, it would be necessary to virtually "dry-up" the reservoir to obtain any appreciable bottom dessication or root freezing.
4. Normal daily summer pumping rate for Chickahominy Reservoir is approximately 22 million gal. If the level of the reservoir were lowered in winter and not refilled by the following summer, dangerously low water supplies would result.

A third alternative which could be employed to control aquatic vegetation involves the use of mechanical devices to harvest weed growth. The regenerative capacity of subdivided fragments of elodea would require the mechanical harvester to not only sever the plant, but also collect all fragments. Incomplete removal of plant fragments would allow conditions prior to treatment to be quickly restored.

A field inspection by Norfolk District personnel of several mechanical cutters and harvesters revealed that most cutters adequately cut aquatic plants, but at the same time they leave a substantial amount of plant fragments unharvested. This floating material can quickly reestablish itself, reinfest controlled regions, and spread to other areas. Thus, it can be assumed that any use of mechanical cutters and harvesters would intensify the present aquatic plant problem, rather than relieve it.

CHEMICAL CONTROL

A combination of two such herbicides, diquat and endothall, was applied to Chickahominy Reservoir, to control the nuisance weed *Egeria densa*. As part of a broader ecological evaluation of the treatment's impact on the reservoir, this study was undertaken to monitor posttreatment herbicide levels in the impoundment.[4,5]

A herbicide's impact, direct or indirect, on any aquatic system is primarily a function of its toxicity to the biota. The effective use of diquat and endothall, respectively, at recommended concentrations up to 40 ppm to control many submerged aquatic macrophytes is well documented. The toxic effects of each herbicide on nontarget organisms is not as well defined. The reported 96-hr median tolerance limit (TLm) of diquat for fresh-water gamefish range from 5 to 150 ppm and higher depending on water hardness. The 96-hr TLms of endothall to fresh-water gamefish falls over much the same range with a tendency to be generally less toxic than diquat.[3,6,8]

The toxicity of the herbicides within the reservoir system is dependent upon the initial concentrations and the contact time of the toxins. In one study diquat, applied at 0.62 ppm to pond water, disappeared after 4 days. A 0.5 ppm treatment in another water system persisted for 14 days. Loss of diquat from water is associated with its rapid absorption by plants and binding in hydrosoils. Endothall, applied at 0.7 ppm, was found to dissipate within 8 days. A 0.5 ppm treatment in another study persisted beyond 20 days. The variability in decline rates for both herbicides is dependent upon the availability of absorbent materials and rates of microbial decomposition.

Materials and Methods

Two study areas at the reservoir, a quiescent bay (SWQ) and a slightly more lotic bay (FWQ), were chosen for intensive study. The reservoir was treated on July 10, 1973, using Aquathol® K (Penwalt Corporation, endothall product) and Diquat (Chevron Corporation) at a rate of 0.75 gal of each herbicide per acre, average depth of 5 ft, calculated to give initial concentrations of 0.17 ppm and 0.23 ppm, respectively, in water. The herbicide mixture was applied from an airboat to the surface of the reservoir.

Pretreatment samples were collected on July 8, and hourly daily samples were collected from then until August 1. A biweekly schedule was then followed until September 1, when a monthly schedule was established. Water samples were collected in glass jars. Three jars for each herbicide were taken at each station. The diquat samples were fixed with H_2SO_4 to lower the pH to 2. One diquat sample for each station was shipped to Chevron Corporation, Richmond, Calif. for analysis. The remainder were put on ice and shipped to Virginia Polytechnical Institute and State University via busline. The samples were then frozen to await extraction. Samples selected for extraction were permitted to thaw overnight or were thawed the day of extraction in an oven at 80°C.

Diquat is a polar cation which will bind to a cation exchange resin when a sample is passed through the column. A series of rinses are made which do not remove the diquat but clean up the column. When the sample is sufficiently cleaned, saturated NH_4Cl, which displaces the diquat, is passed through the column and the eluate is collected.

Table 1
DECLINE OF ENDOTHALL FROM WATER AT CHICKAHOMINY RESERVOIR STUDY STATIONS

Date (July)	FWQ	SWQ	Elapsed time[a]
8	0.00	0.00	0 hr
10	0.05	3.66	4 hr
11	0.00	0.26	12 hr
	0.04	0.16	20 hr
	0.03	0.13	28 hr
12	0.09	0.20	40 hr
13	0.01	0.02	64 hr
14	0.02	0.08	4 days
16	0.02	0.03	6 days
19	0.00	0.03	9 days
22	0.00	0.01	12 days
25	0.00	0.00	15 days
28	0.00	0.00	18 days
3	0.00	0.00	24 days
15	0.00	0.00	36 days

[a] Each value represents the center of a triplet.

The diquat cation reacts with a dithionite indicator to give an intense green color which is compared to a standard colorimetrically. Recovery is around 80% and the lower limit of detectability is 0.01 ppm. The procedure is outlined in detail as Method RM-5, Chevron Chemical Company.[1]

The endothall analysis procedure is described by Carlson,[1] of Pennwalt Corporation. Water samples are boiled in acetic acid until only an endothall-acid solution remains. Endothall is then converted to the N-chloroethylimide derivative by reaction with 2-chloroethylamine hydrochloride. The imide is separated into chloroform and the solution is cleaned with activated charcoal and attaclay, and then taken to dryness. The residue is transferred to a vial with methanol and analyzed on a gas-liquid-chromatograph with a nitrogen detector. Recovery is similar to diquat and the lower limit of detection is also 0.01 ppm.

Results and Discussions

Table 1 represents the decline of endothall observed in water. Each point represents an average of 3 days values. The first point on each graph represents samples collected in "hot spots" and should not be interpreted as initial concentrations in the reservoir. The average value for the first day after treatment in SWQ was 0.18 ppm and in FWQ 0.02 ppm. The herbicide was nondetectable 16 days after treatment in SWQ and 10 days after treatment in FWQ.

Table 2 represents the decline of diquat in water. The average value for the first day after treatment was 0.07 ppm in SWQ and 0.02 ppm in FWQ. The herbicide dissipated within 16 days in SWQ and 10 days in FWQ.

It is apparent that the level of herbicide recovered in FWQ was much less than in SWQ. Chevron Corporation, running duplicate samples, found 102 ppm diquat in the FWQ on the first day of treatment and none thereafter. We attributed this to differences in treatment, environmental conditions, and sampling.

Table 2
DECLINE OF DIQUAT FROM WATER AT CHICKAHOMINY RESERVOIR STUDY STATIONS

Date (July)	FWQ	SWQ	Elapsed time[a]
8	0.01	0.00	0 hr
10	0.00	0.73	4 hr
11	0.03	0.11	12 hr
	0.01	0.07	20 hr
	0.00	0.00	28 hr
12	0.02	0.00	40 hr
13	0.02	0.01	64 hr
14	0.00	0.02	4 days
16	0.01	0.01	6 days
19	0.00	0.00	9 days
22	0.02	0.00	12 days
25	0.00	0.00	15 days
28	0.00	0.02	18 days
3	0.00	0.00	24 days
13	0.00	0.00	30 days
22	0.00	0.00	36 days

[a] Each value represents the center of a triplet.

Chickahominy is part of the Newport News water supply system and several steps were taken to protect the consumers. First, the lake was closed to public use for 10 days following treatment. Samples were taken immediately above the pumping station. In intake pipes in Chickahominy Reservoir, endothall was not detectable after 10 days and diquat was not detectable 16 days after treatment. Second, samples were collected in the impoundments 20 mi down the pipe receiving Chickahominy water and at no time were herbicides detected.

These results of these field studies agree with a review of the literature,[2-10] and demonstrate that the chemical treatment is satisfactory from the standpoint of efficacy and residue tolerance.

Community Metabolism

Additional studies were initiated to study community metabolism. Following application of a 6.7 dihyddrodipyrido (1.2a:2′, 1′C) pyrazidiinium ion (diquat) and 7-oxabicyclo (2.2.1) heptane 2.3-dicarboxylic acid (endothall) herbicide mixture to control the submersed macrophyte egeria (*Egeria densa* planch.) in Chickahominy Reservoir, Va., daily oxygen consumption by the aquatic community regularly exceeded daily community oxygen production in areas of moderate and heavy macrophyte density. Subsequent recovery aided by phytoplankton increases returned these areas to approximate equilibrium within 6 to 7 weeks. Treatment did not alter the pattern of aerobic community metabolism in an area of light macrophyte growth. Community oxygen consumption 6 weeks after treatment indicated that 24 to 76% of the macrophyte biomass lost from the water column in an area of heavy growth, and are in general agreement with a review of the literature.[9-12]

SUMMARY AND CONCLUSIONS

The data of this study demonstrate that herbicide treatments were effective for control of aquatic vegetation and the herbicide residue levels were far below levels toxic to fish or the tolerance levels for potable water. The oxygen levels under conditions of the study were not seriously depleted and there was no fish kill. A delay of one week after treatment, before reopening for fishing, was acceptable to the public.

REFERENCES

1. APHA, American Water Works Association, and Water Pollution Control Federation, in *Standard Methods for the Examination of Water and Wastewater,* 13th ed., American Public Health Association, New York, 1971.
2. **Berry, C. R., Schreck, C. B., and Corning, R. V.,** Control of egeria in a Virginia water supply reservoir, *Hyacinth Control J.*, 13, 24, 1975.
3. **Berry, C. R.and Schrech, C. B.,** Aquatic macroinvertebrate response to field application of the combined herbicides diquat and endothall, *Bull. Environ. Contamination Toxical.*, 11, 374, 1975.
4. **Brooker, M. P. and Edwards, R. W.,** The effects of the herbicide paraquat on the ecology of a reservoir. II. Community metabolism, *Freshwat. Biol.*, 3, 383, 1973.
5. **Daniel, T. C.,** Evaluation of Diquat and Endothall for Control of Water Milfoil *(Myriophyllum exalbescens)* and the Effect of Weedkill on the Nitrogen and Phosphorus Status of a Waterbody. II. Design and Construction of a Shallow Water Sediment Core Sampler, Ph.D. thesis, University of Wisconsin, Madison, 1972.
6. **Davidson, J. F. and Cullen, W. J.,** The determination of diffusion coefficients for sparingly soluble gases in liquids, *Trans. Inst. Chem. Eng.*, 35, 51, 1957.
7. **Fish, G. R.,** Some effects of the destruction of aquatic weeds in Lake Rotoiti, New Zealand, *Weed Res.*, 6, 350, 1966.
8. **Gordon, R. W., Beyers, R. J., Odum, E. P., and Eagen, R. G.,** Studies of a simple laboratory microecosystem: bacterial activities in a heterotrophic succession, *Ecology*, 50, 86, 1969.
9. **Hall, D. J., Cooper, W. E., and Werner, E. E.,** An experimental approach to the production dynamics and structure of freshwater animal communities, *Limnol. Oceanogr.*, 15, 839, 1970.
10. **Hoover, T. B. and Berkshire, D. C.,** Effects of hydration on carbon dioxide exchange across an air water interface, *J. Geophys. Res.*, 74, 456, 1969.
11. **Hosokawa, I. and Ohshima, F.,** An improved method of phosphorus analysis in seawater, *Water Res.*, 7, 283, 1973.
12. **Jewell, W. J.,** Aquatic weed decay; dissolved oxygen utilization and nitrogen and phosphorus regeneration, *J. Water Pollut. Control Fed.*, 43, 1457, 1971.
13. **McConnell, W. J.,** Productivity relations in carboy microcosms, *Limnol. Oceanogr.*, 11, 1, 1966.
14. **McIntire, C. D. and Wulff, B. L.,** A laboratory method for the study of marine benthic diatoms, *Limnol. Oceanogr.*, 14, 667, 1969.
15. **Nichols, D. S. and Keeney, D. R.,** Nitrogen and phosphorus release from decaying water milfoil, *Hydrobiology*, 42, 509, 1973.
16. **Odum, H. T.,** Primary production in flowing waters, *Limnol. Oceanogr.*, 1, 102, 1956.
17. **Quinn, J. A. and Otto, N. C.,** Carbon dioxide exchange at the air-sea interface: flux augmentation by chemical reaction, *J. Geophys. Res.*, 76, 1539, 1971.
18. **Strange, R. J. and Schrech, C. B.,** Response of aerobic community metabolism to chemical treatment of aquatic macrophytes, *J. Aquatic Plant Manage.*, 14, 45, 1976.
19. **Sutherland, E. R.,** A water quality study of the Chickahominy River, Virginia State Water Control Board, Richmond, 1965, 1.
20. **Teal, J. M. and Kanwisher, J.,** The use of pCO for the calculation of biological production, with examples from waters off Massachusetts, *J. Mar. Res.*, 24, 4, 1966.
21. **Walker, C. R.,** Endothall derivitives as aquatic herbicides in fishery habitats, *Weeds*, 11, 226, 1963.
22. **Welch, H. E.,** Use of modified diurnal curves for the measurement of metabolism in standing water, *Limnol. Oceanogr.*, 13, 679, 1968.

Part IV: Research and Development Studies for Control of Eurasian Watermilfoil

Chapter 17

EURASIAN WATERMILFOIL

INTRODUCTION

Eurasian watermilfoil, *Myriophyllum spicatum* L., is an exotic species distinguishable from native species of *Myriophyllum* by its specifically aggressive growth. It is rooted, submersed, and grows principally in quiet water as deep as sufficient light for photosynthetic activity penetrates. It forms a dense floating mat on the water surface. It may reproduce from seed, but more generally, an explosive development results from fragmentation of rapidly growing plant tips.

M. spicatum, native to Europe and Asia, reputedly was released in this country in the 1880s. It spread southward from New Jersey; growth was reported in areas from the Potomac River in 1933. By 1956, it had spread explosively throughout the more saline areas of the lower Potomac and into many embayments of upper Chesapeake Bay. Subsequently, it has spread in all directions and is now found northward into Canada, westward to Washington and southern Canada, southward to Texas and Oklahoma, and across to California.

Eurasian watermilfoil has become a most troublesome aquatic weed. Due to its rapid growth rate, fragmentation, migration, and establishment, milfoil is a serious and real threat to our water resources. In some reservoir areas, heavy infestations have depressed real estate values; stopped recreational water activities such as boating, fishing, skiing, and swimming; clogged municipal water supply intakes; clogged intakes and screens through which raw water passes; hindered small boat navigation; caused commercial fishermen to abandon lines and nets; and provided extensive new mosquito breeding areas.[6-11]

Eurasian watermilfoil, although reported in the U.S. since the 19th century, apparently has become a pest only recently. It has spready widely in the Chesapeake Bay area and in the Northeastern states during the past 20 years. It was allegedly introduced into the Tennessee River watershed by implantation in the Piney River embayment of Watts Bar Lake near the Spring City Boat Dock in east Tennessee about 1953.[12]

M. spicatum was first reported in North Carolina in 1959 when it was observed in one of the fresh-water impoundments on the Pea Island National Wildlife Refuge. This growth was eradicated in 1962 when ocean water overwashed the fresh water during a storm. Small patches of Eurasian watermilfoil were reported by Currituck Sound, N.C. fishermen in 1964, but it was not positively identified until the following year when a localized primary area of infestation found in northern Currituck Sound was estimated to cover 40 ha. Rooted secondary growth of transplants that year were found over an additional 400 ha.

By the following summer (1966), a heavy infestation of watermilfoil covered some 3000 ha of Currituck Sound with another 27,000 ha supporting established growth. Areas of infestation in South Carolina were observed in Back Bay, Albermarle Sound, Kitty Hawk Bay, Croatan Sound, and North River during the fall of 1966. A heavy but limited growth was observed in June 1970 for the first time on the south side of Albermarle Sound in East Lake near Haulover Point and in two isolated patches along Alligatorweed Creek.

The historic pattern of watermilfoil growth, like that of most exotics, seems to be an explosive development following introduction, a stabilized growth at a relatively

* By Gordon E. Smith (retired), formerly Biologist, Tennessee Valley Authority, Muscle Shoals, Ala.

high incidence, and ultimately a gradual subsidence to a lower level in equilibrium with native vegetation. The Eurasian watermilfoil infestation in Currituck Sound and adjacent waters has apparently stabilized.

IDENTIFICATION AND TAXONOMY

Eurasian watermilfoil is closely related to a widespread North American plant, *M. exalbescens*. The latter was not differentiated from *M. spicatum* until plants from Greenland, Canada, and the U.S. were examined and it was separated as a species.[7] Subsequently, *M. exalbescens* sometimes has been considered to be a variety of *M. spicatum*. *M. exalbescens* is presently reported to occur from Greenland, southeastern Labrador, and Alaska south to northern New Jersey, Pennsylvania, Ohio, Indiana, Illinois, Kansas, Arizona, and northwestern Mexico. In much of its range, it grows in glaciated areas. Although both it and *M. spicatum* are very similar in appearance, leaves on well-developed stems of *M. spicatum* have more numerous (14 or more pairs) and closer divisions. Thus, the plant leaves resemble weathered, water-soaked feathers in contrast with the more open appearance of leaves of *M. exalbescens*. Both plants vary in stiffness and curvature of the leaf divisions, in stem color, and in growth habits. Mature *M. spicatum* seeds average considerably smaller in size than those of *M. exalbescens*.

From evidence cited, it appears that Old World *M. spicatum* is a relatively new plant in the U.S. Here its range partially overlaps that of the native *M. exalbescens*, possibly either a variety of *M. spicatum* or sufficiently distinct to be considered a separate but closely related species. Regardless of future changes in nomenclature, it is evident that a plant resembling Old World *M. spicatum* has become a serious pest in numerous areas in the eastern U.S. not occupied by the native *M. exalbescens*.

BIONOMICS AND DISTRIBUTION

Eurasian watermilfoil is a perennial having a wide range of environmental tolerances. The bottom surfaces on which it grows vary from muck to hard packed sand. It is found in fresh waters as well as in those waters ranging in salinity up to 16 parts per thousand (ppt). Exploratory aquarium experiments at Solomons, Md. revealed that the plant grew most rapidly in fresh water, but it still showed vigorous growth at salinities up to 10 ppt. At 20 ppt, it remained alive during the summer, but grew very slowly. Most foliage died except near the growing stem tips. Eurasian watermilfoil can withstand considerable wave action and daily tidal fluctuation of several feet, but grows best in sheltered coves. Although it is most abundant in shallow (4 m) water, it has been found in clear water at depths of 6 m at high tide. Beds also have been found in 2 m of water in which turbidity at the time of the survey prevented sighting of objects below a depth of 1 m.

Reproduction and plant spread occur primarily by stem fragmentation. Small pieces of living stems with leaves are commonly observed drifting in the water, particularly in late fall. Additionally, plant fragments caught on boat motors, fishing gear, or embedded in cakes of drift ice are transported to new areas. If fragments lodge at the bottom, they produce roots and become anchored. Winter buds, separated from the stems, settle to the bottom, sprout with the advent of a new growing season, and comprise another means of reproduction. These buds may be scattered to other areas by tides and currents. Seeds are also formed and likewise distributed, particularly under conditions adverse to vegetative growth.

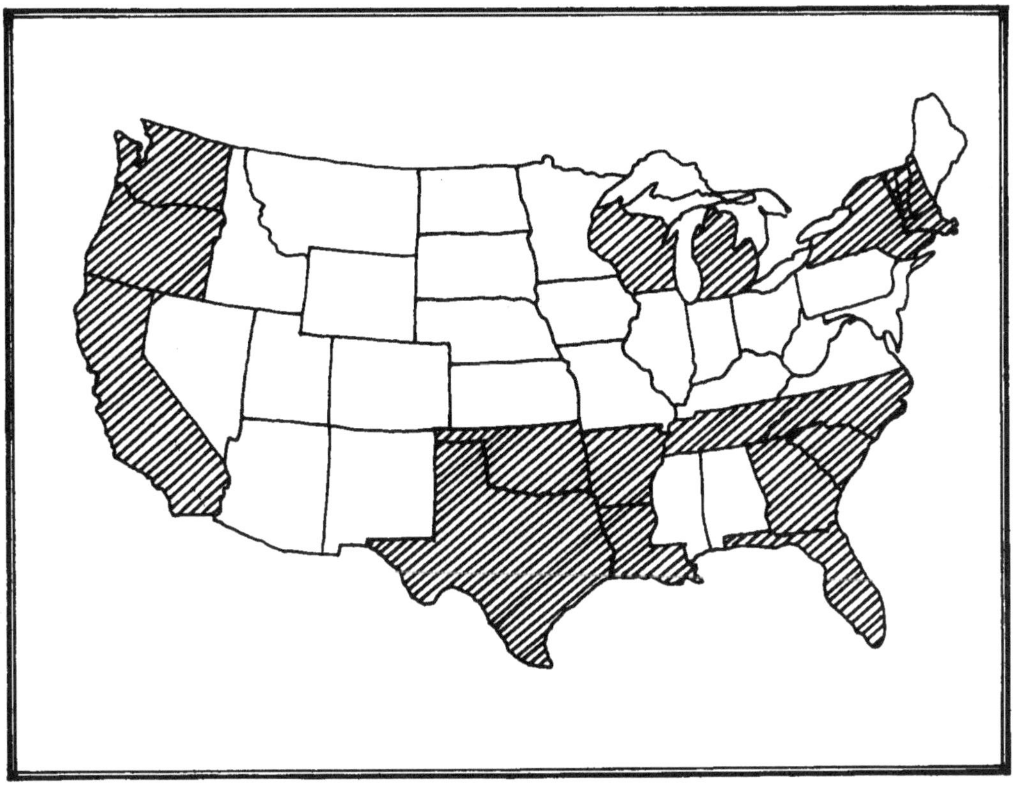

FIGURE 1. Distribution of Eurasian watermilfoil *(Myriophyllum spicatum)* in the U.S. (shaded areas).

Distribution in the U.S.

Eurasian watermilfoil is distributed over many areas of the U.S. in both south and north temperate climates, growing in both fresh and saline waters and rooted in hard-packed sandy soils. The plant tends to be limited to quiescent embayments and waters high in calcium. Growth is most aggressive where nutrient levels are high, but the plant will grow in waters not enriched. During the past 3 decades, it has been gradually distributed over most of the U.S. (see Figure 1).

Distribution in the British Isles

The Agriculture Research Council (ARC), Weed Research Organization (WRO), Begbroke Hill, Kidlington, Oxford, England, was visited on August 10, 1967. The WRO was established by the ARC in 1960 as the official station in Britain for research and technical information about weed control on cropland. It has responsibilities dealing with plant control and management, management of waterways as a flood preventive measure, and of ditches adjacent to agricultural lands to promote good drainage and prevent soil waterlogging. Control methods being studied are mechanical cutting, herbicidal treatments, and biological control. Eurasian watermilfoil in general causes no major problem in England and requires no special program for its control. Inspection of a small tributary of the Thames River near the Oxford Field Station revealed a few small colonies of Eurasian watermilfoil. Inspections of several other potentially favorable areas for this species along the Thames River and in the Slough area gave negative results. Other field observations between London and Oxford concluded that Eurasian watermilfoil was not an important aquatic plant in any of the examined areas. In England there are three species of milfoil: *M. spicatum* (spike milfoil), *M. verticil-*

latum and *M. alterniflorum*. The milfoil was reported to be a native plant occurring in lakes, ponds, and ditches up to an elevation of approximately 400 m above sea level; nevertheless, it is mainly considered a lowland plant. It is locally common in some areas, especially in calcareous waters, and has a wide geographical distribution in England. Spiked milfoil, sometimes a nuisance in English waterways, is not as troublesome as in certain areas of the U.S.

The Biological Records Center was developed from a mapping program of flowering plants and ferns initiated at Cambridge, England, by the Botanical Society of the British Isles (BSBI). The Biological Records Center developed because of demands for services of this type dealing with geographical plant and animal distribution. The system's basis involves determination of presence or absence of species of interest within 10-km squares of a National Grid containing 3500 of these basic units. More than 1½ million records of plant distribution data were collected with the help of amateur botanists for some 1700 plant species. The 40-column, punch card machinery was used for preparation of individual maps showing distribution of a given species. Figure 2 is a distribution map for *M. spicatum* in the British Isles.

The Windermere Laboratory of the Fresh Water Biological Association, Westmoreland, England, was visited on August 15, 1967. Dr. H. C. Gilson, Director of the Station, discussed the Eurasian watermilfoil problem. He pointed out that Lake Windermere was a soft-water lake in a region of Silurian slates and Ordovician outcrops with a minor ring of Carboniferous limestone at its end. Soft-water lakes in general are unfavorable sites for *M. spicatum*, the species being found more abundantly in hard or calcareous lakes in southern England. In any event, *M. spicatum* reportedly occurs only sparsely in Lake Windermere. *M. verticillatum* is apparently more frequent here. Although *M. spicatum* was not seen growing in Windermere, herbarium species specimens collected from this lake were observed. Only three species of milfoil occurred in England, the third being *M. alterniflorum*, an inhabitant of peaty waters in non calcareous rock areas of western and northern England. Several shoreline areas of Lake Windermere between the Ferry House and Ambleside were inspected but Eurasian watermilfoil was not found. Specimens of *Potamogeton crispus, Elodea canadensis, Phragmites communis*, and other littoral plants were seen.

The Grasmere, Thirlmere, and Bassenthwaite Lakes were visited, but they did not reveal any evidence of milfoil *(M. spicatum)*.* This is not to imply that *M. spicatum* does not occur here, but indicates that this milfoil was not a problem plant in these lakes. On August 16 to 17, various waters between Windermere and London were observed — the Manchester Ship Canal, a small reservoir near Cannock, and other streams and pools — but no problem areas of milfoil were encountered.

Distribution in Eastern Europe

On August 18 potential milfoil sites along the Seine River, Somme River, and other waterways between Paris and Antwerp were inspected. No problem areas of milfoil *(M. spicatum)* were found in waterways and ponds observed. Common plants noted at the Somme River included cane *(Phragmites communis)*, cattail *(Typha latifolia)*, white water lily *(Nymphaea alba)*, and coontail *(Ceratophyllum demersum)*. Waterways and pools between Antwerp and Amsterdam were observed on August 19, but milfoil *(M. spicatum)* was not found. On August 20, a trip was made by boat at Amsterdam to inspect the Amstel River, several of the many canals, and the harbor area with seagoing vessels and dry docks. Milfoil *(M. spicatum)* was not found in any of the areas along the boat trip route.

* The above three lakes are located between Ambleside and Cockermouth, England.

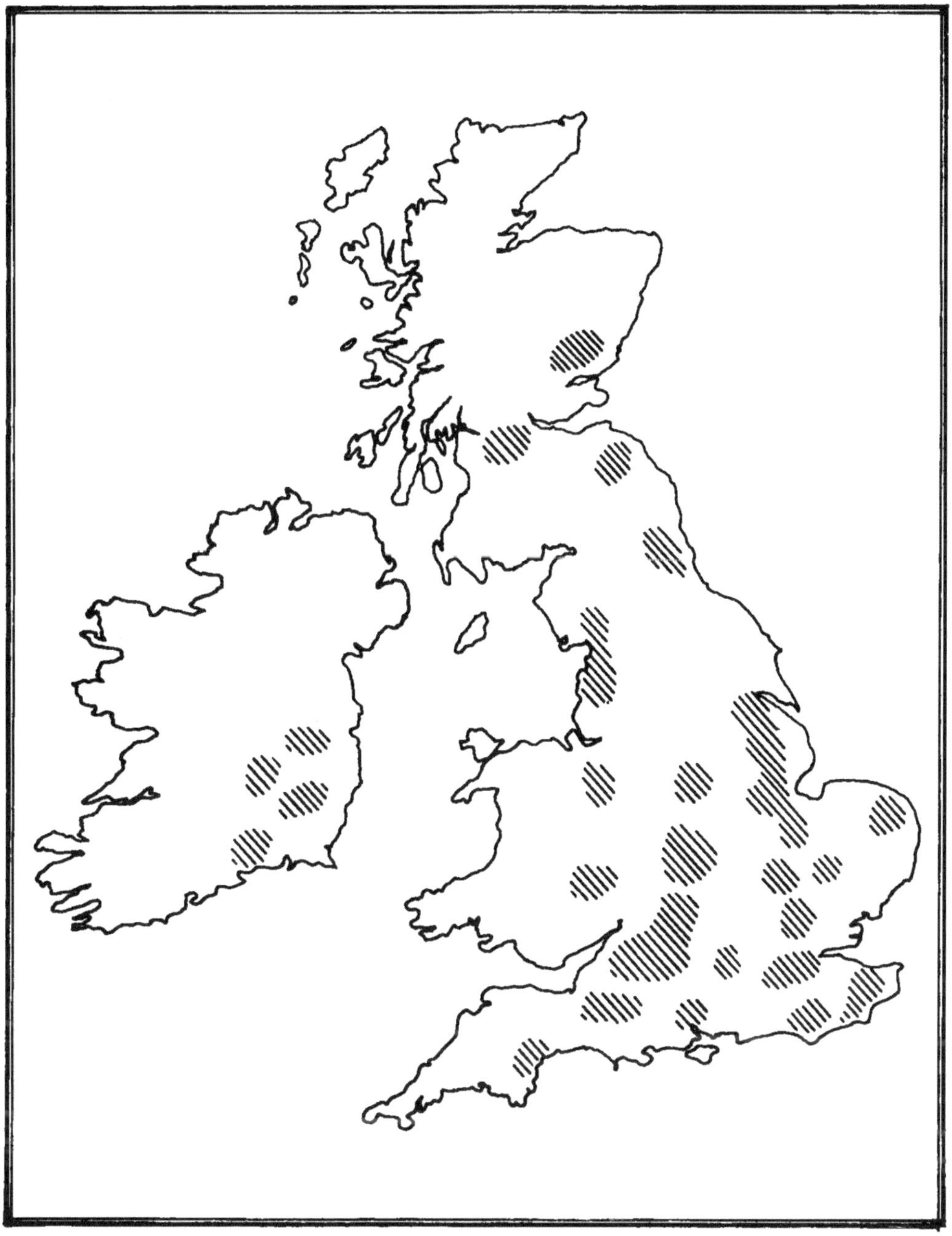

FIGURE 2. Distribution of Eurasian watermilfoil *(Myriophyllum spicatum)* in the British Isles.

The Hugo de Vries Laboratorium, Plantage Middenlaan, Amsterdam, was visted on August 21, 1967. It was thought that the explosive increase in milfoil *(M. spicatum)* in the U.S. was analogous to experiences in Western Europe following the early introduction of *E. canadensis* and of the water fern *(Azolla caroliniana)*. In 1935 (approximately) another aquatic plant *(E. nuttallii)* which caused difficulties was introduced to the continent.

A field trip through Holland via the Zuider Zee Dike was undertaken. The Zuider Zee in deeper parts was reportedly approximately 5 m in depth. Most of the sea is 3 to 4 m deep. Milfoil *(M. spicatum)* was not found growing in the Zuider Zee. Milfoil *(M. spicatum)* was found in isolated spots at Friesland. The milfoil was almost entirely in the vegetative condition with only a single flowering spike found. No fruiting spikes were encountered. Areas from Edam, North Holland, to the vicinity of Meppel in the Netherlands were inspected.

On August 25, several areas in the Northeastern Provinces of Groningen and Drent in the Netherlands were visited. These areas were also checkerboarded with waterways to keep water off agricultural lands. Inspections of several waterways in the Northeastern Provinces were also negative for milfoil *(M. spicatum)*. This milfoil was reported relatively abundant in southern Netherlands and infrequent in the Northeastern Provinces.

On August 26, the Rhein River at Mainz, Germany, was inspected but no milfoil *(M. spicatum)* was found. On August 27 the University of Heidelberg in Germany was visited and areas along the Neckar River were inspected but no milfoil *(M. spicatum)* was found.

Milfoil *(M. spicatum)* was reported to have a wide geographical distribution in Germany, but in southwestern Germany it did not occur in sufficient abundance to require control efforts. We were told that this milfoil occurs in still water and in slowly flowing waters. It thrives best in waters having high lime concentrations like those found in waters of calcereous regions.

Lake Mummelsee (about 1036 m above sea level) in Germany's Black Forest and streams near Baden-Baden were visited on August 29, 1967. Milfoil *(M. spicatum)* was not found at either location. Observation of the Murg River near Gensbach on August 29 was also negative. The plant was not found in areas of the Necar River in Germany.

Distribution in Yugoslavia

Eurasian watermilfoil is found in numerous areas of Yugoslavia and in many different habitats (see Figure 3). Plant overpopulation disables fish production, limits the use of irrigation and drainage canals, disrupts recreation boating activities, and in some areas interferes with hydroelectric power production. It is localized in large bodies of water and forms colonies along river banks or in slow-moving water. Only in artificial habitats wehre there is shallow, quiescent water does it occupy the whole surface. If we separate those habitats in which watermilfoil is in regression, we find that these areas do not fully satisfy the ecological niche as we know it. Moreover, in new artificial lakes, Eurasian watermilfoil is not always found, although conditions are favorable. It is postulated that natural controls operate under these situations and prevent growth of watermilfoil plants. These findings, in light of milfoil's well-established and wide distribution in Britain and Western Europe, support the view that Eurasian watermilfoil *(M. spicatum)* in its native environments does not present special problems as it does in the U.S. In this general area of origin, natural plant enemies are sufficient to provide natural control. To this end, a research project exploring biological control was established with the Institute for Plant Protection at Belgrad, Yugoslavia.[6]

PROBLEMS ASSOCIATED WITH EURASIAN WATERMILFOIL

Dense stands of Eurasian watermilfoil, like those of any other aquatic plant, completely usurp the occupied water area and render all recreation difficult, if not impossible. Boating and swimming are the principal recreational uses most vulnerable to drastic curtailment by dense stands of aquatic vegetation. Commercial fishing also becomes impossible in areas infested with Eurasian watermilfoil. A study in July 1963[10]

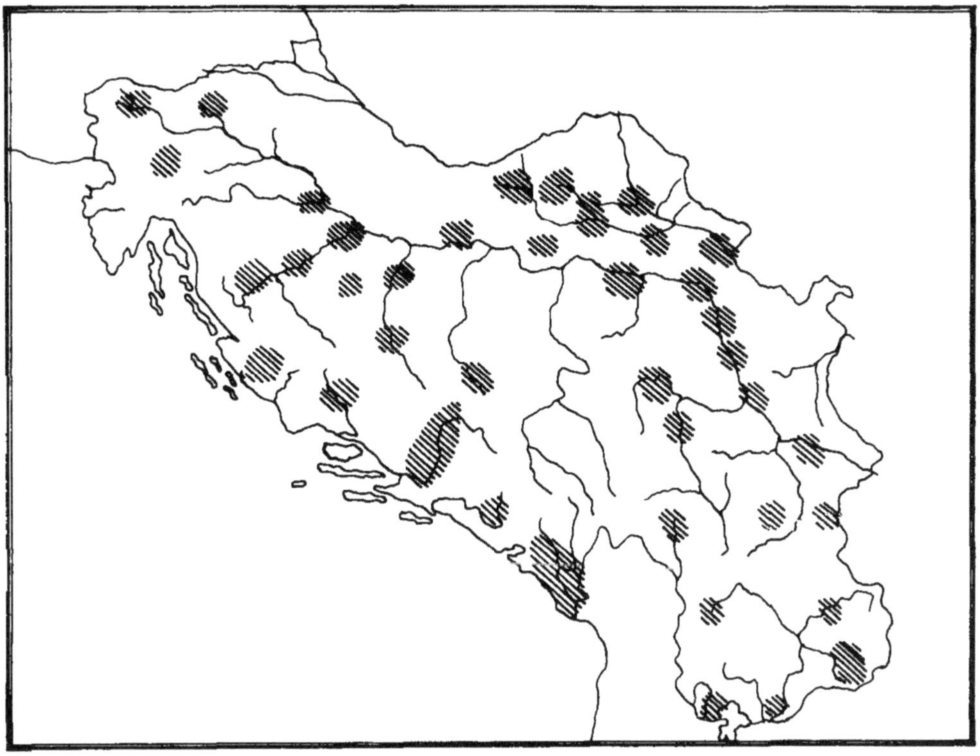

FIGURE 3. Distribution of Eurasian watermilfoil *(Myriophyllum spicatum)* in Yugoslavia.

(before Eurasian watermilfoil appeared) revealed that aquatic plant growth present was composed of 50% widgeon grass, 45% sage pondweed, and 5% redhead grass. In August 1966, the composition had shifted to 90% widgeon grass, 5% wild celery, 1% redhead grass, and traces of naiad and sage. By 1968, Eurasian milfoil comprised an estimated 20 to 25% of the Kitty Hawk Bay plant composition, and in April 1973, it represented approximately 75% of aquatic vegetation. However, there is no evidence that displacement of high food value aquatics by watermilfoil has adversely affected waterfowl populations.

It is suggested that watermilfoil constitutes a human health problem by influencing the increase of disease-causing organisms. There have been charges that milfoil acts as a cultural medium for organisms causing botulism. In addition, it has been implied that milfoil is somehow connected with increases in fecal coliform counts (an indicator of potential disease problems due to water pollution). It has also been stated that dense milfoil mats enhance mosquito breeding by providing habitat for egg laying and hatching.[12-14] The only botulism cases observed have been a few isolated cases of animal botulism affecting a few domestic ducks. The organisms causing these diseases are not harmful to man. There is no basis for charging that increases in fecal coliform are caused by Eurasian watermilfoil; however, the high density of milfoil indicates (as does fecal count) decreasing water quality. This decrease in quality may result from increased nutrients due to run-off from such sources as septic tanks. That milfoil mats do enhance mosquito breeding to some degree seems a valid charge. The magnitude and importance of this is unknown.

Aesthetic revulsion causes part of the opposition to milfoil infestation. Some people consider it unsightly and smelly at certain times of the year. This latter charge is scientifically sound since decay of large milfoil amounts in fall can cause emission of hydrogen sulfide, which has an extremely obnoxious odor. Attempts to put visual or odorous displeasure in dollar terms is extremely difficult. However, it seems logically sound to state that the negative value of milfoil in these terms may make a water area less desirable to some persons who frequent it for recreation, thus providing some degree of negative economic impact on the tourist industry. It is impossible to say whether the economic impact of this is compensated by increased numbers of freshwater fishermen who visit the area.

ENVIRONMENTAL IMPACTS OF WATERMILFOIL CONTROL

The selection of the particular herbicide type and amount to be used for milfoil control (butoxyethanal ester of 2,4-D) is a result of several different studies determining what herbicide has been used in similar situations with negligible environmental damage. Steenis and Stotts[9] stated that granular butoxyethanal ester of 2,4-D proves the most effective herbicide for controlling Eurasian watermilfoil in the upper reaches of Chesapeake Bay. The Tennessee Valley Authority (TVA) has used 2,4-D extensively to treat watermilfoil at rates of 20 to 40 kg ae/ha granular 2,4-D BEE and has found it very suitable for watermilfoil eradication and human safety.[12] Whitney et al.[15] reported that the use of 20 kg ae/ha of 2,4-D resulted in 95% watermilfoil reduction.

The toxicity of 2,4-D to fish has been widely studied. Toxicity, while slight, appears to depend on the formulation used; physical conditions of the test; age, sex, and sensitivity of the species exposed; temperature and exposure time; and other factors. In the 1968 application of 2,4-D (20 kg ae/ha) on 2 A of watermilfoil in Currituck Sound, N.C. no acute effects on fish were observed.[3-5] Other studies of toxicity to fish disclose the acute and chronic concentrations from 500 to 800 ppm depending on conditions. It is anticipated that concentration in the water from the application 20 ae/ha will not exceed 100 ppb almost immediately after application because of rapid herbicide absorption on the vegetation surfaces and dilution caused by water circulation in the treatment vicinity. For this reason, no toxic effect on fish population is observed. Benthic organisms (glass shrimp, damselfly nymphs, scuds, and blue crabs) showed no adverse effects in the 1968 application of 2,4-D in Currituck Sound.[13] In the monitoring of the 1968 watermilfoil treatment, plankton samples indicated no significant variation in numbers or species.

Watermilfoil is extremely susceptible to 2,4-D. The herbicide is absorbed on the plant surface and subsequently absorbed into plant tissues. Observations by Whitney et al.[15] indicate that the herbicide is specifically effective on watermilfoil and is not harmful to most other aquatic species in the area such as southern naiad *(Najas guadalupensis)*, wild celery *(Vallisneria americana)*, redhead grass *(Potomogeton perfoliatus)*, sage pondweed *(P. pectinatus)*, and widgeon grass *(Ruppia maritima)*. Growth of these species increases with decrease in population of dominant watermilfoil. This is beneficial to waterfowl since these are preferred food species. There is some potential damage to shoreline terrestrial vegetation by accidental application. Care must be taken in application to minimize chance of any deposition on terrestrial areas. Meteorological conditions, particularly wind velocity and direction, must be checked to insure no drift of the aerially applied granules, but granule application itself is a safety factor.

Postapplication sampling after the 1968 treatment of 2,4-D on 80 ha in Currituck Sound indicated no change in overall water quality of the treated area and no water samples containing residues of more than 0.10 ppm of the herbicide. The persistence of 2,4-D is determined by a variety of factors: rate of treatment, frequency of application, extent of dilution, biodegradation by microorganisms, metabolism by plants, temperature, and time. In the TVA studies, concentration of the chemical disappeared from water samples within 3 days to 2 weeks. Residue levels in water immediately fell to 100 ppb or less after application and dissipated within 3 weeks after treatment. Residues in fish and benthos also decreased to levels below the detectable limit of 0.10 ppm in the same pattern.[14]

In more than 20 years of extensive use, no recorded incidents of biological herbicide magnification in food chains exist. In summary, there appears to be no significant opportunity for potential long-term effects on the biological accumulation of 2,4-D.

The probability of human exposure to even transitory toxic concentration of 2,4-D during milfoil control activity is extremely remote. Problems associated with acute toxicity of 2,4-D to humans are likely only with purposeful or accidental ingestion of large amounts of undiluted 2,4-D. It is reported that acute fatal poisoning occurs from ingestion of 62 g, and single oral doses of 2 to 4 g are necessary to produce symptoms in humans.[3] The toxicity of 2,4-D to domestic animals varies. The dose required to kill 50% of a population in a given time, LD_{50} for rats, guinea pigs, and rabbits, ranges from 300 to 1000 mg/kg of body weight.[3]

The primary purpose of milfoil control is to alleviate social and economic problems associated with watermilfoil infestation. The areas selected for treatment are those where the greatest need exists to alleviate the social and economic problems of the citizens involved. The primary social impact of milfoil control will be increase in recreational water activity allowed by transformation of the area from weed infested to open water (allowing passage of recreational boating through the infestation) and increased visual aesthetic quality.

Clearing of recreation areas should have positive economic effects since it will stimulate businesses of local persons who derive income from activities associated with water recreation. It will also affect property values in the area, since improved aesthetics will make adjacent property desirable.

REFERENCES

1. **Beaven, G. F.,** Water Milfoil Studies in the Chesapeake Area, No. 60-52, Maryland Department of Research and Education, Annapolis, 1960.
2. **Beaven, G. F., Rawls, C. K., and Beckett, G. E.,** Field observations upon estuarine animals exposed to 2,4-D, *Proc. Northeast. Weed Control Conf.*, 16, 449, 1962.
3. **Dalgaard-Mikkelsen, S. and Poulsen, E.,** Toxicology of herbicides, *Pharmacol. Rev.*, 14(2), 225, 1962.
4. **DeVaney, T. E.,** Chemical Vegetation Control Manual for Fish and Wildlife Management Programs, Resources Publ. No. 48, U.S. Department of the Interior, Bureau of Sport Fishing and Wildlife, Washington, D.C., 1968.
5. **Duggan, R. E.,** Residues in food and feed, *Abstr. Pest. Monit. J.*, 2(1), 5, 1968.
6. **Lekie, M.,** Ecology of the aquatic insect species, *Parapoynx stratiotata* L., *J. Sci. Agric.*, 23, 49, 1970.
7. **Patten, B. C., Jr.,** The status of some American species of *Myriophyllum* as revealed by the discovery of the intergrade material between *M. exalbescens* Fern and *M. spicatum* L. in New Jersey, *Rhodora*, 56(670), 213, 1954.
8. **Pfitzenmeyer, H. T. and Drobeck, K. G.,** Benthic survey for populations of soft-shelled clams, *Mya arenaria*, in the lower Potomac River, Maryland, *Chesapeake Sci.*, 4(2), 67, 1963.
9. **Steenis, J. H. and Stotts, V. D.,** Progress report on control of Eurasian watermilfoil in Chesapeake Bay, *Proc. Northeast. Weed Control Conf.*, 15, 566, 1961.

10. **Steenis, J. H. and Stotts, V. D.,** Progress report on distribution and control of Eurasian watermilfoil in the Chesapeake Bay region, *Proc. Southeast. Weed Control Conf.*, 16, 341, 1963.
11. **Steenis, J. H., Stotts, V. D., Haven, D., and Whipp, A. A.,** Developments of control of Eurasian watermilfoil in the Chesapeake Bay region, *Proc. Southeast. Weed Control Conf.*, 17, 321, 1964.
12. **Smith, G. E. and Isom, B. G.,** Investigations of large scale applications of 2,4-D on aquatic fauna and water quality, *Pest. Monit. J.*, 1(3), 16, 1967.
13. Tennessee Valley Authority, Environmental Control of Eurasian Watermilfoil *Myriophyllum spicatum* L.) in TVA Reservoir, Mimeograph Rep. No. TVA-OHES-72-8, Chattanooga, Tenn., 1972.
14. **Thomas, M. L. H. and Duff, J. R.,** Butoxyethanol ester of 2,4-D in the control of eelgrass (*Zostera marina* L.) and its effects on oysters (*Crassostrea virginica* Gmelin) and other benthos, *Proc. Northeast. Weed Control Conf.*, 22, 186, 1968.
15. **Whitney, E. W., Montgomery, A. B., Martin, E. C., and Gangstad, E. O.,** The effects of 2,4-D application on the biota and water quality in Currituck Sound, North Carolina, *Hyacinth Control J.*, 11, 13, 1973.

Chapter 18

SCREENING TESTS OF 2,4-D*

INTRODUCTION

Environmental factors such as soil, water, air, plants, and animals; forces such as wind, tides, and gravity; conditions such as light, temperature, pollution, and humidity; and processes such as photosynthesis, mineral cycling, and decomposition constitute a complex natural ecosystem. For investigative purposes this concept may include extremes in size of such diverse living/nonliving systems as a drop of water, tidal pool, reservoir, watershed, or estuarine area. Quantitative evaluation must utilize the tools of the physical and biological sciences to measure the ecosystem's development, structure, and function. This discipline is largely an outgrowth of the natural sciences and may be termed "environmental biology." While this science emphasizes the interrelationships of living things and their living physical environment, it leans heavily on past accomplishments of the individual natural sciences. Treatments and modifications of the natural environment of upstream areas may be factors that significantly affect estuarine organisms and are the subject of this chapter.[1-7]

BIOASSAY SCREENING

Effects of several 2,4-D formulations on selected estuarine organisms were studied in acute toxicity tests. The following animals were used: crustacea — brown shrimp *(Penaeus aztecus)*, pink shrimp *(P. duorarum)*, and blue crabs *(Callinectes sapidus)*; mollusk — eastern oyster *(Crassostrea virginica)*, and fish — long-nose killifish *(Fundulus similis)*, mullet *(Mugil cephalus)*, and spot *(Leiostomus xanthurus)*. The animals were tested in flowing sea water and the herbicide was metered into the flowing water with sage pumps. Results of these tests are shown in Tables 1, 2, and 3. The various formulations of 2,4-D were only slightly toxic to the animals tested. The tables give effects at the highest concentrations tested. The no-effect level (NE) for the dimethylamine salt of 2,4-D was 5000 ppb for the blue crab, 1000 ppb for the eastern oyster and 15,000 ppb for long-nose killifish over a 48-hr exposure.

FEEDING TESTS

Twenty-five long-nose killifish *(F. similis)* were placed in each of three compartmented, flowing sea-water aquaria. Two groups were fed 1000 and 100,000 ppb, respectively, of recrystallized N,N-dimethyl-2,4-dichlorophenoxyacetamide in their diet for 2 weeks and uncontaminated food for 2 additional weeks. The third group of control fish received the uncontaminated food for 4 weeks. The food was homogenized fish fillets frozen into small cubes. Recrystallized herbicide was homogenized with the food. Each fish received approximately 31 g of food per week. The fish were fed every other day. Half of the fish from each group were removed at the end of 2 weeks and frozen before residue analysis. The remaining fish from each group were frozen at the end of the 4-week experiment. Results of the tests (Table 4) indicate that the 2,4-D was not detected in fish fed food containing 10,000 ppb of the herbicide. Fish that

* From Duke, T., Effects of 2,4-D Formulations on Estuarine Organisms, Tech. Rep. 12, U.S. Army Engineer Waterways Experiment Station, Vicksburg, Miss.

Table 1
EFFECTS OF VARIOUS FORMULATIONS OF 2,4-D ON THREE ESTUARINE CRUSTACEANS

Animal	Scientific name	Formulation	Concentration (ppb) EC24[a]	Concentration (ppb) EC48[a]
Brown shrimp	*Panaeus aztecus*	Technical acid	10% at 2000	10% at 2000
		Butoxyethanol ester	50% at 5600[b]	50% at 5600[b]
		Dimethylamine salt	2000[b]	10% at 2000[b]
		Propylamine glycol butyl ether ester	1000[b]	1000
Pink shrimp	*P. duorarum*	Butoxyethanol ester	1000	1000
Blue crab	*Callinectes sapidus*	Dimethylamine salt	5.0[b]	5000

[a] EC24 and EC48 = concentrations of 2,4-D that caused mortality or equilibrium loss in 50% of the animals tested after exposure in flowing water for 24 and 48 hr, respectively.
[b] No effect.

From Duke, T., Effects of 2,4-D Formulations on Estuarine Organisms, Tech. Rep. 12, U.S. Army Engineer Waterways Experiment Station, Vicksburg, Miss., 1976.

Table 2
EFFECTS OF VARIOUS FORMULATIONS OF 2,4-D ON THE EASTERN OYSTER *(CRASSOSTREA VIRGINICA)*

2,4-D Formulation	Concentration (ppb) EC96[a]
Technical acid	2000[b]
Butoxyethanol ester	2600
Dimethylamine salt	1000
Propylene glycol butyl ether ester	39% at 1000

[a] EC96 = concentrations of 2,4-D that caused 50% decrease in shell deposition after 96 hr when compared to untreated controls.
[b] No effect.

From Duke, T., Effects of 2,4-D Formulations on Estuarine Organisms, Tech. Rep. 12, U.S. Army Engineer Waterways Experiment Station, Vicksburg, Miss., 1976.

received food containing 100,000 ppb accumulated the herbicide, and all detectable residues were lost after 2 weeks feeding on uncontaminated food.

AVOIDANCE TESTS

Mosquito fish *(Gambusia affinis)* and sheepshead minnows *(Cyprinodon variegatus)* were exposed to untreated water and water containing various concentrations of the butoxyethanol ester of 2,4-D. Fish were tested in an apparatus allowing movement from a holding area into water containing 2,4-D or into clean water. Also, fish movement in relation to various concentrations of 2,4-D was tested. Fish distribution was analyzed statistically (Probability = 0.05). Results are given in Tables 5, 6, 7, and 8.

Table 3
EFFECTS OF VARIOUS FORMULATIONS OF 2,4-D ON THREE SPECIES OF ESTUARINE FISHES

Animal	Scientific name	2,4-D Formulation	Concentration (ppb) EC24	Concentration (ppb) EC48
Long-nose killifish	*F. similis*	Butoxyethanol ester	5000	5000
		Dimethylamine salt	15,000	15,000
		Propylene glycol butyl ether ester	5000[b]	4200[b]
Mullet	*M. cephalus*	Technical acid	50,000[a]	50,000[a]
Spot	*L. xanthurus*	Technical acid	50,000[a]	50,000[a]

[a] No effect.
[b] EC24 and EC48 = concentrations of 2,4-D that caused mortality in 50% of the animals tested after exposure in flowing water for 24 and 48 hr, respectively.

From Duke, T., Effects of 2,4-D Formulations on Estuarine Organisms, Tech. Rep. 12, U.S. Army Engineer Waterways Experiment Station, Vicksburg, Miss., 1976.

Table 4
RESIDUES OF 2,4-D (N,N-DIMETHYL-2,4-DICHOLOROPHENOXYACETAMIDE) FOUND IN THE WHOLE BODY, GILLS, AND LIVER OF *F. SIMILIS*

	2,4-D (ppb)			
	After 2 weeks		After 4 weeks	
Treatment	Whole body	Gills and liver	Whole body	Gills and liver
Untreated controls	0	0	0	0
10,000 ppb in food	0	0	0	0
100,000 ppb in food	8400	3000	0	0

Note: The treated fish were fed contaminated food for the first 2 weeks and uncontaminated food for the second 2 weeks.

From Duke, T., Effects of 2,4-D Formulations on Estuarine Organisms, Tech. Rep. 12, U.S. Army Engineer Waterways Experiment Station, Vicksburg, Miss., 1976.

C. variegatus avoided 100 and 1000 ppb of 2,4-D but not 10,000 ppb (Table 5). In experiments with different herbicide concentrations, they avoided the highest concentrations tested (Table 6).

G. affinis sought water free of 1000 and 10,000 ppb of 2,4-D but did not seek water free of 100 ppb (Table 7). When given a choice between two concentrations of 2,4-D, *G. affinis* avoided water with 10,000 ppb but failed to discriminate between water with 100 and 1000 ppb (Table 8).

GROWTH AND PHOTOSYNTHESIS

Four species of marine unicellular algae (*Chlorococcum* sp., *Dunaliella tertiolecta, Isochrysis galbana,* and *Phaeodactylum tricornutum)* were exposed to the technical

Table 5
PERCENT OF SHEEPSHEAD MINNOWS *(CYPRINODON VARIEGATUS)* SEEKING WATER FREE OF THE BUTOXYETHANOL ESTER OF 2,4-D

Concentration (ppb)	In "clean" water (%)	x^2 Value
10	54.3	NS[a]
100	64.8	9.15
1000	67.0	11.56

[a] Not significant.

From Duke, T., Effects of 2,4-D Formulations on Estuarine Organisms, Tech. Rep. 12, U.S. Army Engineer Waterways Experimental Station, Vicksburg, Miss., 1976.

Table 6
RESPONSES OF SHEEPSHEAD MINNOWS TO TWO CONCENTRATIONS OF THE BUTOXYETHANOL ESTER OF 2,4-D

Concentrations (ppb)			
High	Low	In low concentration (%)	x^2 Value
10,000	1000	94.6	116.74
10,000	100	92.9	93.55
10,000	100	95.0	113.40
1000	1000	43.7	NS[a]
1000	10	47.8	NS[a]
100	10	46.7	NS[a]

[a] Not significant.

From Duke, T., Effects of 2,4-D Formulations on Estuarine Organisms, Tech. Rep. 12, U.S. Army Engineer Waterways Experiment Station, Vicksburg, Miss., 1976.

acid and butoxyethanol ester of 2,4-D. Effects upon growth and photosynthesis were measured.

Effects of 2,4-D on growth were studied by growing the algae in artificial sea water supplemented with vitamins and trace elements in optically matched tubes. Growth was measured spectrophotometrically at 525 nm after 10 days. Data, expressed as concentrations of herbicide that reduced growth by 50% (LD^{50}) and 100% (LD^{101}) when compared to untreated controls, are given in Table 9.

Effects of herbicides upon evolution of oxygen by marine unicellular algae were measured in a Gilson® differential respirometer. Untreated cultures in the logarithmic growth phase were centrifuged and resuspended in growth medium containing 2,4-D. Evolution of oxygen was measured at 10-min intervals for 90 min. The herbicide concentrations that reduced the rate of oxygen evolution by 50% (LD^{50}) and 100% (LD^{100}) compared to untreated controls are given in Table 10.

Table 7
AVOIDANCE OF THE BUTOXYETHANOL ESTER OF 2,4-D BY MOSQUITO FISH *(G. AFFFINIS)* WHEN GIVEN A CHOICE BETWEEN 2,4-D-FREE WATER AND WATER CONTAINING 2,4-D

Concentration (ppb)	Avoidance (%)
100	51.1[a]
1000	79.3[b]
10,000	93.8[b]

[a] Not significant.
[b] $p = 0.05$.

From Duke, T., Effects of 2,4-D Formulations on Estuarine Organisms, Tech. Rep. 12, U.S. Army Engineer Experiment Station, Vicksburg, Miss., 1976.

Table 8
RESPONSES OF MOSQUITO FISH (*G. AFFINIS*) TO TWO DIFFERENT CONCENTRATIONS OF THE BUTOXYETHANOL ESTER OF 2,4-D

Concentrations (ppb)		
High	Low	In low concentration (%)
1000	100	48.7[a]
10000	100	92.6[b]
10000	1000	92.4[b]

[a] Not significant.
[b] $p = 0.05$.

From Duke, T., Effects of 2,4-D Formulations on Estuarine Organisms, Tech. Rep. 12, U.S. Army Engineer Experiment Station, Vicksburg, Miss., 1976.

TRANSLOCATION TESTS

Turtle grass *(Thalassia testudinum)* and widgeon grass *(Ruppia maritima)* in sand that contained 0, 1, 10, 20, and 50 kg ae/ha 2,4-D in the form of butoxyethanol ester adsorbed on attaclay wall were placed in wooden boxes 0.25 m in area and 10 m deep. Five replicates of each concentration were made. The boxes were placed in Santa Rosa Sound Estuary (Florida) and sprigged with the sea grasses, *T. testudinum* and *R. maritima*. After 6 weeks, the sea grasses were harvested and weighed. Animals were collected from the sand and counted. No effect of 2,4-D upon *T. testudinum* and *R. maritima* was found. Resulting plant biomass in experimental boxes was similar to that of the controls, and there were no significant differences in the faunal contents of the boxes. Results of residue analyses are given in Table 11.

Table 9
CONCENTRATIONS OF 2,4-D (ppb) THAT DECREASED GROWTH OF MARINE UNICELLULAR ALGAE BY 50% (LD^{50}) AND 100% (LD^{100}) AFTER 10 DAYS

	Chlorococcum sp.		*D. tertiolecta*		*I. galbana*	
Formulation	LD^{50}	LD^{100}	LD^{50}	LD^{100}	LD^{50}	LD^{100}
Technical acid	50,000	75,000	75,000	100,000	50,000	75,000
Butoxyethanol ester	75,000	150,000	75,000	300,000	75,000	250,000

From Duke, T., Effects of 2,4-D Formulations on Estuarine Organisms, Tech. Rep. 12, U.S. Army Engineer Experiment Station, Vicksburg, Miss., 1976.

Table 10
CONCENTRATIONS OF 2,4-D (ppb) THAT DECREASED PRODUCTION OF OXYGEN BY MARINE UNICELLULAR ALGAE BY 50% (LD^{50}) AND 100% (LD^{100}) AFTER 90 MIN

	Chlorococcum sp.		*D. tertiolecta*		*I. galbana*	
Formulation	LD^{50}	LD^{100}	LD^{50}	LD^{100}	LD^{50}	LD^{100}
Technical acid	60,000	90,000	90,000	100,000	60,000	85,000
Butoxyethanol ester	100,000	200,000	100,000	400,000	100,000	300,000

From Duke, T., Effects of 2,4-D Formulations on Estuarine Organisms, Tech. Rep. 12, U.S. Army Engineer Experiment Station, Vicksburg, Miss., 1976.

Table 11
RESIDUES (ppb) OF 2,4-D IN *T. TESTUDINUM* AND *R. MARITIMA* AFTER EXPOSURE TO "AQUA-KLEEN®" FOR 6 WEEKS IN SANTA ROSA SOUND ESTUARY

	T. testudinum		*R. maritima*	
Treatment	Roots	Leaves	Roots	Leaves
Control	0	0	0	0
1 kg ae/ha	0	0	0	0
10 kg ae/ha	690	180	100	100
20 kg ae/ha	1400	290	470	790
50 kg ae/ha	3400	620	780	260

Note: Each value is the average of five replicate field samples.

From Duke, T., Effects of 2,4-D Formulations on Estuarine Organisms, Tech. Rep. 12, U.S. Army Engineer Waterways Experimental Station, Vicksburg, Miss., 1976.

Table 12
BIOLOGICAL EFFECTS OF HERBICIDES ON AQUATIC ORGANISMS

Composition	Species	Concentration	Effect
2,4-D Acid (technical)	Mullet	50,000	No effect (48 hr)
	Long-nose killifish	50,000	No effect (48 hr)
	Phytoplankton	100,000	No decrease in productivity (4 hr)
	Oysters	200,000	No reduction in shell growth
	Bluegill	350,000	24-hr TLm in Louisiana River water
	Large-mouth bass	350,000	24-hr TLm in Louisiana River water
	Bluegill	390,000	24-hr TLm in Bayou de Siard, La.
	Large-mouth bass	375,000	24-hr TLm in Bayou de Siard, La.
		375,000	48-hr TLm in Bayou de Siard, La.
2,4-D Acid (with emulsifiers)	Bluegill	8,000	24-hr TLm distilled water
		8,000	48-hr TLm distilled water
2,4-D (Technical 70% ae) butoxyethanol ester	Long-nose killifish	75,000	100% mortality 24 and 48 hr
		5,000	50% mortality 24 and 48 hr
		2,500	0% mortality 24 and 48 hr
	Eastern oyster	5,000	57% decrease in shell growth in 96 hr
		5,000	33% decrease in shell growth in 96 hr
		50	6% decrease in shell growth in 96 hr
	Pink shrimp	1,000	10% mortality 24 and 48 hr
	Brown shrimp	10,000	60% mortality 24 and 48 hr
		3,200	40% mortality 24 and 48 hr
		1,000	0% mortality 24 and 48 hr
2,4-D (Aqua-Kleen® 20) butoxyethanol ester	Eastern oyster blue crab, soft shell clam, pumpkinseed sunfish	(26 kg ae/ha)	No direct toxic effects
	Blue mussel	(26 kg ae/ha)	General increase in numbers and biomass
2,4-D Butoxyethanol ester	*(Mytilus edulis)*		

From Duke, T., Effects of 2,4-D Formulations on Estuarine Organisms, Tech. Rep. 12, U.S. Army Engineer Waterways Experiment Station, Vicksburg, Miss., 1976.

ACUTE AND CHRONIC BIOLOGICAL EFFECTS

Although herbicidal effectiveness may often be adequately expressed in terms of the acid equivalent (as in the case of phenoxy compound), the composition of the herbicide (substituents as amines, esters, or salts) or formulation (carrier, wetting agent, and other so-called inert ingredients) is quite important due to its effects on many organisms. Chlorinated phenoxy compounds are not particularly persistent or cumulative in the food chains. They have relatively low acute toxicity and few discernible chronic effects and persistent ecological changes are apparent except for the target species. A summary of these effects is given in Table 12.

REFERENCES

1. **Steenis, J. H. and Stotts, V. D.,** Progress report on control of Eurasian watermilfoil in Chesapeake Bay, *Proc. Northeast. Weed Control Conf.*, 15, 566, 1961.
2. **Steenis, J. H. and Stotts, V. D.,** Progress report on distribution and control of Eurasian watermilfoil in the Chesapeake Bay region, *Proc. South. Weed Conf.*, 16, 341, 1963.
3. **Steenis, J. H., Stotts, V. D., Haven, D., and Whipp, A. A.,** Developments of control of Eurasian watermilfoil in the Chesapeake Bay region, *Proc. South. Weed Control Conf.*, 17, 321, 1964.
4. **Smith, G. E. and Isom, B. G.,** Investigations of large scale applications of 2,4-D on aquatic fauna and water quality, *Pest. Monitor. J.*, 1(3), 16, 1967.
5. Tennessee Valley Authority, Environmental Control of Eurasian Watermilfoil *(Myriophyllum spicatum* L.) in TVA Reservoir, Chattanooga, Tenn., 1972.
6. **Thomas, M. L. H. and Duff, J. R.,** Butoxyethanol ester of 2,4-D in the control of eelgrass *(Zostera marina* L.) and its effects on oysters *(Crassostrea virginica* Gmelin) and other benthos, *Proc. Northeast. Weed Control Conf.*, 22, 186, 1968.
7. **Whitney, E. W., Montgomery, A. B., Martin, E. C., and Gangstad, E. O.,** The effects of 2,4-D application on the biota and water quality in Currituck Sound, North Carolina, *Hyacinth Control J.*, 11, 13, 1973.
8. **Duke, T.,** Effects of 2,4-D Formulations on Estuarine Organisms, Tech. Rep. 12, U.S. Army Engineer Waterways Experiment Station, Vicksburg, Miss., 1976.

Chapter 19

CHESAPEAKE BAY STUDY*

INTRODUCTION

Eurasian watermilfoil *(Myriophyllum spicatum* L.) will grow in clear waters up to 9 ft deep at mean low tide in the Chesapeake Bay and its tributaries, and exists in fresh to brackish waters of up to 15‰ salinity. Its dense surface mat, normally present from June to October, seriously hampers or even prevents fishing, swimming, and boating; it chokes out valuable waterfowl foods; provides ideal conditions for mosquito breeding; causes lowered real estate values; and in many places makes the harvest of shellfish impossible or even smothers these animals by trapping sediments.

A cooperative research-control program of Eurasian watermilfoil has been conducted between the U.S. Fish and Wildlife Service, the Maryland Game and Inland Fish Commission, the Maryland Department of Chesapeake Bay Affairs, Baltimore, the Virginia Institute of Marine Science, Gloucester Point, and the Chesapeake Biological Laboratory of the University of Maryland's Natural Resources Institute, Solomons. This program has involved screening a large number of herbicide formulations for their effectiveness in controlling watermilfoil, and in addition, the Chesapeake Biological Laboratory has investigated the feasibility of underwater mowing of milfoil with and without the subsequent application of herbicides. Thus far, it appears that the butoxyethanol ester of 2,4-dicholorophenoxyacetic acid (2,4-D BEE) holds the most promise for consistent control of watermilfoil in tidal areas.[5-9]

Herbicide use in waters yielding animal food for human consumption poses two immediate considerations — toxicity to organisms in the treated areas, and herbicide residues occurring in edible fauna. Extensive experiments showed that 2,4-D BEE at recommended rates in Chesapeake Bay waters had no primary lethal effects on oysters, soft-shelled clams, blue crabs, *Callinectes sapidus,* and various species of test fish (mostly pumpkinseed, *Lepomis machrochirus).* No adverse effects were noticed among adult oysters held in treated plots when using new oyster shell growth as a bioindicator in the manner described by Butler.[2,3]

Once primary toxicity proved a negligible factor in field applications of 2,4-D BEE, it became necessary to determine the herbicide's accumulation and loss rate (if any) in animals found in treated locations. Other than accumulation and loss rates of 2,4-D BEE in oysters and clams, this study was designed to determine the length of time a treated shellfish-producing area must be closed before its animals may be harvested free of detectable residues (precluding illegal residues from milfoil control treatments).[1,5]

METHODS AND MATERIALS

In setting up the 1966 field study for 2,4-D BEE residues in exposed test animals, several site criteria had to be considered. Milfoil had to be present in stands dense enough to simulate growth conditions requiring 2,4-D BEE application. Salinity had to be high enough to support oysters and clams. A natural population of clams had to be present. Accessibility by outboard boat and clam dredge had to be relatively easy. Extensive field surveys pointed to one area, apparently ideal. Above Chaptico

* By Charles K. Rawls, University of Maryland, Natural Resources Institute, Chesapeake Biological Laboratory, Solomons, Md.

Bay, Md. on the east side of the Wicomico River (a tributary of the Potomac), vigorous milfoil stands grew so profusely that they offered adequate vegetative containment for the herbicide to be applied. Plots were laid out in the Chesapeake Bay where no previous 2,4-D application had been made, thus eliminating possible direct herbicide contamination of endemic fauna (Figure 1).

Plot Establishment

In late June of 1966, plots (about 65 m to a side) were measured with a floating polypropylene line. Three were to be treated and one would serve as a control. Plot corners were marked with 5-m partially seasoned bamboo poles pushed about 1 m into the bottom. Poles were also placed 45 and 90 m from perimeters of plots chosen for treatment. Another pole was situated near the center of each plot. All poles carried red flags for easy sighting and fewer navigation hazards. Supplementing each pole (in case of loss) were weighted buoys made from empty plastic quart oil bottles painted international orange. Oysters and caged clams would be placed in the center of each plot and at the 45- and 90-m yard stations. Each plot was separated from the other by a 400-m buffer zone of milfoil. Throughout the study location, redhead grass, *Potamogeton perfoliatus* L. and widgeon grass, *Ruppia maritima* L. grew among the milfoil. Near the shoreline naiad, *Najas guadalupensis* (Spreng.) Magnus, was common. The average water depth at low tide in the plots was 1.3 m.

Test Animals

Clams and oysters were used for test animals since they remain in relatively fixed postions. Crabs and fish can leave a treated area if it becomes unsuitable. Adult oysters, taken above Highway Bridge 301 from Beacon Bar in the Potomac River, were placed on strandard oyster trays. Clams dredged from the Potomac near Blakiston Island, were placed in hardware cloth cages and positioned next to the trayed oysters. Ropes, bridled to trays and cages, were fastened to the marker poles. The buoys, in turn, were fastened to the trays. Herbicide application was delayed a week to allow time for bioassay animals to adjust to their new environment.

Herbicide Treatment

Before herbicide treatment on July 5, 1966, each tray and cage of test animals was examined for mortality resulting from natural causes or from mechanical injury incurred during dredging. All oysters were living, but most caged clams were dead either from predation by eels, *Anquilla rostrata,* who had forced themselves through cage seams, or presumably from exposure to extremely hot weather from the time they were dredged until the time they were placed in plot positions.

The herbicide used in this experiment was Amchem's Aqua-Kleen®, a low volatile 20% acid equivalent (ae) 2,4-D BEE impregnated for slow release on baked attaclay granules. Plot 1 was treated at 33 kg ae/ha, Plot 2 at 11 kg ae/ha, Plot 3 at 67 kg ae/ha, and Plot 4 (the farthest upstream) served as the control. A single application of 2,4-D BEE granules on plots to be treated was made by forced air distribution using a Sol-Fi air blower attached to a flexible hose connected to a spreader head. The spreader head was mounted on the bow of an outboard hull powered by an 18-hp Johnson® outboard motor. The granular herbicide was fed by gravity flow to the air blower.

Sampling Design

Bedded clams and oysters from trays were used for residue analysis. At each station on each collection date 15 to 20 adult oysters and a similar number of bedded clams were taken. All samples were taken to the Chesapeake Biological Laboratory, Solo-

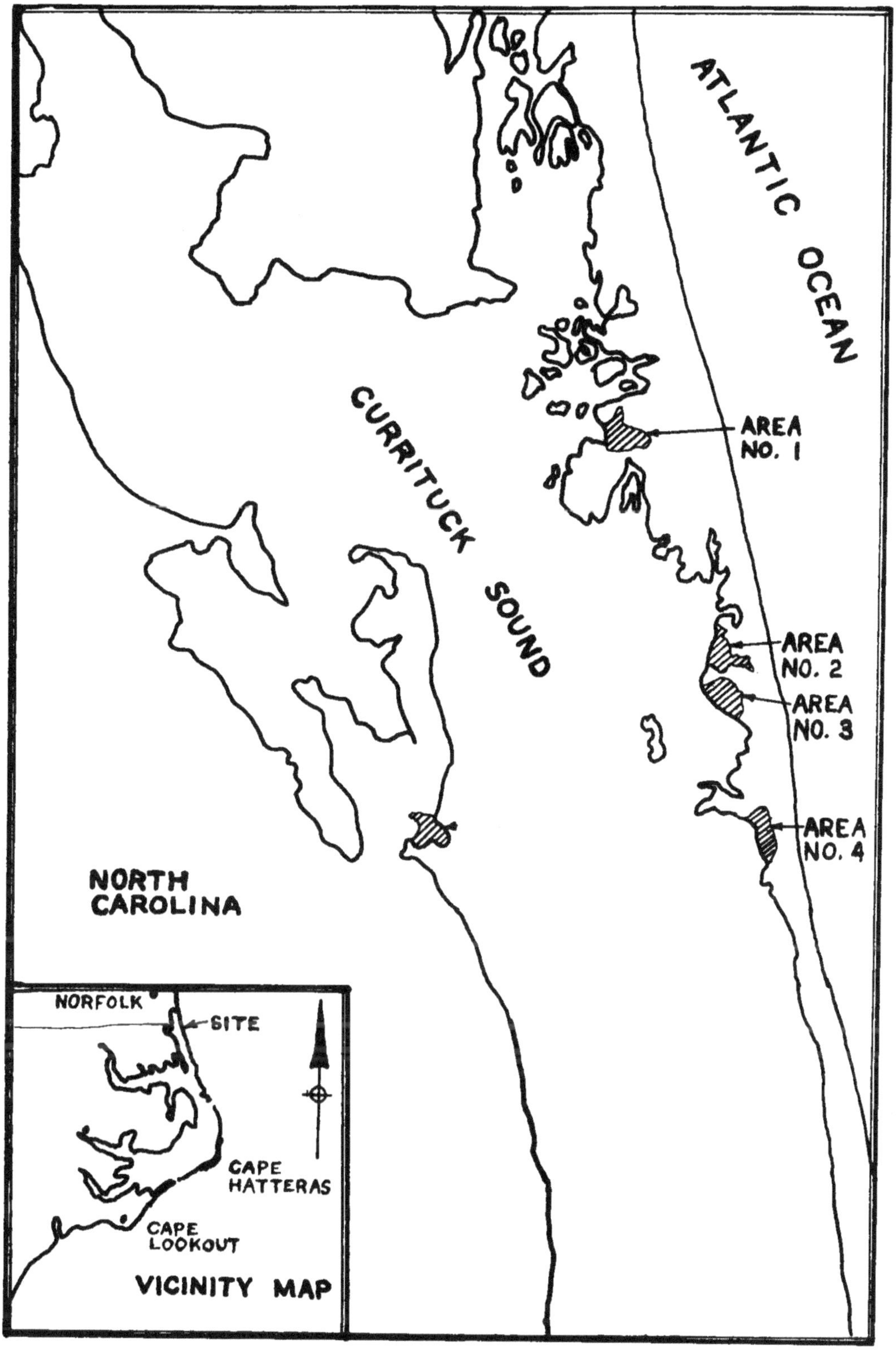

FIGURE 1. Chesapeake Bay study area for treatment of Eurasian watermilfoil with 2,4-D BEE granular herbicide.

mons, Md. for preparation and labeling. Oysters and clams were shucked. During preparation for market, about one half the clam siphone and all periostracum were discarded. Samples were washed separately for 5 min under flowing tap water, drained, placed in new polyethylene containers, and covered with just enough water to prevent freezer burn.

DISCUSSION OF RESULTS

Data for the accumulation and loss of 2,4-D BEE residues for oysters and soft-shelled clams are summarized in Tables 1 and 2, respectively. A background level of 225 ppb 2,4-D was indicated as the zero treatment, zero time check level. All treatment levels were less than 1000 ppb except the highest rate 67 kg ae/ha the 3rd day of sampling in both oysters and soft-shelled clams; all samples taken at 45 m or more from the treatment area had less than 500 ppb residue for either oysters or soft-shelled clams.

SUMMARY OF EFFECTS

Crabs, fish, clams, and oysters will accumulate and lose 2,4-D BEE residues in varying amounts depending on the species of bioassay animal, the distance of animals from treated areas, the herbicide dosage applied, the water movement through study plots, and the degree of vegetative containment within and around herbicide application sites. Clams and oysters acquire greater residues than do fish and crabs. In most instances, it appears that residues become undetectable by the 35th day of exposure. The level of residue depends upon the distance animals are from treated areas, the dosage of herbicides applied, water movement through study plots, and the degree of vegetative containment within and around sites of herbicide application.

Residue peaks are reached 3 days after treatment and decline rapidly thereafter.

If shell fish are to be harvested immediately near an area of 2,4-D BEE application, there should be a minimum 500-yard buffer zone of vegetation separating the harvest area from the treated area; if the rate of application is limited to 22 kg ae/ha and a tolerance level of lppm 2,4-D residue is permitted, there need be no restrictions for time and distance.

Table 1
SUMMARY OF 2,4-D BEE RESIDUES (ppb) IN OYSTERS OF 0, 11, 33, AND 67 KG AE/HA TREATMENT OF 2,4-D BEE FOR 2, 7, 14, 21, AND 35 DAYS AFTER TREATMENT

Rate (kg/ha)	Time (days)	Distance (m) 0	45	90
0	0	225	225	225
11	3	425	350	300
33	3	300	175	200
67	3	1750[a]	350	375
11	7	100	350	250
33	7	500	350	200
67	7	750	225	200
11	14	175	175	175
33	14	175	50	200
67	14	425	225	50
11	21	225	175	200
33	21	325	300	300
67	21	150	300	200
11	35	225	175	225
33	35	225	200	225
67	35	225	200	100
	Treatment Means			
11		238	245	238
33		385[a]	215	225
67		660	260	145
	Treatment Means			
	3	825[a]	292	292
	7	463	308	217
	14	258	130	142
	21	233	258	233
	35	225	192	183

[a] Significant difference at the 5% level of probability, using the pooled error term of rate by distances as a test of significance.

Table 2
SUMMARY OF 2,4-D BEE RESIDUE (ppb) IN SOFT SHELLED CLAMS FOR 0, 11, 33, AND 67 KG AE/HA FOR 0, 3, 14, 21, AND 35 DAYS AFTER TREATMENT IN PARTS PER BILLION

Rate (kg/ha)	Time (days)	Distance (m)		
		0	45	90
0	0	225	225	225
11	3	800	100	100
33	3	300	200	300
67	3	1500[a]	200	200
11	7	225	100	100
33	7	600	200	200
67	7	400	250	200
11	14	150	100	75
33	14	400	150	200
67	14	700	150	225
11	21	50	100	50
33	21	250	225	100
67	21	500	100	225
11	35	50	100	50
33	35	225	200	250
67	45	225	225	225
	Treatment Means			
11		225	100	75
33		335	205	210
67		665[a]	185	430
	Treatment Means			
	3	866[a]	183	200
	7	408	183	166
	14	416	133	166
	21	266	142	125
	35	166	175	175

[a] Significant difference at the 5% level of probability using the pooled error term of rates by distances as test of significance.

REFERENCES

1. **Burchfield, H. P., Johnson, D. E., and Storrs, E. E.,** Guide to the Analyses of Pesticide Residues, Public Health Service, Bureau State Services, U.S. Department of Health, Education and Welfare, Atlanta, 1965.
2. **Butler, P. A.,** Effects of pesticides on commercial fisheries, *Annu. Sess. Gulf Caribb. Fish Inst.*, 13, 168, 1960.
3. **Butler, P. A.,** Monitoring pesticide pollution in estuaries, *Proc. Northeast. Weed Control Conf.*, 22, 7, 1968.
4. **Coakley, J. E., Campbell, J. E., and McFarren, E. F.,** Determination of butoxyethanol ester of 2,4-Dichlorophenoxyacetic acid in shellfish and fish, *J. Agric. Food Chem.*, 12(3), 262, 1964.
5. **Rawls, C. K.,** Aquatic plant nuisances, *Proc. Interstate Comm. Potomac River Basin*, 1, 51, 1964.
6. **Rawls, C. K.,** Field tests of herbicide toxicity to certain estuarine animals, *Chesapeake Sci.*, 6(3), 150, 1964.
7. **Rawls, C. K.,** Providing estuarine fauna for residue analyses of field-applied herbicides, *Proc. 19th South. Weed Conf.*, 19, 397, 1966.
8. **Rawls, C. K.,** Opinions on the Status of Certain Marine Vegetation Species in the Chesapeake Bay and its Tributaries, No. 66-96, University of Maryland Research Institute, College Park, 1966.
9. **Rawls, C. K.,** The accumulation and loss of field applied butoxyethanol ester of 2,4-dichlorophenoxyacetic acid in eastern oysters *Crossostrea viginian* and soft-shelled clams, *Mya arenarian*, *Hyacinth Control J.*, 9, 62, 1971.

Chapter 20

EURASIAN WATERMILFOIL STUDIES IN NEW JERSEY*

INTRODUCTION

Eurasian watermilfoil, *Myriophyllum spicatum*, a native of Europe, Asia, and parts of Africa, has been present in this country since at least the turn of the century. Recently, it has increased rapidly and is now an important pest in numerous fresh and brackish waters of New Jersey and other states where it chokes out waterfowl food plants; interferes seriously with boating, fishing, and swimming; and lowers real estate values. Granular formulations of esters of 2,4-dichlorophenoxyacetic acid are the most effective chemical herbicides. Cooperative studies are continuing on various aspects of the biology and control of Eurasian watermilfoil.

ROCKAWAY PARK LAKE STUDY

Formulations of 2,4-dichlorophenoxy (2,4-D) are used extensively and effectively for control of aquatic vegetation in lakes, ponds, and reservoirs. There are, however, the questions of the effect of this aquatic herbicide on the general chemical quality of these potable water supplies and of the occurrence and persistence of 2,4-dichlorophenol (2,4-DCP) and 2,4-D.

Aquatic Plant Treatment

A granular formulation of an isooctyl ester of 2,4-D was added to Rockaway Park Lake, Rockaway, N.J. for the control of watermilfoil. The immediate and prolonged effects of 2,4-D on water quality were observed for a period of 132 days after treatment to determine the secondary effects of decaying vegetation on chemical water quality.[1,2]

Sampling and Chemical Analysis

Rockaway Park Lake covers a surface area of 4 ha and has an average depth of 2 m. Samples were taken at two sites at the edge and at two sites in the center of the lake. No water escaped from the lake spillway during the sampling period. Most of the chemical analyses were made in accordance with the *Standard Methods* manual.[10] The granular formulation contained 20% by weight acid equivalent (ae) of the isooctyl ester of 2,4-D. This material was added at the rate of 18 kg ae/ha or a concentration of 1.34 mg/ℓ. This formulation was found also to contain 0.65 mg/g of 2,4-DCP or a concentration of 4.36 μg/ℓ in the lake. The chemical quality of the lake water is shown in Table 1 just before addition of the 2,4-D. The high pH value of 9.4 was attributed to a tributary carrying drainage from a limestone quarry. The bicarbonate (total) alkalinity and the total hardness were of low to medium concentrations. The color value of 20 is fairly low for surface waters. Organic matter was measured indirectly through the chlorine demand and biological oxygen demand (BOD) analyses. The values of 0.4 mg/ℓ and 3.5 mg/ℓ, respectively, indicate small amounts of organic matter. The natural or background threshold odor number was quite low and was qualitatively identified as ''earthy'' or ''musty''.

* From cooperative studies of the Department of Sanitation, Rutgers University and the National Institutes of Health, Bethesda, Md.

Table 1
CHEMICAL WATER QUALITY OF ROCKAWAY PARK LAKE[a]

Analyses	Result[b]
pH	9.4
Total alkalinity	42.5 mg/ℓ as CaCO
Total hardness	54.5 mg/ℓ as CaCO
Color	20 Color units
Chlorine demand[c]	0.40 mg/ℓ as Cl^2
Biochemical oxygen demand[d]	3.5 mg/ℓ as O_2
Temperature °F	73
Dissolved oxygen	119.5% saturation
Threshold odor number	2
2,4-DCP	0.0
Phenol	9.0 g/ℓ as phenol
2,4-D	8.0 g/ℓ as 2,4-D

[a] Sample taken 6/27/61 and analyzed 6/28/61.
[b] Average of four samples.
[c] Contact time = 15 min; chlorine residual (OT) = 0.1 mg/ℓ.
[d] 5-Day, 20°C, BOD.

From Faust, S. D. and Aly, O. M., *Proc. Northeast. Weed Control Conf.*, 17, 460, 1963. With permission.

Significant Effects

No significant effects on the chemical characteristics of pH, total hardness, and total alkalinity (Table 1) by the 2,4-D were observed throughout a sampling period. Likewise, there were no significant deviations from the original values for color, chlorine demand, BOD, and dissolved oxygen. These latter analyses were made primarily to observe any secondary effects on water quality by decaying vegetation. Apparently the release of organic matter was too slow to register an appreciable effect. The threshold odor number remained at the value of 2 throughout the sampling period. The occurrence and persistence of 2,4-D, 2,4-DCP, and phenol at various time intervals up to 132 days are given in Table 2. The maximum concentration of 2,4-D (49.5 g/ℓ) was observed 13 days after treatment whereupon there was a gradual decrease. These data show that very little 2,4-D was detected in the free-flowing portion of the lake despite the fact that the herbicide was applied at a concentration of 1.34 mg/ℓ. The 2,4-DCP exhibited the same pattern of occurrence and persistence as 2,4-D, but at lesser concentrations. There appeared to be a slight increase of 2,4-DCP to the 10 to 11 g/ℓ range after 70 days. A spot check for 2,4-DCP was made at 132 days where the concentration was observed to be 13 g/ℓ. A very low level, 5.0 to 15.0 g/ℓ, of background phenols was observed.

Results of this study indicate that any 2,4-DCP present in a lake treated with 2,4-D would be there primarily as an impurity from the formulation. There is no evidence to suggest that an appreciable concentration of 2,4-DCP would arise from the biological degradation of 2,4-D.

The levels of 2,4-DCP shown in Table 2 can be compared with threshold taste and odor concentrations of five chlorinated phenols as given in Table 3. The odor level of Rockaway Park Lake water was not affected by the observed concentrations of 2,4-DCP in Table 2 since the threshold value of two remained constant through 70 days of examination. A spot check was made for the typical medicinal taste and odor of chlorinated phenols at 132 days since a concentration of 13.0 g/ℓ of 2,4-DCP was found. No medicinal taste or odor was observed.

Table 2
OCCURRENCE AND PERSISTENCE OF 2,4-D, 2,4-DICHLOROPHENOL, AND PHENOL

Determination-Average of Four Sampling Areas

Time	2,4-D (ppb)	2,4-DCP (ppb)	Phenol (ppb)
Before[a]	8.0	0.0	9.0
1 hr	38.5	1.0	5.0
2 days	45.0	3.0	11.0
13 days	49.5	2.0	15.0
22 days	30.8	.5	7.8
27 days	25.8	2.0	7.8
48 days	25.8	0.0	9.0
55 days	20.0	11.3	7.5
70 days	24.5	10.5	9.3
132 days[b]	13.0	13.0	0.0

[a] June 27, 1961.
[b] November 16, 1961.

From Faust, S. D. and Aly, O. M., *Proc. Northeast. Weed Control Conf.*, 17, 460, 1963. With permission.

Table 3
THRESHOLD TASTE AND ODOR CONCENTRATION OF VARIOUS PHENOLIC COMPOUNDS

Component	Geometric mean threshold (μg/ℓ) Taste	Odor
Phenol	1000	1000
2-CP	4	2
4-CP	1000	250
2,4-DCP	8	2
2,6-DCP	2	3
2,4,6-TCP	1000	1000

From Faust, S. D. and Aly, O. M., *Proc. Northeast. Weed Control Conf.*, 17, 460, 1963. With permission.

WATER TREATMENT STUDY

Organic esters of 2,4-D are used effectively for the control of some species of aquatic weeds in lakes, ponds, and reservoirs. These compounds are also sometimes discharged as waste waters from the manufacturing process and can persist for long time periods in surface waters. Most commercial herbicide formulations contain 2,4-DCP as an impurity. In this study, the effect of various conventional treatment processes on the removal of 2,4-D (Na salt); the isopropyl, butyl, and isooctyl esters of 2,4-D; and 2,4-DCP from surface water were examined.[2]

Table 4
ADSORPTION PARAMETERS OF 2,4-DICHLOROPHENOL AND 2,4-D COMPOUNDS

Compound	1/n	K
2,4-Dichlorophenol	0.32	0.122
2,4-D	0.56	0.118
2,4-D, Isopropyl	0.63	0.283
2,4-D, Butyl ester	0.52	0.200
2,4-D, Isooctyl ester	0.33	0.117

From Faust, S. D. and Aly, O. M., *Proc. Northeast. Weed Control Conf.*, 17, 460, 1963. With permission.

Limestone Treatment

The high solubility of calcium and magnesium salts of 2,4-D determines the fate of the herbicide in lakes and reservoirs, and the possible use of limestone products for its chemical removal at potable water treatment plants. The solubilities and activity coefficient for Ca $(2,4\text{-D})^2$ and Mg $(2,4\text{-D})^2$ were determined in distilled water in KCl solutions ranging from 0.001 *M* to 0.1 *M*. The solubility of CA $(2,4\text{-D})^2$ was found to be 9.05×10^{-3} (4000 mg/ℓ) and the solubility of Mg $(2,4\text{-D})^2$ was found to be 2.51×10^{-2} (11,100 mg/ℓ) in distilled water. The activity solubility products constants were determined as 1.096×10^{-6} for Ca $(2,4\text{-D})^2$ and 1.55×10^{-5} for Mg $(2,4\text{-D})^2$. Consequently, the usual aquatic herbicide dosages of 1 to 5 mg/ℓ of 2,4-D are not affected by normal calcium and magnesium ion concentrations of natural surface waters, and limestone products cannot be used for the chemical removal of 2,4-D from surface waters.

Coagulation of one mg/ℓ of the sodium salt of 2,4-D and of the isopropyl, butyl, isooctyl esters of 2,4-D with 100 mg/ℓ of either aluminum sulfate or ferric sulfate failed to show any significant removal of these compounds.

Activated Carbon Treatment

Various amounts of activated carbon were added to fixed concentrations of several 2,4-D compounds and of 2,4-DCP. These suspensions were stirred rapidly at room temperature for 30 min filtered, and the amount of 2,4-D or 2,4-DCP absorbed per weight of carbon was determined. These data were used to construct adsorption isotherms in accordance with the Freundlich equation: $\log x/m = \log k \div 1/n \log c$, where x = amount of 2,4-D adsorbed − mg/ℓ; m = amount of carbon − mg/ℓ; and c = residual concentration − mg/ℓ.

By plotting x/m vs. c on log-log paper, a straight line is obtained where the slope is $1/n$ and the intercept at $c = 1$ is K. K and $1/n$ are parameters for the particular system and are used to calculate the carbon dose required to reduce the concentration of the contaminant to a desired level. These parameters are given in Table 4.

Chlorination

A concentration of 0.75 mg/ℓ of chlorine was added to solutions containing 3 mg/ℓ of the 2,4-D compounds. Residual chlorine was determined at various time intervals and the residual 2,4-D compound was determined after 3 hr. Residual chlorine was not appreciably reduced by the Na salt of 2,4-D. In the case of the organic esters, some reduction was observed. The latter may be attributed to the phenolic impurity found in the formulations of these compounds.

Table 5
ION EXCHANGE REMOVAL OF 2,4-D AND 2,4-DCP

Compound	Resin	Breakthrough[a] (Vol mls)	Amount removed (mg)
2,4-D (Na salt)	Dowex 1 Cl^-	240	56.0
120 mg/ℓ	IRC-50 H^+	0	2.0
	IR-120 H^+	0	2.0
2,4-D butyl ester	Dowex 1 Cl^-	0	3.0
10 mg/ℓ	IRC-50 H^+	0	2.9
	IR-120 H^+	0	1.0
2,4-Dichlorophenol	Dowex 1 Cl^-	100	41.5
100 mg/ℓ	IRC-50 H^+	0	31.1
	IR-120 H^+	0	26.2

[a] Solution volume = 500 mℓ.

From Faust, S. D. and Aly, O. M., *Proc. Northeast. Weed Control Conf.*, 17, 460, 1963. With permission.

Potassium Permanganate

Various 2,4-D compounds at concentrations of 1 mg/ℓ were not affected when treated with 3 mg/ℓ of $KMNO^4$ for 1 hr. On the other hand, 1 mg/ℓ 2,4-DCP was found to be completely oxidized with a $KMNO^4$ dosage of 1.25 mg/ℓ, and pH was found to be a factor influencing the efficiency of phenol oxidation. A pH range of 0.50 to 6.0 was found to be optimum when buffered solutions of 1.0 mg/ℓ 2,4-DCP at various pH values were subject to 1.0 mg/ℓ of $KMNO^4$.

Ion Exchange Resins

Aqueous solutions of 1 mg/ℓ 2,4-DCP, of 120 mg/ℓ 2,4-D (Na salt), and of 10 mg/ℓ 2,4-D butyl ester were passed through beds of the following ion exchange resins at a flow rate of (1) Dowex 1 (HCl^-) = strongly basic anion exchanger, (2) IR-120 (H^+) = strongly acidic cation exchanger, and (3) IRC-50 (H^+) = weakly acidic cation exchanger.

Table 5 shows the breakthrough volumes and the amount of each compound adsorbed on the different resins. The 2,4-D (Na salt) was completely removed on the anion resin by ion exchange. In all other cases, removals were due to adsorption.

BIODEGRADATION KINETICS OF 2,4-D

Biological degradation of organic pesticides, when provided as the sole substrate, has seldom been examined in aquatic systems. It would be extremely desirable, therefore, to develop a model system from which two pertinent questions could be answered: (1) "What is the biological availability of various organic pesticides to degradation?" and (2) "What is the rate of this availability?"

An aquatic and terrestrial herbicide, 2,4-D, was selected as the refrence or model compound because it is manufactured, sold, and applied in large quantities each year, its sodium salt is soluble in water which climates some experimental difficulties, and a heterogeneous community of aquatic microorganisms can be developed or acclimatized to degrade it.[3-5]

The term "biodegradation" may often cause confusion and misunderstanding. Biodegradation is the conversion of potentially harmful chemical compounds to innocuous substances by microorganisms. There are, however, a number of extents to which biodegradation may occur. Primary biodegradation is considered the minimum extent necessary to change molecular identity of the compound. Acceptable biodegradation is carried to the point at which some undesirable property is lost. Ultimate biodegradation is carried to water, carbon dioxide, and inorganic materials. The latter definition is assumed herein. Verification of the ultimate biodegradation of 2,4-D was obtained from the loss of light energy adsorption in the ultraviolet region and from oxygen uptake measurements.

Research into the persistence of 2,4-D in aquatic environments has been limited. Okey and Bogan[8] using an activated sludge system, found that 2,4-D underwent considerable metabolism. They found, however, that cultures adapted to degrade 2,4-D could not degrade 2,4,5-trichlorophenoxyacetic acid.

Aly and Faust[2] used bottom mud samples collected from two lakes. The first lake had been treated the previous year with 2,4-D while the second lake had never been treated with 2,4-D. It was observed that when 2,4-D was applied to the mud from the lake previously treated, the herbicide disappeared in 35 days. 2,4-D persisted in muds from the untreated lake for 65 days. These authors emphasized importance of times involved in degradation (35 to 65 days) of the two samples. When compared with soil samples requiring approximately 2 weeks, these times show that the biodegradation in lakes is much slower.

Schwartz[9] found that only a small fraction of 2,4-D was degraded by a mixed microbial population in a dilute medium of mineral salts. Presence of larger amounts of nutrient broth as a supplemental source of organic carbon had no appreciable effect on the decomposition rate.

DeMarco et al.[3] investigated several environmental conditions under which 2,4-D would persist or be biodegraded in natural lake waters. In warm aerobic waters, 2,4-D was apparently biodegraded within 6 days, whereas, 2,4-D persisted for as long as 80 days in cold deoxygenated waters. These warm and cold conditions were selected to simulate stratified and impounded reservoir waters.

It becomes apparent from the above literature review that 2,4-D is degraded quite readily in terrestrial environments, whereas, conditions in aquatic environments may not be as favorable for rapid disappearance. It also appears that kinetics of 2,4-D biodegradation have not been extensively examined.

Enrichment Cultures

Domestic activated sewage sludge was used as a source of microorganisms. Procedures were employed to acclimatize this heterogeneous population of aquatic microorganisms to 2,4-D. This technique involves the sequential addition of increasing amounts of 2,4-D to the microorganisms over a period of 12 to 14 days. The acclimatized microorganisms were maintained by periodic additions of the 2,4-D and a combination of 2,4-D plus synthetic sewage. An aerobic environment was maintained by a continuous shaking device.

Manometric Techniques

Determination of the mass of acclimatized cells was obtained by measuring the total dry solids. A specific quantity of acclimatized cells was added to 100-mℓ volumetric flasks. The following nutrients were then added to the flasks: $MgSO^4$, $FeCl^3$, $CaCl^2$, and PO^4 buffer. To the control flasks, distilled water was added to bring the volume to 100 mℓ. To the enrichment flasks, the desired amount of 2,4-D was added and the volume brought to 100 mℓ by distilled water.

Aliquots (50 mℓ) from the control and enrichment flasks were then placed in separate Warburg flasks. In the center well of each flask was placed 1.0 mℓ of potassium hydroxide. Oxygen uptake subsequently was measured for the appropriate time period. All tests were conducted at 20 ± 0.5°C. The pH was maintained at 7.0 ± 0.2.

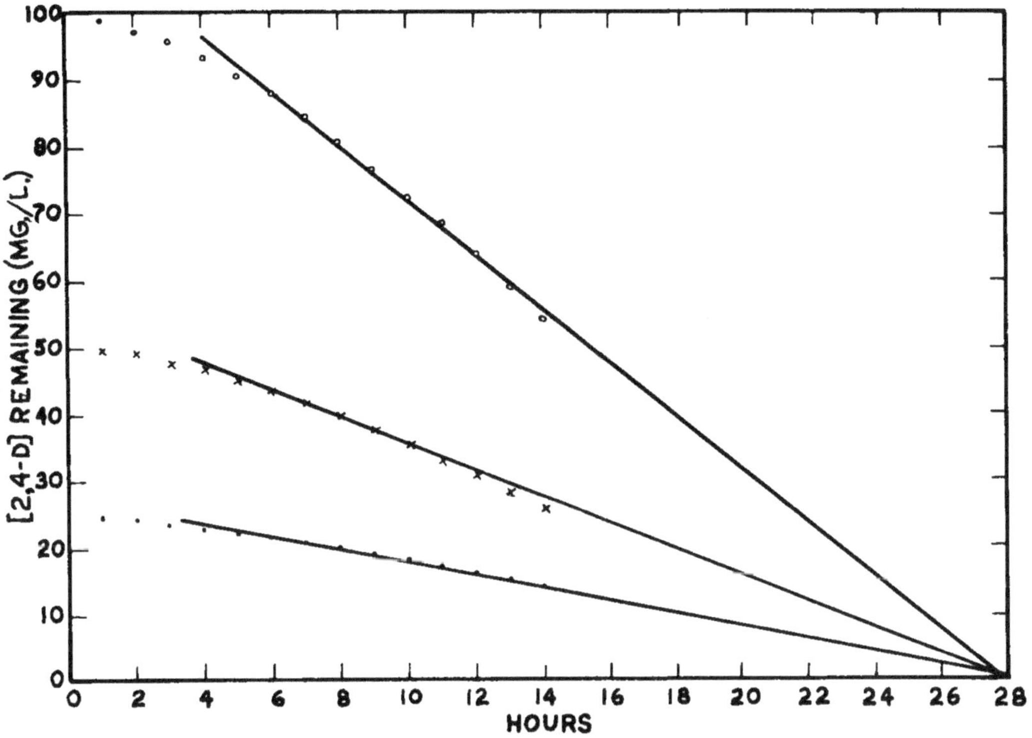

FIGURE 1. Biological degradation of 2,4-D for different concentrations, holding microorganisms constant. (Redrawn from Hemmet, R. B., Jr. and Faust, S. D., *Residue Rev.*, 29, 192, 1969. With permission.)

Analytical Methods

2,4-D residuals were determined at 284 nm[1]. This analytic information confirmed disappearance of the 2,4-D molecule in the aquatic microorganic ecosystem.

Materials

Eastman Kodak®, white label grade, 2,4-D was used.

Kinetic Model

The biological degradation of 2,4-D was assumed to follow the stoichiometry suggested by this reaction:

$$2,4\text{-D} + 7.5O^2 = 8CO^2 + 2H^2O = 2HCl\,(A.M.) = K$$

Thus, 1.00 mg of 2,4-D should require 1.09 mg of oxygen for complete conversion to carbon dioxide.

Microorganism Concentration Held Constant

Figure 1 shows that as the herbicide concentration increases, the oxidation rate also increases. The k values are 0.885, 2.24, and 3.98 mg for herbicide concentrations of 24, 50, and 100 mg/ℓ, respectively. Confirmation of zero-order kinetics, in this case, is to see if the percent of 2,4-D oxidized per unit time is the same. For example, the percent of 2,4-D oxidized between the 9th and 10th hr was 4.0, 4.0, and 4.5 for herbicide concentrations of 25, 50, and 100 mg/ℓ, respectively. Therefore, the oxidation rate was independent of initial 2,4-D concentration.

Table 6
SUMMARY OF ZERO-ORDER RATE CONSTANTS FOR THE BIODEGRADATION OF 2,4-D BY AQUATIC MICROORGANISMS

	k Value (mg/ℓ^{-hr-1})			
(A.M.) (mg/ℓ dry weight)	(2,4-D) (mg/ℓ)			
	25	50	75	100
500[a]	—	1.20	—	2.10
1000	—	1.89	—	4.11
500[b]	0.47	1.08	—	1.89
1000	0.885	2.24	—	3.98
500[c]	—	1.25	—	—
750	—	—	2.42	—
1000	—	—	—	3.37

[a] (2,4-D) held constant, microorganisms varied.
[b] Microorganisms held constant, (2,4-D) varied.
[c] Ratio of (A.M.) to (2,4-D) held constant at 10:1.

From Hemmett, R. B., Jr. and Faust, S. D., *Residue Rev.*, 29, 192, 1969. With permission.

Microorganism and Herbicide Concentrations Varied

For this analysis, microorganism concentration and 2,4-D were varied, but their weight ratios were held constant. Apparent zero-order kinetics were observed again since the percentages of oxidation per hour were 3.0, 3.3, and 3.5 for herbicide concentrations of 50, 75, and 100 mg/ℓ, respectively. The rate constants are summarized in Table 6.

Significant Effects

Results suggest that biodegradation of 2,4-D follows zero-order kinetics. The most important constituent of the reaction is, therefore, the size and nature of the microbial population. It has been observed that approximately 68% of the 2,4-D concentration is used in the respiration process as measured by manometric techniques, whereas, 32% is incorporated into the microbial cells as an energy reserve.

Rates of biodegradation depend on several variables. The first is the time necessary for the enzyme systems to acclimatize to the herbicide. The second is the effect that age of the acclimatized microorganisms has on the rate. It has been shown that the rate decreases as the acclimatization period increases until it reaches a stabilization point. Therefore, each new addition of 2,4-D should be degraded faster than the previous one provided the stabilization point has not been reached. The third variable is the natural condition of the aquatic environment.

The biodegradation appeared to follow zero-order kinetics, with the oxidation rate independent of the substrate concentration. When the ratio of microorganism concentration to 2,4-D concentration was held constant, constant oxidation rates were observed suggesting the enzyme systems were saturated with substrate. Thus, substrate concentration was not a limiting factor in these kinetic studies, further substantiating the zero-order kinetic model. Acclimatization time also appeared as a factor. The

the acclimatization period, the shorter the length of effect of ultimate biodegradation of 2,4-D.

SUMMARY OF EFFECTS

The effect of a 2,4-D granular formulation applied at the rate of 1.32 ppm (20 kg ae/ha granular material) on water quality of Rockaway Park Lake, N.J. was studied for 132 days after application. During this time, there was no adverse effect on pH, total hardness, total alkalinity, dissolved oxygen, BOD, chlorine demand, or change in odor or taste. Although there was a level of 0.65 mg/g of 2,4-DCP/g of 2,4-D in the original formulation, there was no accumulation of this material due to biological degradation.

Calcium and magnesium salts of 2,4-D have relatively high solubilities in water at 25°C. Therefore, ordinary concentrations of calcium and magnesium ions in surface water do not remove 2,4-D through precipitation reactions and limestone products cannot be used to remove 2,4-D at water treatment plants. However, it is found that 2,4-D compounds can be removed from surface waters by activated carbon treatment. Chlorination and potassium permanganate oxidation do not remove 2,4-D compounds from water, but do remove 2,4-DCP from water. Usual treatments with ferrous sulfate and aluminum coagulation do not remove 2,4-D compounds from surface water, but 2,4-D (Na salt) and 2,4-DCP can be removed from surface water by strongly basic anion exchange resins.

The activity of microorganisms in biodegradation of 2,4-D definitely influences the residue levels following application. Their ability to degrade 2,4-D, however, depends on the microorganism adaptation to utilize the 2,4-D as a carbon source. The degradation rate is increased substantially after the adaptation of microorganisms from the use of 2,4-D. Initial applications to an aquatic system will therefore result in longer period of time with detectable residues than will subsequent applications.

REFERENCES

1. **Aly, O. M. and Faust, S. D.,** Determination of 2,4-dichlorophenoxy acetic acid in surface waters, *J. Am. Water Works Assoc.*, 55, 639, 1963.
2. **Aly, O. M. and Faust, S. D.,** Studies on the fate of 2,4-D and ester derivatives in natural surface waters, *J. Agric. Food Chem.*, 12, 541, 1964.
3. **Demarco, J., Symons, J. M., and Robeck, G. G.,** Behavior of synthetic organics in stratified impoundments, *J. Am. Water Works Assoc.*, 59, 965, 1967.
4. **Faust, S. D. and Aly, O. M.,** Some effects of 2,4-D and 2,4-DCP on drinking water quality, *Proc. Northeast. Weed Control Conf.*, 17, 460, 1963.
5. **Hemmett, R. B., Jr. and Faust, S. D.,** Biodegradation kinetics of 2,4-dichlorophenoxyacetic acid by microorganisms, *Residue Rev.*, 29, 192, 1969.
6. **Burttschell, R. H.,** Chlorine derivatives of phenol causing taste and odor, *J. Am. Water Works Assoc.*, 51, 205, 1959.
7. **Lee, F. G. and Morris, C. J.,** Kinetics of chlorination of phenol-chlorophenolic tastes and odors, *Int. J. Air Water Pollut.*, 6, 419, 1962.
8. **Okey, R. W. and Bogan, R. A.,** Apparent involvement of electronic mechanism in limiting the microbial metabolism, *J. Water Pollut. Control Fed.*, 37, 692, 1965.
9. **Schwartz, H. J.,** Microbial degradation of pesticides in aqueous solutions, *J. Water Pollut. Control Fed.*, 39, 1701, 1967.
10. **Burchfield, H. P., Johnson, D. E., and Storrs, E. E.,** Guide to the Analyses of Pesticide Residues, U.S. Department of Health, Education and Welfare Public Health Service, Bureau State Services 1 (I.E.3):1, 1965.
11. **Butler, P. A.,** Effects of pesticides on commercial fisheries, *Annu. Sess. Gulf Carib. Fish. Inst.*, 13, 168, 1960.

Chapter 21

EURASIAN WATERMILFOIL STUDIES IN FLORIDA*

INTRODUCTION

Eurasian watermilfoil *(Myriophyllum spicatum)* has been established for several decades in Florida. Extensive growth is largely limited to the Crystal-Homosassa River Basin on the west coast extending into the Chassakowitzka Bay, just north of Tampa Bay. In March 1967, it spread over approximately 1200 ha and seriously threatened such famous tourist attractions as Weeki Wachee, Homosassa, and Crystal Springs. During recent years, this area has been partly overgrown with hydrilla *(Hydrilla verticillata),* and the problem is greatly compounded by the plant's aggressive nature. Eurasian watermilfoil is a submersed vascular aquatic plant that produces a mat of vegetation several meters thick and tends to dominate the water surface where it grows. It can severely damage water resources by preventing fishing, water skiing, boating, swimming, navigation, irrigation, and drainage.

Effective control of Eurasian watermilfoil by herbicide use has been of great value. A high degree of susceptibility to the phenoxy herbicides has been demonstrated. However, there is a need to more closely define the relationship between herbicide concentrations and their phytotoxicity to the plant. This study was designed to determine the optimum herbicide concentrations required for effective and economical Eurasian watermilfoil control.

HERBICIDE EVALUATION

Eurasian watermilfoil was not added to the U.S. Department of Agriculture's evaluation (USDA) program until late 1967[1] when growth tanks of Eurasian watermilfoil were established at Fort Lauderdale, Fla. for these studies. The plant was found easily cultured in outside growth tanks. Results of several commonly used aquatic herbicides on Eurasian watermilfoil are shown in Table 1. All herbicides evaluated were highly active on this plant at parts per million weight (ppmw). Test plots of ½ acre each were set out in the Crystal River area and in the Homosassa River area (Figure 1) to evaluate selected herbicides under field conditions. Applications were made July 8 and 9, 1968, at low tide and were evaluated at 2-week intervals by three visual estimates of the percent control. Results given in Table 2 indicate that most commonly used herbicides were effective. Regrowth was evident by the end of the season so that a program of repeated application would be necessary for good control.[2]

TIME CONCENTRATION STUDIES

The purpose of these experiments was to obtain information of the relationship between concentration of 2,4-D butyl ether ester (BEE) in water and phytotoxicity to watermilfoil for various periods of exposure to herbicide concentrations.[6]

Apical cuttings of watermilfoil were planted in pots containing a 1:1 sand-peat soil mixture. Each pot held three 7.5-cm cuttings. In the first test the pots were placed in

* From cooperative studies of the Aquatic Plant Management Laboratory, U.S. Department of Agriculture, Agricultural Research Service, Ft. Lauderdale cooperating with the U.S. Army Environmental Hygiene Agency, Edgewood Arsenal, Md., and U.S. Army Corps of Engineers, Office of the Chief of Engineers, Washington, D.C.

Table 1
HERBICIDAL ACTIVITY OF SELECTED CHEMICALS EVALUATED ON EURASIAN WATERMILFOIL IN STILL-WATER TESTS[a]

Chemical	Rate (ppmw)	Weeks after treatment (% control)		
		2	4	6
Silvex® (PGBE)	1	75	83	100
	2.5	90	100	100
	5	95	100	100
Silvex® (BEE)	1	70	80	100
	2.5	87	100	100
	5	95	100	100
Silvex® (IOE)	1	83	98	100
	2.5	95	100	100
	5	95	100	100
2,4-D Amine	1	43	68	88
	2.5	48	100	100
	5	57	100	100
2,4-D (PGBE)	1	58	85	98
	2.5	60	100	100
	5	75	100	100
2,4-D (BEE)	1	55	83	99
	2.5	68	100	100
	5	80	100	100
2,4-D (IOE)	1	70	78	99
	2.5	77	100	100
	5	85	100	100
2,4-D (tertiary fatty acid amine)	1	78	100	100
	2.5	85	100	100
	5	93	100	100
Endothall (k salt)	1	62	99	100
	2.5	80	100	100
	5	86	100	100
Diquat	0.5	100	100	100
	1	100	100	100
	2.5	100	100	100

[a] Each value is the average of three replications.

From Blackburn, R. D. and Weldon, L. W., Weed Investigations for Aquatic and Noncropland Areas, Annu. Rep., Plantation Field Laboratory, U.S. Department of Agriculture, Fort Lauderdale, Fla., 1967.

20-ℓ aquaria in the growth room. The temperature was 25°C, light intensity was 80 to 100 fc on a 12-hr photoperiod. Water in the aquarian was replaced every 3 days. After 8 days, appropriate volumes of 2,4-D BEE stock solution were added to the aquaria to bring the concentration to 0.5, 1.0, 2.5, or 5.0 ppmw 2,4-D. After 0-, 1-, 2-, 4-, 8-, 16-, 24-, 48-, or 96-hr contact, plants were removed from the aquaria, rinsed in running water for 30 min, and placed in gallon jars in the growth room. Exposure times were replicated three times and consisted of two pots per jar. Because of the number of jars involved, this test was conducted in two parts, 11 days apart. The first part consisted of 1- to 8-hr, and the second part, 16- to 96-hr contact times. A second test was con-

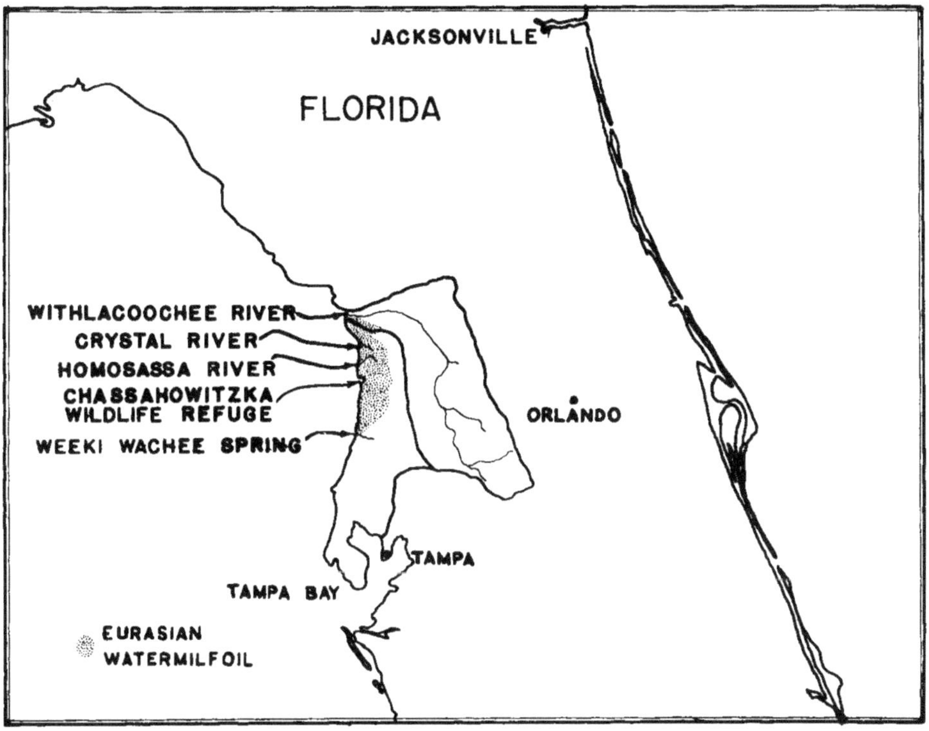

FIGURE 1. Test plot study area in the Homosassa River and infestation of Eurasian watermilfoil.

ducted in which treatments were applied in outside tanks. Potted milfoil plants were placed in the tanks, and after 2 weeks, treatments of 0.5, 1.0, or 2.5 ppmw 2,4-D were applied as previously described. Water temperature at the time of treatment was 18°C. Treatment of the plants, exposure times, and number of replicates were the same as in the first test. Observations were made at weekly intervals to determine percent herbicidal injury.

DISCUSSION OF RESULTS

Test 1

As might be predicted, herbicide injury to the plants increased with increasing concentration or exposure time. Eurasian watermilfoil was destroyed after a 1-hr exposure to a 5 ppmw concentration of 2,4-D (Tables 3 and 4). A 4-hr exposure period was required at the 2.5 ppmw rate. In one part of the 1.0 ppmw treatment rate test, an 8-hr exposure produced adequate control, but in the second part, no exposure time was adequate. None of the exposure times were adequate at the 0.5 ppmw concentration in these tests.

Test 2

In this test (Table 5), as in the one previous, plant injury increased with both increasing herbicide concentration and/or exposure time. Maximum injury to watermilfoil within the 0.5 ppmw treatment was observed 6 weeks following a 96-hr exposure to the treatment. The injury rating decreased at later observation periods indicating that regrowth was occurring. The slow rate of increase in injury rating with increasing exposure time is evidence that the 0.5 ppmw concentration was too low to be effective.

Table 2
HERBICIDE ACTIVITY OF SELECTED CHEMICALS ON EURASIAN WATERMILFOIL IN FIELD TEST PLOTS AT HOMOSASSA, FLORIDA[a]

Herbicide	Rate (ppmw)	Weeks after treatment (% control[b]) 2	4	6
2,4-D IOE	2.5	75	81	86
(isoactyl ester)	2.5	95	76	91
	5.0	72	83	85
	5.0	95	95	90
2,4-D BEE	2.5	51	76	71
(Butoxyethyl ester)	2.5	93	90	96
	5.0	70	83	71
	5.0	71	63	69
Diquat	0.5	80	86	93
	0.5	85	77	92
	1.0	90	93	96
	1.0	85	90	91
Hydrothal	2.0	70	75	95
	2.0	85	81	90
	4.0	70	83	85
	4.0	71	81	95

[a] Each value is the mean of three estimates.
[b] 0 = no control; 100 = complete kill of plants. Each value is the average of three replications.

From Blackburn, R. D. and Weldon, L. W., Weed Investigations for Aquatic and Noncropland Areas, Annu. Rep., Plantation Field Laboratory, U.S. Department of Agriculture, Fort Lauderdale, Fla., 1969.

The 1.0 ppmw treatment rate produced a maximum injury to and effective control of watermilfoil 7 weeks after a 48-hr exposure period. The 24-hr exposure period (although producing approximately 90% injury to the plants after 8 weeks) was not adequate because increase in injury rating beyond this period was too gradual to be significant.

The 2.5 ppmw treatment rate produced 100% injury to the plants 7 to 8 weeks after an 8-hr exposure period. Under the condition of the second test, 48 hr-contact to 2,4-D BEE applied at the rate of 1.0 ppmw acid equivalent (ae) is required for complete kill of the test plants.

HERBICIDE RELEASE STUDIES

The conventional method of applying liquid formulations of herbicides for submersed aquatic weed control is to apply a phytotoxic herbicide concentration in the entire volume of water. Injury to nontarget organisms (both plant and animal) caused by this type of treatment can be quite severe, depending on toxicity of the herbicide.[3,4]

This injury may be reduced, however, by applying herbicides in inert carrier to give controlled release, rather than treating the entire volume of water. It may be possible, with selected carriers, to release the herbicide near the soil-water interface where plant

Table 3
THE EFFECT OF 2,4-D ON EURASIAN WATERMILFOIL GROWN AND TREATED UNDER CONTROLLED ENVIRONMENTAL CONDITIONS

Part One of Test 1

Rates of 2,4-D (ppmw)	Exposure time (hr)	Weeks following treatment (% control[a]) 2	4	6	8	10
Control		0	2	12	18	20
0.5	1	5	13	40	42	40
	2	5	18	55	60	57
	4	12	28	58	57	52
	8	13	30	68	73	63
1.0	2	7	12	50	82	77
	4	10	20	67	87	88
	8	15	32	78	93	93
		18	38	80	99	100
2.5	1	13	48	87	98	97
	2	15	65	90	93	90
	4	18	77	96	100	100
	8	20	87	97	100	100
5.0	1	17	90	95	100	100
	2	32	95	98	100	100
	4	28	97	98	100	100
	8	37	98	99	100	100

[a] 0 = No control; 100 = complete kill of the plants. Each value is the average of three replications.

From Elliston, R. A. and Steward, K. K., *Hyacinth Control J.*, 10, 38, 1972. With permission.

growth originates and propagating structures are located. Proper application timing may prevent dense growths of aquatic plants by inhibiting or preventing growth early in the growing season. Controlled release of a herbicide from a carrier may also prevent regrowth that emerges from propagating structures buried in the bottom muds.

The investigation was comprised of laboratory determinations of the rate of herbicide release from experimental formulations and of bioassay of herbicide formulations.[5]

Release Rate Determinations

Butoxyethanol ester (BEE) of 2,3-D-14 C (radioactive carbon of 2,4-D) herbicide-polyvinyl chloride complexes were prepared from a premixed plastisol by the following general procedure. A PVC plastisol was prepared by mixing 50 g of Geon 135 PVC resin wth 50 g of dioctylphthalate (DOP) plasticizer. Herbicide-plastisol formulations containing 10, 20, and 30% herbicide were prepared by adding 0.20, 0.40, and 0.60 g herbicide to 1.80, 1.60, and 1.40 g plastisol, respectively. Five microcuries of 2,4-D-1-14 C herbicide (16 mg) in 1.0 mℓ of plasticizer were added to each formulation. The mixtures were stirred, poured into 6-mm ID glass tubes, and cured at 125 to 130°C for 30 min. The rod-like products were removed from the glass tubes and cut into pellets weighing approximately 82 mg. Five pellets were placed in 400 mℓ of tap water in a screw-capped, glass jar and stored at 25°C. Each concentration of 2,4-D-PVC

Table 4
THE EFFECT OF 2,4-D ON EURASIAN WATERMILFOIL GROWN AND TREATED UNDER CONTROLLED ENVIRONMENTAL CONDITIONS

Part Two of Test 1

Control		0	1	1	1	1
0.5	16	10	32	67	62	55
	24	10	28	70	62	60
	48	12	42	85	73	63
	96	13	37	82	65	60
1.0	16	13	47	72	55	53
	24	17	42	75	65	60
	48	20	67	88	85	85
	96	20	75	90	87	90
2.5	16	25	92	99	100	100
	24	25	92	99	100	100
	48	48	97	100	100	100
	96	50	99	100	100	100
5.0	16	37	98	100	100	100
	24	67	100	100	100	100
	48	80	100	100	100	100
	96	87	100	100	100	100

[a] 0 = No control; 100 = complete kill of the plants. Each value is the average of three replications.

From Elliston, R. A. and Steward, K. K., *Hyacinth Control J.*, 10, 38, 1972. With permission.

formulation was replicated three times. Aliquots of 1.0 $m\ell$ were taken at 1, 3, 5, and 7 days and at weekly intervals thereafter through 20 weeks. Each 1.0-$m\ell$ sample was placed in 14 $m\ell$ of scintillating fluid for analysis by liquid scintillation counting. Samples were analyzed for radioactivity using a liquid scintillation counter, which had a counting efficiency of 95.6% (unquenched standard).

Preparation of Experimental Formulations for Bioassays

PVC was evaluated as a plastic carrier by preparing herbicide-PVC formulations from a premixed plastisol. Formulations were prepared according to the following procedure. A PVC plastisol was prepared by dispensing 100 g of Geon 135 resin in 150 g of DOP plasticizer. Herbicide was added to the plastisol and the mixture stirred and poured into aluminum pellet molds. Thermal curing in an oven at 125 to 135°C for 30 min produced pellets weighing approximately 75 mg.

Bioassays

In September 1969, field-collected cuttings of Eurasian watermilfoil were planted in outside culture tanks which contained 3700 ℓ of water. These tanks were 2.4 m in diameter and 2 m deep. Treatments of 2,4-D BEE formulated in PVC, 30% active ingredients (a.i.), were applied in June 1970 at concentrations of 1.0, 2.0, and 4.0 ppmw. The 1.0 and 2.0 ppmw rates were compared with a field-tested commercial clay-granule formulation of 2,4-D BEE.[12,13,14] Each randomized treatment was replicated three times. Herbidical effects were observed and recorded 1 week after treatment and at 2-week intervals thereafter. Injury was rated on a 0 to 100% scale which indicates no effect to complete kill.

Table 5
THE EFFECT OF THREE CONCENTRATIONS OF 2,4-D ON EURASIAN WATERMILFOIL WHEN APPLIED IN OUTSIDE TANKS AT VARIOUS EXPOSURE TIMES[a]

Test 2

Control		0	1	1	1	1
0.5	1	5	10	23	35	28
	2	5	12	32	45	38
	4	8	15	37	45	40
	8	10	27	53	63	53
	16	10	33	58	68	58
	24	12	43	68	73	63
	48	15	45	77	77	70
	96	17	52	88	78	75
1.0	1	5	13	42	53	47
	2	8	17	43	53	48
	4	12	18	43	55	50
	8	13	33	67	67	68
	16	15	48	83	88	92
	24	17	55	88	90	92
	48	18	68	97	100	100
	96	23	82	99	100	100
2.5	1	10	25	78	68	53
	2	12	33	85	82	77
	4	15	53	88	88	90
	8	17	70	97	100	100

[a] Plants were placed under controlled conditions after exposure to the 2,4-D.

[b] 0 = No control; 100 = complete kill of the plants. Each value is the average of three replications.

From Elliston, R. A. and Steward, K. K., *Hyacinth Control J.*, 10, 35, 1972. With permission.

Residue Analysis

Samples for residue analysis were taken before herbicide application and then after 1, 4, 7, 14, 28, 42, and 56 days. Three 1.0-ℓ samples were taken at equal intervals around the circumference of the pools. The three samples were composited, mixed, and 1.0-ℓ subsample of the composite taken for extraction. The pH of the composite sample was adjusted to ph 12 with sodium hydroxide, shaken, and allowed to stand for approximately 30 min to hydrolyze the ester. The pH of the sample was then adjusted to pH 1.5 with concentrated hydrochloric acid, and a 500-mℓ aliquot was extracted twice with two 25-mℓ volumes of chloroform. The chloroform extracts were then evaporated to remove the chloroform, and the 2,4-D acid was converted to the methyl ester using a Boron trifluoride methylation procedure[6] and analyzed by gas chromatography.

A Micro-Tek® Model 150 gas chromatograph with a Ni electron capture detector was used for analysis of water extracts. The instrument was equipped with a 2 m by 6 mm glass column packed with 80/90 mesh Chromosorb W coated with 3% OV-1. The operating parameters were oven temperature, 200°C; inlet temperature, 235°C; outlet temperature, 240°C; detector temperature, 235°C; outlet temperature, 240°C; detector temperature, 250°C; carrier gas N; column flow rate, 60 cc/min; and scavenger flow

Table 6
PERCENT CONTROL AND HERBICIDE RESIDUE OF ATTACLAY AND PVC FORMULATIONS OF 2,4-D BEE FOR CONTROL OF EURASIAN WATERMILFOIL IN OUTSIDE GROWTH POOLS

Weeks after Treatment	Attaclay control (%)[a]	Residue (ppmw)	PVC pellet control (%)	Residue[b] (ppmw)
	1 ppm weight of herbicide			
1	38	0.560	8	1/m[c]
2	96	0.103	17	1/m
4	97	1/m[c]	18	1/m
6	95	1/m	33	1/m
8	90	1/m	20	1/m
	2 ppm weight of herbicide			
1	17	0.850	10	0.250
2	43	0.412	28	1/m[c]
4	99	1/m[c]	30	1/m
6	99	1/m	57	1/m
8	100	1/m	37	1/m

[a] Attaclay granular 2,4-D BEE herbicide (Aqua Kleen®), Amchem Products, Ambler, Pa.

[b] PVC (polyvinyl chloride) pellets containing 2,4-D BEE (Experimental herbicide prepared by U.S. Army Environmental Health Agency, Edgewood Arsenal, Md.).

[c] Herbicide residue less than 0.1 ppmw in water.

From Steward, K. K. and Nelson, L. L., *Hyacinth Control J.*, 10, 35, 1972. With permission.

rate, 20 cc/min. The retention time for the methyl ester of 2,4-D relative to aldrin as unity was 0.30. The percent recovery of known amounts of the methyl ester was calculated to be 54.4%. All analyses were corrected for recovery.

Phytotoxicity and Herbicide Residue

The results of growth pool tests to evaluate attaclay granular (Aqua Kleen®) and PVC pellets containing the BEE of 2,4-D are summarized for percent control and herbicide residue in the water after treatment in Table 6. Attaclay granular at 1 and 2 ppm provided satisfactory control of Eurasian watermilfoil. The residue in the water was below 0.1 ppm, 2 weeks after treatment. PVC pellets applied at 1 ppm rate did not supply sufficient herbicide release.

SUMMARY OF EFFECTS

Field application of 2,4-D granules [2,4-D BEE and 2,4-D isoactyl ester (IOE)], diquat and hydrothal resulted in significant control of Eurasian milfoil for the summer season, applied in early July.

Time-concentration treatments in the laboratory and growth pools resulted in 99% dieback at an optimum concentration of 1 ppm of herbicide for 4 days exposure.

Herbicide-release studies indicated 99% dieback from attaclay granules at 1 ppm. Residues in the water of the growth tanks persisted above 0.1 ppm for 2 weeks. In moving water, this period would be less depending on the rate of movement or exchange.

REFERENCES

1. **Blackburn, R. D. and Weldon, L. W.,** Weed Investigations for Aquatic and Noncropland Areas, Annu. Rep., Plantation Field Laboratory, Fort Lauderdale, Fla., 1967.
2. **Blackburn, R. D. and Weldon, L. W.,** Weed Investigations for Aquatic and Noncropland Areas, Annu. Rep., Plantation Field Laboratory, U.S. Department of Agriculture, Fort Lauderdale, Fla., 1969.
3. **Stanley, R. A.,** Studies on Nutrition, Photosynthesis, and Respiration in *Myriophyllum spicatum* L., Ph.D. thesis, Duke University, Durham, N.C., 1970.
4. **Wajtalik, T. A., Hall, T. F., and Hill, L. O.,** Monitoring ecological conditions associated with wide-scale applications of DMA, 2,4-D to aquatic environments, *Pest. Monitor. J.*, 4, 184, 1971.
5. **Steward, K. K. and Nelson, L. L.,** Evaluations of controlled release PVC and attaclay formulations of 2,4-D on Eurasian watermilfoil, *Hyacinth Control J.*, 10, 35, 1972.
6. **Elliston, R. A. and Steward, K. K.,** The response of Eurasian watermilfoil to various concentrations and exposure periods of 2,4-D, *Hyacinth Control J.*, 10, 38, 1972.

Appendixes

Appendix A

DEFINITIONS AND EQUIVALENTS

cfs = cubic feet per second; cfs × 448.8 = gallons per minute; cfs × 5.4×10^6 = pounds per day

g/ℓ = grams per liter; $g/\ell \times 1{,}000$ = parts per million

g/m^2 = grams per square meter; $g/m^2 \times 8.922$ = pounds per acre

ha = hectare; ha × 2.471 = acres

kg = kilograms; kg × 2.2 = pounds

kg/ha = kilograms per hectare; kg/ha × 0.8922 = pounds per acre

metric tons × 2,204 = pounds

μg-at P/g = microgram atoms phosphorus per gram = 31 parts per million phosphorus

μg-at P/ℓ = microgram atoms phosphorus per liter = 31 parts per billion phosphorus

μg/g = micrograms per gram = ppm

$\mu g/\ell$ = micrograms per liter = ppb; $\mu g/\ell \times 10^{-3}$ = parts per million

$\mu g/m^2$ = micrograms per square meter; $\mu g/m^2 \times 8.92 \times 10^6$ = pounds per acre

mg/g = milligrams per gram; mg/g × 10^3 = ppm

micromoles P per 100 grams × 0.31 = parts per million

mg/ℓ = milligrams per liter = parts per million

mg/m^3 = milligrams per cubic meter; $mg/m^3 \times 0.00272$ = pounds per acre-foot

mg/m^2 = milligrams per square meter; $mg/m^2 \times 8{,}922$ = pounds per acre

mg % = milligrams percent = milligrams in 100 grams = 1 part in 100,000 parts wet weight; mg % × 0.1 = ppm

mgd = million gallons per day; mgd × 1.547 = cubic feet per second

ppb = parts per billion

ppm = parts per million

P_2O_5 = phosphorus pentoxide; $P_2O_5 \times 0.436$ = P

PO_4 = phosphate; $PO_4 \times 0.326$ = P

NO_3 = nitrate; $NO_3 \times 0.226$ = N

Parts per million × cubic feet per second × 5.4 = pounds per day (gallons per minute × 2.228×10^{-3} = cubic feet per second)

Parts per million × 8.34 × gallons per day × 10^{-6} = pounds per day (gallons per minute × 1440 = gallons per day)

Lineal Measure (Length)

1 inch (in.)	= 2.54 centimeters = 25.4 millimeters = 25,400 microns
1 foot (ft)	= 12 inches = 30.48 centimeters = 0.3048 meter
1 yard (yd)	= 3 feet = 0.9144 meter
1 rod (rd)	= 5.5 yards = 16.5 feet = 5.029 meters
1 mile (mi)	= 160 rods = 1760 yards = 5280 feet = 1.6094 kilometers
1 micron (μ)	= 0.001 millimeter = 0.00004 inch (1/25,400)
1 millimeter (mm)	= 0.0394 inch
1 centimeter (cm)	= 10 millimeters = 0.3937 inch
1 meter (m)	= 100 centimeters = 3.281 feet = 39.37 inches
1 kilometer (km)	= 1000 meters = 0.6214 miles = 3281 feet

Square Measure (Area)

1 square inch	= 6.45 square centimeters
1 square foot	= 144 square inches = 0.0929 square meter = 929 square centimeters
1 square yard	= 9 square feet = 0.8361 meter
1 square rod	= 272.25 square feet = 30.25 square yards = 25.293 square meters
1 acre	= 43,560 square feet = 4,840 square yards = 160 square rods = 0.4047 hectare
1 square mile	= 640 acres = 259 hectares
1 square meter	= 1550 square inches = 10.764 square feet = 1.196 square yard
1 hectare	= 2.471 acres = 10,000 square meters
1 square kilometer	= 0.3861 square mile
1 square centimeter	= 0.155 inch

Volume or Capacity Measure (Cubic)

1 cubic inch	= 16.387 cubic centimeters
1 cubic foot	= 1728 cubic inches = 29.922 U.S. liquid quarts = 25.714 U.S. dry quarts = 0.80357 U.S. bushel = 28.316 liters = 28.320 cubic centimeters
1 cubic yard	= 27 cubic foot = 0.7646 cubic meter
1 cubic millimeter	= 0.001 cubic centimeter
1 cubic centimeter	= 0.001 cubic decimeter = 0.061 cubic inch
1 cubic decimeter	= 0.001 cubic meter
1 cubic meter	= 35.31 cubic feet = 1.308 cubic yard = 264.2 gallons

Volume or Capacity Measure (Liquid)

1 level tablespoonful	= 3 level teaspoonfuls
1 fluid ounce (U.S.)	= 2 tablespoonfuls = 29.57 milliliters
1 cupful	= 8 fluid ounces
1 pint (pt)	= 2 cupfuls = 16 fluid ounces = 0.4732 liter
1 quart (qt) (U.S.)	= 2 pints = 0.9463 liter
1 gallon (gal) (U.S.)	= 4 quarts = 0.8333 gallon (Imperial or Br.) = 231 cubic inches = 0.1337 cubic foot = 3.785 liters
1 gallon (Br.)	= 1.2009 gallons (U.S.) = 4.546 liters
1 liter	= 1,000 milliliters = 1.057 quart = 0.2642 gallon

Volume or Capacity Measure (Dry)

1 quart (U.S.)	= 2 pints = 1.1012 liters
1 bushel (bu) (U.S.)	= 32 quarts = 4 pecks = 1.244 cubic feet
1 bushel (Br.)	= 1.2843 cubic feet = 36.368 liters
1 liter	= 0.9081 dry quart (U.S.)

Weight or Mass Measure

1 grain (gr)	= 64.7989 milligrams
1 ounce (oz) (avoirdupois)	= 437.5 grains = 28.3495 grams
1 pound (lb) (avoirdupois)	= 16 ounces = 7000 grains = 453.59 grams

1 ton (U.S. short)	=	2000 pounds = 907.185 kilograms
1 ton (U.S. long)	=	2240 pounds = 1016.047 kilograms
1 microgram	=	1 gamma = 0.001 milligram (mg)
1 gram (gm)	=	1000 milligrams = 15.432 grains = 0.03527 ounce
1 kilogram (kg)	=	1000 grams = 2.205 pounds
1 ton (metric)	=	1000 kilograms = 2204 pounds

At maximum density at 39.164°F., one gallon of pure water weighs 8.345 pounds.
At all temperatures up to 100°F., one gallon of water weighs approximately 8.3 pounds.

GLOSSARY OF BIOLIMNOLOGICAL TERMS

Some of these words have several meanings. Those given here are the ones that relate to weed control as used in this text.

Absorption — The process by which a chemical is taken into plants, animals, or minerals. Compare with adsorption.

Abyssal depth (lakes) — In a limnological sense, that depth at which the water remains uniform in temperature or is "stagnant".

Accretion — A term denoting the process by which alluvion is made. However, the two words, accretion and alluvion, are often used synonymously.

Active ingredient — The part of the pesticide which will kill pests or prevent damage by them. Usually it is the same as the "technical" material in the formulated product.

Acute toxicity — Any relatively brief illness caused by a poison; the degree of acute toxicity may vary from mild to fatal depending on the amount of poison absorbed.

Adsorption — The process by which chemicals are held on the surface of a mineral or soil particle. Compare with absorption.

Aerobic — Living in air (oxygen). The opposite of anaerobic.

Alga, Algae (i.e.) — Simple aquatic plants without true leaves, stems, or roots, sometimes having brown or reddish pigments; oval or spherical one- or multiple-celled green plants with a ribbon-like thallus (body).

Alluvion (lake) — In its legal meaning alluvion is an accretion to land, made gradually, composed of detritus deposited by streams or of deposits accumulated by the action of waves and currents. On lakes, accretions are made: by alluvial deposition on the advancing front of a delta; by filling of shoreline lake bottom by erosion detritus carried in by affluents, gullies, and superficial surface run-off or rainwash; by shifting dune sand; and by waste disposal such as tailings from mine operations. Also natural accretions may be made to islands, bars, and beaches of lakes. The deposits are realities, but whether or not, in specific instances, they constitute legal alluvion which involves land ownership is, ultimately, a matter for court decisions.

Amoeba — One-celled animals consisting of a protoplasmic mass with protruding cytoplasm providing means of locomotion, found commonly in ponds.

Amorphous — Substance, without form, in biology, describing especially lower forms of life that have no specialized structure; in geology, masses that have no continuity or stratification; in chemistry, chemicals that have no crystallinity.

Anabolism — A plant or animal process in which food is converted to living tissue; constructive metabolism as opposed to destructive metabolism or catabolism; a building from the simple to the complex.

Anaerobic — Living in the absence of air (oxygen). The opposite of aerobic.

Annual — A plant that grows from seed, produces seed the same year, then dies. New plants are produced from the seed the following year.

Antagonism — The loss of activity of a chemical when exposed to another chemical.

Aquatic weeds — Weeds that grow in water, either on top or under the surface.

Attractive nuisance — A legal term for any object which might attract children or other persons to it and then might injure or hurt them as a result. Examples: sprayers and empty pesticide containers.

Aufwuchs — Aufwuchs includes all those organisms that are attached to a submerged substrate but do not penetrate into its surface. This term is broader and more inclusive than periphyton; aufwuchs includes crustaceans, insects, and other forms.

Authochtonous detritus — Particulate matter originating in the lake.

Bacteria — Microscopic one-celled organisms that do not have chlorophyll; they usually multiply by simple division; very important for decay-causing functions in the soil and for disposal of sewage; disease-causing organisms in man and animals.

Biochemical oxygen demand (BOD) — The amount of dissolved oxygen in parts per million, required by organisms for the aerobic biochemical decomposition of organic matter present in water. This determination may be made by any of several acceptable laboratory techniques. A test for biochemical oxygen demand is standard procedure for public health, pollution control, and conservation agencies. Sewage treatment plant effluents are commonly tested for biochemical oxygen demand; industrial effluents are also tested by pollution control agencies in order to establish required levels of treatment.

Biological control — Pest control without the use of chemicals, i.e., parasites, predators, or diseases, etc. used to control pests.

Bitter lakes — Those containing large amounts of sulfates such as magnesium sulfate in their waters. In usage, restricted to the arid regions in the U.S. In the glaciated region of the northern U.S., and especially in the Lake States, the bogs are commonly peat bogs, and most frequently occur in basins, representing the sites of former lakes, or occur as formations partly or completely surrounding existing lakes. Bogs are further differentiated according to the characteristic surface vegetation such as sphagnum moss, leather leaf, and the conifers black spruce, tamarack, and white cedar. However, where in the ecologic succession the tree vegetation becomes dominant the term swamp often replaces "bog" in usage. Likewise marsh may replace bog for very large treeless wetland areas characterized by grasses and sedge. The term quaking bog is commonly applied to sedge-grass mats on the margins of lakes, and other semifloating types of peat in lake basins.

Carrying capacity (recreational) — The measure of the capacity of a lake for boating, skiing, bathing — recreational use in general — and residential occupation of the shore and shore border land without patent overcrowding, pollution, and consequent danger to health and safety. Carrying capacity may be limited if a single use is given priority; also, it may be expanded if the surface area of the lake is zoned for particular uses and the time for use in each zone is specified.

Catabolism — Destructive metabolic changes; process of breaking down tissue into simpler substance; opposed to anabolism; also spelled katabolism.

Caution — A signal word on labels of pesticides to alert users that the pesticide is toxic.

Chelating agents — Organic compounds as citric and tartaric acid which react with metal ions to give a chemically inert product due to the way in which the metal ions are bound into multiple ring arrangements; water softeners are chelating compounds.

Chlorosis — The yellowing of a plant's green tissue.

Chronic poisoning — Poisoning which occurs as a result of small, repeated doses of pesticide over a long period of time.

Chronic toxicity — How poisonous a pesticide is to an animal (or man) after small, repeated doses over a period of time.

Common name — A well-known, simple name of a pesticide accepted by the Pesticide Regulation Division of the Environmental Protection Agency.

Closed ecological system — An isolated environmental situation, as in a spacecraft chamber, in which organisms, animals, and humans have to depend on self-contained biochemical chemical cycles.

Closed system — A situation in science that assumes that all variables relevant to a problem are present and correctly represented.

Contaminate — Pollute; the addition of an unwanted material (often a pesticide) where it could do harm or damage.

Cryolimnion — A proposed term for a body of open water in a nearly completely frozen over, or ice-covered, inland lake. On some lakes the open water may represent either the location of the deeper holes in the basin, or the location of springs on the bottom or shore.

Cyptodepressions (lakes) — Lake basins whose deep parts are below sea level.

Diatoms — Microscopic plants which occur abundantly as floating forms in plankton. They sometimes occur abundantly enough to produce a water "bloom" and give a color tint, yellowish or brownish, to the water. Diatoms are notable for their shells of silica, and the siliceous character imparted to bottom deposits by their remains.

Diffused water — A legal term for vagrant water on the surface of land not yet in natural watercourses or lakes. It includes overland sheet runoff and temporary puddles, or splash, occurring after a rainstorm. Vagrant water collected and stored in artificial ponds or other enclosures becomes the property of the land owner. Diffused water has no precise meaning as a hydrogeologic term.

Diluent — Any liquid or solid material used to dilute or carry an active ingredient.

Dilute — To make thinner by adding water, another liquid, or solid.

Disciplined waters — A shop term applied to a stream or other waters which have been modified by engineering measures and structures and thereby brought under control to a degree to make it better serve some particular use.

Dismal — A term sometimes applied to fresh-water swamps in eastern parts of the U.S. especially in the flat coastal plain region — from Virginia southward.

Disposal area (lake improvement) — An area of land, usually in a marsh or swamp, where dredged material from a lake improvement project is deposited.

Drawdown — An engineering term for a decrease in the levels of reservoirs or other impoundments, caused by intensive use for hydroelectric power production, or to withdrawls made as a safety precaution against flooding, ice damage, or for other management practices.

Dry lakes — This term is applied to the sites of former lakes. In usage the bed of the former lake is not necessarily literally dry, but may support a marsh type of vegetation or, in the lower part of the basin, even aquatic vegetation. The desiccation, or loss of surface water, is not due mainly to loss by evaporation as in arid regions, but by drop in level of a local water table, filling by aquatic vegetation, or by land drainage operations.

Dam — An engineering structure, or man-made structure of any kind, designed to hold back a flow of water and create an impoundment, or raise the level of a preexisting body of water. Dam may be used in a figurative sense for a body of standing water. Where DAM is a place name, the name usually implies the impoundment of water as well as the engineering structure. Impoundments may be formed by natural dams, such as those by lodgment of driftwood across a stream channel, by alluvial deposition, by landslides, and those made by the beaver.

Dam lake — One created by a dam. The term is usually applied to a water body formed back of a man-made dam.

Dam pond — An impoundment back of a man-made dam. It is not restricted in size, and can be the equivalent of reservoir.

Decontaminate — To remove or break down the unwanted material (usually a pesticide) so it cannot do any harm or damage.

Degree of exposure — The amount or extent to which a person has been in contact with a toxic pesticide.

Dehydrogenation — Removal of hydrogen, particularly in the cell's energy transfers, involving oxidation-reduction processes.

Desiccation (of lakes) — Loss of water by direct evaporation. Desiccation usually implies thorough dryness of the whole lake bed area. Siccation of lakes may result primarily from other causes than desiccation or evaporation. Other causes are (1) drainage by cutting outlets or by dredging natural outlet streams; (2) escape of water through opening of subterranean outlets; and (3) removal or destruction of dams.

Destratification (of a lake) — Artificial circulation or mixing of the water. A device using compressed air may be used to force the cold stagnant water of the hypolimnion up through the thermocline thus mixing it with the warmer oxygenated water of the epilimnion.

Detention reservoir — A basin (above a dam) constructed for the temporary storage of a streamflow and superficial surface run-off. Also called retarding reservoir.

Dystrophic lakes — Those lakes associated with acid peat bogs, and largely filled with sediments consisting of unhumified or peaty organic matter. Typically the water is yellowish or brownish, low in calcium, often acid, and low in fish productivity.

Ecology — Study of the interrelationships between living organisms with each other with their environment.

Effluent — (1) The water discharged after use in the treatment of sewage; (2) water discharged from storm and sanitary sewers; (3) water discharged after the treatment of industrial products, or from tanks or containers after any use.

Egress (lake) — Egress is legally restricted to the act of departing from a lake. Where public access to lake waters is provided, egress from the waters, by the public, is implied or stipulated to avoid trespass on private property.

Embankments (lake) — Depositional features along shorelines, such as spits and bars.

Embayment — A descriptive word for a recess in a shoreline. Rarely used in place names.

Enriched lake — A eutrophied lake that has received inputs of nitrates, phosphates, and other nutrients, thereby greatly increasing the growth potential for algae and other aquatic plants. Most frequently, enrichment results from the inflow of sewage effluent but may sometimes be related to the movement of commercial agricultural fertilizer into the lake basin.

Emphermeral (lakes, ponds) — Short-lived lakes and ponds. Examples: beach pools and lagoons; deflation lakes; ponds in dune hollows; water bodies in river flood plains left by one flood and destroyed by succeeding floods; some lime sink ponds that have intermittent subterranean outlets.

Epilimnion — In a thermally stratified lake, the turbulent layer of water that extends from the surface to the thermocline or metalimnion.

Equalizing reservoir — Banks Lake, a large irrigation reservoir in the Grand Coulee near the Grand Coulee Dam on the Columbia River in the state of Washington, is called an equalizing reservoir. During the season of abundance of flow, in the river, water is lifted at a pumping station at the Dam to a feeder canal in the Grand Coulee which fills the reservoir. The supply in Banks Lake thence makes it unnecessary to pump water for irrigation from the river in the spring and fall when the stage is low.

Evaporation loss — Water lost from a lake surface by combined effect of solar radiation and wind circulation. Evaporation losses may be as little as 24 in. but can be as high as 70 in./year depending upon the climatic environment.

Evaporation suppression — Various fatty alcohols, which create monomolecular films are in use in Australia and southwestern U.S. to reduce evaporation losses from reservoirs. Commonly, these materials are either frequently applied or constantly applied by distributors mounted on rafts.

Evapotranspiration losses (lakes) — The combined losses from a lake surface due to evaporation, sublimation, and transpiration; exact measurements are extremely difficult. Commonly, evapotranspiration losses are stated as estimates.

Evolution — Development implying a slow, gradual process. In biology, organic evolution supposes gradual modification of the simplest or unicellular life forms through long time periods leading to their present existence as various species with all their differentiating characteristics; another significant aspect of the theory of evolution suggests that heredity enables animals to retain structural similarities and that variations account for making individuals different and adaptive to their environment.

Extralegal access — Extralegal access is involved when a person gains access to a lake over land not owned by himself, over private property without permission, or over public property that is riparian in location but is devoid of riparian rights, such as the right of way of a public road. Extralegal implies that the right of the person to use the water is unsettled or moot.

Fall overturn — A phenomenon that may take place in a lake in early autumn. Beginning with a cooling of the surface waters and changes in density, a general circulation or mixing of the water from top to bottom takes place, resulting in physical and chemical uniformity, which may be referred to as the initial phase of winter stagnation. (The opposite of spring Turnover.)

False bottom (lake) — A colloquial term usually applied to a semisuspended or soupy organic mass through which a weight easily sinks. Soft marl is also referred to very frequently as a false bottom.

False ponds — Mirage effects in desert regions, which resemble water bodies.

False shoreline — The line of contact between the open water of a lake and the front, or edge, or a "floating" mat of vegetation built out from the shore. Here the mat does not represent a former and higher stage of the lake and is therefore technically a part of the lake, even though its surface may be above water level.

Federal Waters, inland — The designation Federal applies to inland waters, including lakes, that are used for commercial navigation. Boating on Federal Waters is subject to the navigation rules of the U.S. Coast Guard. Some of the inland lakes, connected with or tributary to the Great Lakes that are navigable by federal test, may be subject to Coast Guard rules and control.

Fen — A British term sometimes used in the U.S. and often loosely as the equivalent of marsh, bog, and swamp. Technically the term fen may be restricted to a marsh or bog that has an alkaline mucky or nonpeaty soil. The landward, or peripheral zone of an acid bog surrounding a lake, may be called "marginal fen" because the soil is more humified.

Fertile lake — Lakes are often described as being fertile or infertile. A "fertile lake" is one that has a large standing crop, that is, a profile growth of aquatic plants and usually accompanying profile aquatic fauna. The degree of fertility cannot be measured or determined solely by chemical analysis of the waters or bottom soil. Chemical tests of the water may reveal extreme acidity, alkalinity, salinity, or presence of toxic matter and afford an explanation why a lake is unproductive, or has only a meager biomass.

Filter dams — Previous dams of loose stones, or stones and brush, placed in outlets of lakes to prevent fish from moving out, or at the mouths of inlet streams to prevent fish from entering.

Finger lakes — Long, narrow lakes occupying deep troughs in deeply eroded straight preglacial valleys in glaciated regions. "Finger" is frequently used as a descriptive term and in place name for lakes, regardless of origin, on the basis of a close, or only a faint, resemblance in shape to a finger of the human hand.

Fish kill — Destruction of fish in lakes or ponds due to prolonged ice and snow cover and resulting oxygen deficiency; or in very shallow lakes due to freezing of the water down to the bottom; or in summer due to oxygen deficiency caused by excessive amounts of organic matter in suspension. Also, fish may be killed by toxic pollutants, disease, or oxygen lack due to thermal additions from effluent of power plants.

Fish ladder — Any device designed to facilitate the movement of migrating fish over a dam. A conventional fish ladder consists of a stair-like series of small ponds connected by flowing water. The fish may swim from pond to pond and gain access to the lake or reservoir. Varying degrees of success are claimed by those involved with the administration of fish ladders.

Fishery — The term fishery of an inland lake is used in a special sense pertaining to the total game fish population and its environment. On the Great Lakes and larger inland lakes the term is ordinarily used to include the economic aspects of commercial fishing.

Floating island — A free-floating mat of vegetation. These floating mats usually represent a mass detached from a shore of quaking bog or marsh by storms or by a rise in lake level.

Floating sand — A patch of floating sand is a phenomenon occasionally observed on the calm water of a lake. The dry sand particles float because they are unable to break the surface tension of the water.

Flood zone — Some large reservoirs, impoundments, or streams are subject to a rise in level during a season of high precipitation. The area of shore land inundated by the rise in level above the normal operating pool level is the flood zone.

Floodplain (of a lake) — That part of a lake basin plain, lying between the shoreline and the shore cliff, and subject to submergence during a high stage of the lake.

Floodwater (in impoundments) — A word sometimes applied as a proper name to the flooded area back of a dam, similar to reservoir, lake, pond, and flowage.

Floor (of a lake) — The lower or deeper lying part of the bottom, especially of a deep lake. The term floor is often used interchangeably with bed and bottom. However, for many particular kinds of standing water bodies it seems appropriate to make a distinction. In some instances of usage flatness or horizontality is implied, and it is also limited to the firm surface lying beneath soft autogenic deposits, but commonly these interpretations are not observed.

Flowage easement — The right to flow a tract of land to a designated level; usually sold voluntarily. The land owner may reserve the right to enter upon and use the entire flowage for riparian purposes but he always retains ownership to the bottomland located on his tract. Ingress and egress may be across an exclusive or semiexclusive right of way between the uppermost contour elevation established by the easement and the waterline. The right-to-flow land usually embodies the right to manipulate the water level.

Freshlet (lake) — Small affluent streams which have a high rate of flow during the spring season and often carry a heavy silt load during their peak flow. (Overflowing of shallow stream.)

Groin — A low narrow wall-like structure extending out into the lake normal to the shore. It may be constructed of timber, stone, concrete, or steel and is usually impermeable. Its purpose is to catch littoral drift, or to trap sand, and to protect a beach from destructive erosion. It is sometimes called a wing dam.

Ground water lake — One whose water represents the level of the ground water or the upper boundary of the zone of saturation.

Growth regulator — A pesticide chemical which increases, decreases, or changes the normal growth or reproduction of a plant.

Habitat (lake) — The total of the environmental conditions which affect the life of plants and animals. Each lake or part of a lake has its own peculiar habitat. The natural environment in which a population of organisms lives.

Harbor line — On the Great Lakes, a line (set by Army Corps of Engineers), beyond which no wharfs can be constructed.

Harbor of refuge — A name given to havens on the shores of the Great Lakes located in between commercial harbors, and designated primarily to be a place of refuge for small craft during storm periods.

Harbors — A place of safe anchorage for ships and pleasure craft. Harbors are geographic features on the shore lines of the Great Lakes, but havens on the inland lakes are rarely called harbors.

Head (of lake) — The influent (stream) end, or where nonsingle influent is present the end opposite the outlet. The term has no application to lakes in enclosed basins. Where an influent enters a lake very near the outlet the term is inapplicable, also in numerous other instances where a number of small affluents enter a lake, or where a lake is nearly a perfect circle in shape.

Head of outlet — The place where the water leaves the lakes and enters the outlet stream, or effluent.

Gabion — A specially designed basket or box of corrosion-resistant wire used to hold rock and other coarse aggregate. These gabions may be locked together to form groins, sea walls, revetments, deflectors, and other structures. Their flexible construction permits minor adjustments of alignment resulting from undercutting, filling, and settling.

Gap — A passage or narrow channel between shore land and an island. In Lake of the Woods, Canada gap is used as a generic, Devils Gap.

Glacial lakes — The term glacial applies to those lakes formed as a result of glacial action during the Pleistocene. The sites of extinct glacial lakes may be determined by old beach and shoreline features and lacustrine sediments. The term glacial is also applied to lakes associated with existing, or present-day, glaciers.

Glade — Among its several connotations, glade may mean a water-covered marsh or wet prairie. However, geographic instances of glades of this kind, other than the Everglades of Florida, are very infrequent. Glade has been used as a term for a stretch of open water in the ice of a lake or river, but instances of usage in this sense are not common.

Grass lakes — Grass as an element in a place name for a lake is common, but it is also applied to very shallow marshy lakes, supporting a grassy vegetation consisting of true grasses and "false grasses", the latter mostly sedges. "Grass lake" is sometimes applied to a marsh, not water covered, occupying the bed of a former lake.

Gravel pit lakes — In the mining of gravel from the glacial drift in Michigan and other states, excavations often extend far below the level of the water table, and when mining operations are abandoned, bodies of open water, often of considerable size, are left. Usually the gravel pit lake is bordered by steep bare slopes which descend into deep water at the shoreline, and the environment is unsightly debris from mining operations. However, in favorable locations, gravel pit lakes have been successfully modified and developed for recreational uses. Bordering land may also be improved as a residential subdivision. Abandoned ponds are often used as swimming holes by youngsters but constitute a serious hazard because of deep, cold water.

Hole — In usage in the U.S. the term has been applied to several widely different kinds of depressional features, both very small and very large, both wet and dry. In numerous locations, it is a local name for a small lake or pond, occupying a sink, closed pit or hollow which may be either deep or shallow. Hole is also a name for a deep hollow and a deep water in a lake basin, and for either shallow or deep pockets containing water set in the terrain of a marsh or swamp. "Holes" are often identified by a geographic name, and occasionally in such names the generic Hole Lake is used. Also locally a bay of a lake, especially if the bay is "landlocked", is sometimes called a "hole."

Holomictic lake — One in which the circulation, or mixing, of the waters takes place throughout the entire depth. This is in contrast to a meromictic lake in which circulation is only in the upper layers, as an occasional occurrence.

Humus — Dark-colored topsoil, formed in part by the decay of organic matter.

Hydrogen-ion concentration — A measure of acidity or alkalinity, expressed in terms of the pH of the solution. For example, a pH of 7 is neutral, from 1 to 7 is acid, and from 7 to 14 is alkaline.

Hydrophilic — Referring to colloids which are easily hydrated, and therefore not readily coagulable; opposite of hydrophobic.

Hydrophobic — Opposite of hydrophilic; refers to colloids which are not easily hydrated and therefore coagulate readily.

Hypolimnion — The deep layer of a lake lying below the metalimnion and removed from surface influences.

Ice lake — One from which ice is harvested in winter, and stored under insulation for summer, or any later use. Cutting and storage of ice from lakes, once important, is now a vanishing industry. The right to harvest ice from an inland lake in Michigan is the exclusive right of the riparian.

Ichthyotoxin — Any substance which kills fish. Frequently used to eliminate or control rough fish populations and to manage lakes with overstocked or stunted populations of pan fish. Rotenone, a natural derivative of the South American derris root, has been widely used throughout the world as an ichthyotoxin but today there are many synthetic materials used with variable success. Proper use of ichthyotoxins requires technical knowledge of the lake involved and most frequently a permit is required from the conservation agency having jurisdiction over fishing.

Impoundment — A body of water ponded, or held back by a dam dike, floodgate, or any other barrier. The word impoundment is generally a common noun, and only rarely has an attached specific to make it a place name.

Impoundment, reservoir — One in which water is held for a considerable period of time and released to increase or maintain stream flow.

Influent (of lake) — Any surface stream flowing into a lake in contrast to an outflowing stream, or effluent. Often used synonymously with inlet and outlet — in a hydrologic sense; ground water may be influent in the form of springs or seepage.

Ingredients statement — The part of the label on a pesticide container which gives the name and amount of each pesticide chemical and the amount of inactive material in the mixture.

Inhalation toxicity — How poisonous a pesticide is to man or an animal when breathed in through the lungs.

Integrated control — A system in which two or more methods are used to control a pest. These methods may include cultural practices, natural enemies, and selective pesticides.

Interval — Period of time. The time period between two pesticide applications or between the last pesticide application and harvest.

Invert emulsifier — An agent or additive which allows water to remain suspended in oil rather than settling out. The usual emulsifier allows suspension of oil in water.

Krebs cycle — One last phase in the breaking down of glucose in animal and plant cells, producing carbon dioxide; also known as citric acid cycle.

Krumatology — The science which is concerned with the phenomena of water waves.

Lac — French for lake. The word appears as a generic in names of lakes (for example Lac Vienux Desert in the Upper Peninsula of Michigan) and infrequently as a common in parts of the U.S. where the influence of early French exploration and settlement remains.

Lacology — That branch of science which is concerned with the study of lakes, ponds, and all other lentic water bodies — their geologic origin and evolution; hydrography; geomorphology; biologic aspects; economic and social significance; development use; and management and conservation of water and shoreland. Lacology, the science of lakes, would be analogous to Potamology, science of rivers.

Lacuster — The central part of a lake.

Lacustrine — The person dwelling in a house on or built over the water of a lake. The term applies mainly to prehistoric lake dwellers inhabiting a lake.

Laurential lakes — A geographic name for the group of great lakes drained by the St. Lawrence River.

Limited access (lake) — Access to the water surface for all riparians at a single place. Except for this limited access location the lake shore remains natural and unchanged.

Limnetic zone — Open water of a lake, pond, or marsh.

Limnology — Study of fresh waters in terms of their chemical, biological, and physical aspects.

Littoral — The word pertains to the shore, either or both the shore land and shore water and near shore bottom of a lake. Littoral is also a name for the shore land for an indeterminate distance back from the water but as such is pretty much restricted to sea coasts. Littoral, in a limnological sense, refers to shallow water in which aquatic vegetation can be present on the bottom, as contrasted with deep water or the profundal.

Littoral currents — Pertaining to lakes, the currents alongside, or parallel, to the shore.

Littoral drift — The material that moves in the littoral zone under the influence of waves and currents.

Littoral plants — Littoral plants are those growing on or near the shore.

Littoral shelf — A shallow submerged shelf or terrace, extending lakeward from the water line of a beach. Where a cliff or high bank rises from the water line and is subject to the erosive action of waves, the cliff recedes under wave attack and a terrace of cutting by waves, and of deposition by currents, is gradually formed and often extends outward a considerable distance from shore, ending in deep water with a step front or "drop-off." The beach may be regarded as the part of the shelf which lies above the water level. Also called: subaqueous terrace, submerged terrace; submerged platform; cut and built terrace; wavecut terrace; wave-built terrace; and sublacustrine shelf.

Littoral zone — The shoreward region of a body of water in which light penetrates to the bottom; in lakes or ponds from shoreline to the lake-ward limit of rooted aquatic plants; in oceans, from shoreline to a depth of 200 m.

Lobe (of a lake) — The word is sometimes used as a common noun in descriptions and also as a proper noun, a part of a geographic name, for a longish rounded segment in the nature of a bay or arm.

Loch, logh — Respectively, Scottish and Irish words for lake.

Marine — Having to do with animals and plants which live in the ocean.

Metabolism — The chemical activities that take place in the cells of living organisms involving two fundamental procedures, catabolism and anabolism, simultaneously at work; the former refers to the breaking up of substances into constituent parts, the latter, building up of complex substances from simpler ones.

Metric — A system of measurement which is used throughout the world in scientific work. It uses meters, grams, and liters as units. (Note: the key word here is "scientific". It is standard in science.)

Metalimnion — The layer of water in a lake between the epilimnion and hypolimnion in which the temperature exhibits the greatest difference in a verticle direction. Used interchangeable with thermocline.

Meromictic lakes — Those lakes that at the time of winter cooling undergo only partial circulation down to a depth determined by a density stratification.

Milligram (mg) — A unit of weight in the metric system; about 28,500 mg equals 1 oz.

Monitoring system — A regular system of keeping track of and checking up on whether pesticides are escaping into the environment.

Multipurpose — Doing more than one job; a pesticide which kills more than one pest.

Multiple use lake — When applied to water bodies, multiple use has traditionally implied that management could include more than one major category of use; flood control, hydro-electric power generation, navigation, irrigation, recreation, and water supply. Due to space demands, conflicts of interest, and the intense competition of various recreational uses for lake waters, multiple use is being applied in a somewhat different sense. A lake may be classified as a multiple use lake if it is managed for fishing and water skiing. (Refer to economic classes of lakes and lakelands for other recreational uses.) Lakeshore activities are included as lake uses. On many lakes, the surface water has been zoned on a time or spatial basis to facilitate multiple use.

Natural enemies — The predators and parasites which exist in the environment and attack pest species.

Natural resource — Any material that occurs in nature and is useful to man; coal forests fertile land, oil, and minerals of many kinds are wealth-producing materials.

Navigable (inland lakes) — The meanings of navigable have been determined pretty largely by statutes and court decisions and, therefore, have variations in time and place, accordingly. Navigability is not necessarily restricted to deep, open, unobstructed waters and passage of large vessels engaged in trade, but on the contrary in some instances very shallow lakes, with water barely deep enough to float a saw-log, a canoe, or a row boat may have been declared navigable.

Nitrogen trap — Lakes, marshes, and swamps effectively remove a large percentage of the dissolved nitrogen from inflowing water. As a result of the nitrogen being fixed by aquatic vegetation, the outflowing water has a lower concentration of nitrogen and other nutrients.

Nitrogenous wastes — Wastes of animal or plant origin which contain a significant concentration of available nitrogen. Excessive enrichment of lakes by nitrogenous wastes (for example: sewage effluent), called eutrophication, represents one of the most serious threats to these valuable water resources in the U.S.

Nonconsumptive use — Nonconsumptive uses are those that do not reduce the supply, such as recreational (bathing, fishing, boating, and hunting), navigation, and aquaculture.

Nontarget — Any plant, animal, or other organism at which a pesticide application is not aimed, but may accidently be injured by the chemical.

Nonvolatile — A pesticide chemical that does not evaporate (turn into a gas vapor) at normal temperatures.

Oligotrophic — Waters with a small supply of nutrients and hence a small organic production.

Pelagic — On an inland lake a term sometimes applied to the deeper bottom zones, 10 to 20 m or more, characterized by the absence of aquatic vegetation and the presence of ooze or mud deposition.

Plephyte — A term for lake bottom deposits consisting mainly of fine nonfibrous plant remains.

Penstock — A closed tube used to conduct water under pressure from a reservoir to the turbine house.

Perennial lake — One that has water in its basin throughout the year and not subject to extreme fluctuations in level.

Periglacial lakes — Lakes formed during the Pleistocene, on unglaciated land, near the borders of ice sheets.

Periphyton — A term for the assemblage of the microorganisms which form a coating on the stems and leaves of submerged plants, stones, and other objects on lake bottoms.

Peroxisomes — Membrane-bounded particles found in the cytoplasm of plant cells. Peroxisomes contain oxidases and catalase and are probably the site of photorespiration.

Permissive riparian rights — Temporary or limited riparian rights obtained by either verbal or written request. Sometimes dependent upon the payment of a fee for the use or enjoyment of a lake. Permissive rights normally cover short periods of time and may be terminated without prior notice or warning. No prescriptive may envolve from permissive use of a lake.

Phosphate — Essential nutrient (see macronutrients) for plants, supplied naturally from the weathering of rocks and artificially from the mining of rock phosphate, an insoluble evaporite. Rock phosphate is processed by treatment with sulphuric acid to produce "superphosphate," containing some soluble phosphorus pentoxide (P_2O_5) fertilizer, and superphosphate can be treated further, with phosphoric acid, to increase the P_2O_5 content as "triple superphosphate." Gypsum is a by-product. Phosphates are also used industrially. They can cause eutrophication when discharged, mainly in sewage. Phosphate is fairly immobile in soils and contamination of water from phosphate fertilizer is not severe. See phosphorus cycle.

Phosphorus cycle — The circulation of atoms of phosphorus, brought about mainly by living organisms. Phosphorus is an essential constituent of DNA and RNA and is involved in the ADP-ATP (see ATP) process whereby energy is made available to cells. The death and decomposition of living tissues returns phosphorus to the soil in a form available to plants, so continuing the cycle.

Photic zone — The surface waters of a sea or lake, penetrated by sunlight. This zone includes the euphotic and dysphotic zones.

Photochemical — Applied to chemical reactions brought about by energy supplied to substances by light.

Photosynthesis — The synthesis by living cells of organic compounds from simple inorganic compounds, using light energy. In green plants the absorption of light by chlorophyll initiates photochemical reactions ("light reactions") in which oxygen is released from water and light energy is converted into chemical energy by the formation of ATP. This is photophosphorylation. Subsequent "dark reactions" result in the reduction of carbon dioxide (using hydrogen which originated in the water) to carbohydrates, utilizing the energy stored in ATP.

Potable water — Water suitable for drinking purposes. Detailed laboratory tests by competent public health agencies are necessary to determine the safety of drinking water. Generally, untreated lake water is not potable water, but in wilderness areas lake water is easily rendered potable by use of chlorination kits.

Potash lakes — A term applied to closed and enclosed lakes, in semiarid and arid regions, whose brine contain a high proportion of salts of potassium. In the U.S. such lakes occur in the western part of the Sand Hills regions of Nebraska.

Pothole (lake, pond) — The term pothole is in current use in Michigan for small pit depressions, generally circular or elliptical, from a few feet to 40 to 50 ft deep. They are constructional in origin and most common in outwash plains and recessional moraines but are also recognized in till plains. Many have dry bottoms, but many also contain lakes, intermittent ponds, and bogs of various kinds.

Phytotoxicity — Injury to plant life caused by a chemical or other agent, or toxicity produced by another plant.

Plant regulator — Growth regulator; a chemical which increases, decreases, or changes the normal growth on reproduction of a plant.

Plankton — Microscopic aquatic organisms which are free-floating and capable of limited movement.

Plant — An organism characterized chiefly by a holophytic mode of nutrition dependent upon the possession of chlorophyll. Some plants, however, are parasitic (see parasitism) or saprophytic. Most have cellulose cell walls, are sedentary, and have branching bodies. The distinction between plants and animals becomes blurred in some groups (e.g., slime moulds, unicellular algae, protozoa), which are now often regarded as part of a third kingdom, the Protista.

Plasmagenes — Cytoplasmic particles or substances which can reproduce and pass on inherited qualities to daughter cells. Unlike genes, they are not inherited through the chromosomes of the gametes in a Mendelian way. Plasmagenes may be present in bodies such as plastids (plastogenes) or mitochondria.

Plasmalemma — The cell membrane. In plant cells the term is applied only to the external membrane (ectoplast) and not to the one surrounding the vacuole. See tonoplast.

Posion control center — An agency (usually a hospital) in all the major cities which is informed of the proper first aid and antidotes for poisoning emergencies — including pesticide poisonings.

Precautions — Safeguards; safety measures; warnings as stated on the pesticide label.

Primary production — The production or organic matter from inorganic materials within a certain period of time by autotrophic organisms with the help of radiant energy.

Productivity (of a lake) — The rate of production at which organisms are grown; expressed with reference to total, kinds, time, and rate.

Profundal bottoms — The deeper bottoms of a lake in contrast to shallows shoals and littoral shelves. Also a limnological concept, the profoundal is a life zone lying at greater depths than the metalimnion. Emergent and submerged plants of the littoral are absent, and the communities of flora and fauna are those adapted to a low content of oxygen, or oxygen-less conditions.

Pumped storage — A system for generating peak-load electricity by turbines turned by a fall of water from a reservoir filled by pumps lifting water from a lower level. During base-load periods, when demand for electrical power is low, water is lifted from the lower to the higher level, and at times of peak demand the water is released.

Recreational pool — Residual storage over and above the required capacity of a flood control reservoir. The designed capacity of the flood control structure can be increased to provide for a permanent recreational pool without impairing its emergency flood capacity. Also called a conservation pool.

Registration — Approval by the Environmental Protection Agency of a pesticide for uses as stated on its label.

Regulatory officials — Those persons working for the federal or state government who enforce the rules and laws.

Respirator — A face mask which filters out poisonous gases and particles from the air so that a person can breathe and work safely.

Riprap — Coarse stones, natural boulders, and cobbles or artifically broken rock fragments laid (either loosely or cemented) against the basal slope of a bank for the purpose of preventing wave cutting.

Riverine lakes — An inclusive term for impoundments created by dams across rivers. Riverine includes impoundmented water bodies variously called: reservoirs, ponds, basins, storage basins, flowages, flooding, and floodwaters.

Salinity (of lake water) — Salinity commonly signifies water containing a high concentration of salt or sodium chloride. Technically the term is not so restricted and may refer to water containing chlorides, sulfates and carbonates of sodium, potassium, calcium and magnesium, or to total dissolved solids.

Salt flat — Salt-encrusted bottom of a dessicated lake.

Salt lakes — The term is commonly applied to those whose waters contain a high concentration of sodium chloride (NaCl).

Saprophyte — Organism (plant or Protista) which obtains food in solution from the dead or decaying bodies of other organisms. Many fungi and bacteria are saprophytes, so are a few Protozoa (e.g., Euglenia, see Euglenophyta) and flowering plants (e.g., Bird's Nest Orchid, Neottia). Saprophytes carry out the essential process of breaking down organic matter into simple substances such as carbon dioxide and nitrates, which are then available for synthesis by autotrophic organisms into new organic matter.

Saturation — A relative humidity of 100%, measured by comparing the difference in readings between a dry bulb and wet bulb thermometer. In unsaturated air, water evaporating from the wet bulb will lower its reading. In saturated air there will be no evaporation from the wet bulb; the water on the bulb will be at the same temperature as the surrounding air.

Settling basin — An artificial basin for collecting the sediment of a river before it flows into a reservoir, and thereby preventing the rapid siltation of the latter. The basin is designed with weirs for drawing off the clear water. The term is also applied to industrial sedimentation structures used for the removal of pollutant materials from factory effluents.

Severed riparian frontage (lake) — Highways and railroads built across bays or arms of lakes frequently sever riparian lands from the lake if the culverts are not large enough to provide for the passage of boats. The severed riparian frontage has little value unless adequate access to the lake can be obtained.

Stabilisation (Biol.) — The increase of dominance that ends in a stable climax, produced by the invasion of species leading to the establishment of a population most completely fitted for prevailing conditions. This degree of adaptation to conditions ensures that the climax is permanent and can be changed only by a change in the conditions to which the population is fitted.

Stoichiometric — A chemical is stoichiometric when its component elements are present in the precise proportions represented by its chemical formula. A stoichiometric mixture is one that will yield a stoichiometric compound on reaction (i.e., it consists of elements in exactly the proportion required to yield a stoichiometric compound utilizing all the components of the mixture).

Surface water — Rivers, lakes, ponds, and streams, etc. which are located above ground.

Surfactant — A chemical or agent used in a pesticide formulation to make mixing easier and help the material to spread over and completely wet the surface to be sprayed. Examples: detergent, emulsifier, wetting agent.

Synecology — The study of the relationships between communities of species and their environment.

Systemic — A pesticide that is taken up by one part of a plant or animal and moved to another section where it acts against a pest.

Taxis — A locomotory response of a cell or organism in which the direction of movement is oriented with relation to the stimulus (cf. kinesis). In chemotaxis the stimulus is a chemical (e.g., a pheromone emitted by a female moth to attract a male; cane sugar exuding from the female sex organ (archegonium) of a moss plant to attract spermatozoids). In phototaxis (heliotaxis) the stimulus is light, in geotaxis it is gravity, in the thermotaxis it is heat, and in rheotaxis it is flowing movements in the immediate environment (e.g., water currents).

Thermocline — In thermally stratified lakes, the layer below the epilimnion. It is the stratum in which there is a rapid rate of decrease in temperature with depth; a minimum of 1°C/m in depth.

Tolerance — The amount of a pesticide that can remain on any food (plant or animal) that is to be eaten by livestock or humans. The tolerance is set by the Environmental Protection Agency.

Tolerant — Not susceptible to (injured by) a pesticide application.

Toxicant — A poison. The chemical in a pesticide formulation that can injure or kill the pest as well as humans, animals, or plants.

Trade name — A brand name. The name given to a pesticide by a manufacturing company to identify it as their product.

Turbidity — The degree of opaqueness of the water due to the amount of fine (particles) in suspension. The particles that cause turbidity may also determine apparent color. Carp, suckers, and sturgeon commonly extract food from the bottom deposits of lakes. Their feeding involves sucking, digging, and rooting which creates localized turbidity, and this in turn may adversely affect the growth of certain aquatic plants. Their feeding action may also be considered as a form of cultivation that occurs in shallow water that is otherwise protected from wave and current action.

Underground water — Waterways which are located beneath soil surface which wells get their water from.

USDA — U.S. Department of Agriculture.

Wetlands — The term wetlands, for a broad group of wet habitats, is in common use by specialists in the field of wild life, especially waterfowl management. It includes features that are permanently wet, or intermittently water covered, such as swamps, marshes, bogs, muskegs, potholes, swales, glades, slashes, and overflow land of river valleys. Large open lakes are commonly excluded, but many kinds of ponds, pools, sloughs, holes, and bayous may be included.

Index

INDEX

D

E

F

G

H

I

N

Q

R

S

T

U

V

W

X

Y

Z